Moths, Myths, and Mosquitoes

# Moths, Myths, and Mosquitoes
## *The Eccentric Life of Harrison G. Dyar, Jr.*

Marc E. Epstein

# OXFORD
UNIVERSITY PRESS

Oxford University Press is a department of the University of
Oxford. It furthers the University's objective of excellence in research,
scholarship, and education by publishing worldwide.

Oxford   New York
Auckland   Cape Town   Dar es Salaam   Hong Kong   Karachi
Kuala Lumpur   Madrid   Melbourne   Mexico City   Nairobi

New Delhi   Shanghai   Taipei   TorontoWith offices in
Argentina   Austria   Brazil   Chile   Czech Republic   France   Greece
Guatemala   Hungary   Italy   Japan   Poland   Portugal   Singapore

South Korea   Switzerland   Thailand   Turkey   Ukraine   VietnamOxford is a registered
trademark of Oxford University Press

in the UK and certain other countries.Published in the United States of America by
Oxford University Press
198 Madison Avenue, New York, NY 10016

© Oxford University Press 2016

All rights reserved. No part of this publication may be reproduced, stored in
a retrieval system, or transmitted, in any form or by any means, without the prior
permission in writing of Oxford University Press, or as expressly permitted by law,
by license, or under terms agreed with the appropriate reproduction rights organization.
Inquiries concerning reproduction outside the scope of the above should be sent to the
Rights Department, Oxford University Press, at the address above.

You must not circulate this work in any other form
and you must impose this same condition on any acquirer.

Library of Congress Cataloging-in-Publication Data
Names: Epstein, Marc E., author.
Title: Moths, myths, and mosquitoes : the eccentric life of Harrison G. Dyar, Jr./Marc E. Epstein.
Description: Oxford [UK] ; New York : Oxford University Press, 2016. |
Includes bibliographical references and index.
Identifiers: LCCN 2015014631 | ISBN 9780190215255 (hardback)
Subjects: LCSH: Dyar, Harrison G. (Harrison Gray), 1866-1929. | Entomologists—United
States—Biography. | Lepidoptera—Catalogs and collections. | BISAC: SCIENCE / Life Sciences /
Zoology / Entomology. | SCIENCE / History.
Classification: LCC QL31.D89 E67 2016 | DDC 595.7092—dc23 LC record available at http://lccn.
loc.gov/2015014631

1 3 5 7 9 8 6 4 2
Printed in the United States of America
on acid-free paper

For William E. Miller, Wallace J. Dyar, and fellow entomologists

# CONTENTS

*List of Figures*  xi
*Acknowledgments*  xix
*Introduction*  xxiii
*Maps*  xxvii

**PART I. Preparatory Stages**
1. The Dyars and the Hannums: 1805–1875    3
   The Hannums and President Lincoln    5
   Marriage to Eleonora Hannum, Birth of Dyar, Jr.,
      and Death of Dyar, Sr.    6
2. Dyar, Jr.: Early Growth Stages and Development: 1875–1889    8
   Life in Boston: School, Mother, and Future Wife    11
3. Collecting and Rearing Lepidoptera, and Dyar's Law: 1882–1891    16
   Collecting and the Catalogue    17
   Early Mentors and Publications    19
   Dyar's Law of Geometric Growth    22
4. Long Collecting Trips and Sawflies: 1889–1897    26
   The Fifteen-Month Trip    28
   Sawflies: A Nine-Year Interlude    31
5. Postgraduate Education on Classification of Moths and Bacteria at
   Columbia: 1892–1895    34
   Neumoegen and the Myth of the Name "Dyaria"    36
   Reclassification of Lepidoptera    38
   Talks on Lepidoptera Classification    41
   Doctorate in Bacteriology    43
6. Genealogy of the Limacodidae and the Dyars    45
   The New York Slug Caterpillars    46
   Dyar's Genealogy and Connection with the Hannums    50
7. Last Days in New York, L.O. Howard, and a Move to
   the U.S. National Museum    51
   L.O. Howard and Bringing Dyar to Washington    52

## PART II. Beginning a New Life at the USNM and in Washington

8. Life in the District of Columbia and Wellesca Pollock   *59*
   Stony Man and Wellesca Pollock   *61*
   Zella Dyar, Her Mother, and Place in Washington Society   *62*
9. Beginning a Career Building the National Collection of Lepidoptera   *65*
   Dyar's "List" and Expanded Lepidoptera Interests   *67*
   Dyar Joins the Entomological Society of Washington   *71*
10. Battle of the Titans Smith and Dyar, and Their New Love—the "Skeets"   *74*
    Battles between Dyar and Smith   *76*
    Dyar and Smith's Rule of the Mosquitoes   *82*
11. Collecting and Travels: 1898–1909   *84*
12. Literature Wars and Last Battles with Smith: 1904–1911   *95*
    Battles with Mosquito Workers and the Mitchell Lawsuit   *99*
    Last Battles with Smith   *102*
13. Gains and Losses as a Professional Lepidopterist: 1907–1914   *106*
    Paid Positions and Problems with Doctor Barnes   *112*

## PART III. Scandal, Divorces, and Their Aftermath

14. Wellesca's Bahá'í Faith, New Wealth, and Growing Concerns by Zella as Dyar's Life Begins to Unravel: 1906–1908   *117*
    Involvement with Dyar in Washington Real Estate and Loans to George F. Pollock   *119*
    Zella's Ultimatum and Return of the Stones from 'Abdu'l Bahá   *121*
    Dyar's Decline in Collecting   *123*
15. Marriage Troubles: 1909–1913   *125*
16. The Separation: 1914–1915   *131*
17. Divorce Wrangling in Reno: 1915   *137*
    "A Fool and His Friend"   *142*
18. Divorces, Appeals, and Bigamy: 1916–1921   *144*
    Dyar v. Dyar   *146*
    Allen v. Allen   *147*
    Wellesca's Fictional Marriage and Fictional Divorce   *150*
    Bigamy: An Allen Marriage Rather Than a Dyar Divorce   *150*
    Invention of an Alter Ego: Wilfred P. Allen   *151*
19. After the Scandal, Dismissal, and Friend Knab: 1917–1918   *153*
    Recovering from Breakdown in Connecticut   *158*
    Dyar's Partnership with Frederick Knab   *159*

## PART IV. The Final Decade: Attempts at Reinstatement

20. The National Collection of Lepidoptera, Its Workers, and Their Tiffs: 1920s   *165*
    Schaus and the Problem with Honorary Titles   *168*
    Tiffs between Dyar and Schaus with Their Helpers   *171*
21. Mosquitoes and Dyar's Pursuit of Reinstatement   *174*
    Dr. Clara S. Ludlow and the Army Medical Museum   *176*
    Travels for Mosquitoes: 1916–1924   *179*

22. Dyar and His Tunnels: Dupont Circle & B Street     *184*
    The B Street Tunnel     *187*
    The 1917 Collapse of the Dupont Circle Tunnel     *189*
    Why Dyar Dug the Tunnels     *189*
23. Personal Life and Bahá'ís in the 1920s     *191*
    Life on B Street     *192*
    Dyar's Heterodox Bahá'í Beliefs     *195*
24. Unity in the USNM Lepidoptera Section and Acquiring the Barnes Collection     *201*
    William Barnes's Concern over Dyar's Influence on the Purchase of His Collection     *205*
25. Final Travels, Projects, and Days as Custodian of Lepidoptera     *207*
    Dyar and Howard on Issues Related to Mosquito Control     *208*
    Dyar Revisits His "Pet" Group, the Limacodidae     *211*
    Final Days as Custodian of the USNM Collection of Lepidoptera     *213*
26. Financial Collapse and Final Push for Reinstatement: 1925–1929     *215*
    The Last Push for Reinstatement: January 1929     *218*
    The Tunnels after Dyar's Death     *223*

Epilogue     *225*
    Wellesca's Admission to the Fictitious Marriage     *225*
    It Took Two Entomologists to Replace Dyar     *225*
    Final Days of Zella P. Dyar and Wellesca P.A. Dyar     *226*
    Final Resting Places     *227*
    Grandchildren Find Each Other at the Same School     *228*

*Chronology     229*
*Notes     237*
*Selected Bibliography     285*
*Index     297*

# LIST OF FIGURES

1.1 Harrison Gray Dyar, Sr., from Harrower, C. S. (1875). *In memoriam: Harrison Gray Dyar.* New York, Jones.   4

1.2 Parthenia Hannum (*a*) and (*b*) Hannum with Nettie Colburn and President Lincoln, from Maynard, N. C. (1891). *Was Abraham Lincoln a spiritualist?* Philadelphia, Rufus C. Hartranft.   5

2.1 Dyar's childhood home at Linwood Hill near Rhinebeck, New York, courtesy Drawings & Archives, Avery Architectural and Fine Arts Library, Columbia University.   9

2.2 Floor plan for Dyar's fictitious "Woodville Inn" (*a*), courtesy of OSIA and (*b*) floor plan of the Linwood Hill cottage by Alexander Jackson Davis, courtesy of Drawings & Archives, Avery Architectural and Fine Arts Library, Columbia University.   9

2.3 Roxbury Latin School (*a*), courtesy of Wikipedia Commons; Westchester Park brownstone (*b*), courtesy of Google maps; location of brownstone on map (*c*) courtesy of LOC.   12

2.4 Mrs. Eleonora Rosella Dyar with her spirit guide Christal (*a*), courtesy of Marcus Lisle; Mrs. Dyar (note "Ayers" misspelling) (*b*); Mr. Marcellus S. Ayer (*c*), courtesy of American Antiquarian Society (rights and reproduction).   13

2.5 Wedding announcement for Harrison G. Dyar and Zella Peabody, courtesy of the Library of Congress (LOC).   14

3.1 Earliest entry in Dyar's "Blue Book," courtesy of the Office of the Smithsonian Institution Archives (OSIA).   17

3.2 Measurements used in development of Dyar's Law of Geometric Growth, courtesy of OSIA.   23

4.1 Stoneman House, courtesy of the National Park Service, Yosemite, CA.   27

4.2 Denver's Union Station (*a*), courtesy of the Denver Public Library, Western History Collection, z-2533; John Lembert portrait (*b*), courtesy of ANSP Archives Collection 457.   28

(xii)  List of Figures

4.3   Adult of *Ameteastegia pallipes* (a), courtesy of Ken Wolgemuth and *Craesus latitarsus* (b), copyright John Pickering, Discover Life (www.discoverlife.org).   31

5.1   Dyar, Jr. with professor and classmates at Woods Hole, 1893 (a,b), courtesy of the library of the Marine Biological Laboratory, Woods Hole.   35

5.2   Berthold Neumoegen (a), and his publication in *The Canadian Entomologist* describing *Dyaria singularis* (b), courtesy of National Archives, L.O. Howard Photo Collection (NARA-LOH), and the Entomological Society of Canada (ESC).   37

5.3   Dyar portrait in December 1893 while attending Columbia (a), courtesy of ANSP Archives Collection 457; Dyar's thesis (b), courtesy of BHL and the New York Academy of Sciences; Dyar's comparative figures on Lepidoptera evolution (c), courtesy of BHL and the New York Academy of Sciences   39

6.1   Geneaology of the New York slug caterpillars, 1899 (a); and Dyar genealogy (b), courtesy of NYES and LOC.   46

6.2   Miss Emily L. Morton (a); the first page (b) and plate (c) of the New York slug caterpillar series (1895) (b), courtesy of AES and NYES.   47

7.1   Sixteen West Ninety-fifth Street, the last of the New York City properties owned by Harrison and Perle Dyar, which remain today (a), courtesy of Google Maps; detail of the entrance (b), courtesy of Carolyn Rogers.   52

7.2   Dyar, Jr. (a) and L.O. Howard (b), courtesy of OSIA; Howard with cigar (c), courtesy of and NARA-LOH.   53

8.1   Dyar's home in an exclusive neighborhood in Dupont Circle (a) and the Indonesian Embassy (b), both courtesy of the author.   60

8.2   George Freeman Pollock, his father George H. Pollock (a), and Wellesca Pollock (b) (1888), courtesy of Shenandoah National Park Archives.   62

9.1   Old National Museum building (now Arts and Industries) (a) and entomology workers inside (b) (c. 1905), courtesy of OSIA.   66

9.2   Title page of "A List of North American Lepidoptera" (a) and Dyar's description of *Calidota zella* within (b), courtesy of Biodiversity Heritage Library, Smithsonian Institution (BHL).   68

10.1  Dyar panning for mosquito larvae at Herzog Island, Maryland (a). courtesy of NARA-LOH and Smith in his typical field attire (b), courtesy of OSIA.   75

10.2  Rev. George D. Hulst (a), courtesy of OSIA and Dyar's paper (b), courtesy of Biodiversity Heritage Library, Smithsonian Institution (BHL).   80

List of Figures (xiii)

10.3 Smith's report on New Jersey mosquitoes (1904) (*a*) and illustration of mosquito from Smith's New Jersey report by L.H. Joutel (*b*), courtesy of the New Jersey Department of Agriculture; P.N. Knopf's illustrations of mosquito larvae (*c*), courtesy of the NYES.   83

11.1 H.G. Dyar, Jr., with his son Otis, courtesy of the Handy family.   85

11.2 Royal Victoria Hotel in Nassau, Bahamas, courtesy of OSIA   86

11.3 Andrew Caudell with his wife Penelope ("Poodle") (*a*) and Wellesca Pollock (*b*), courtesy of the Cundiff family and Wallace J. Dyar.   86

11.4 Moyie and Kuskanook racing on Kootenay Lake (1908), courtesy of the Royal BC Museum, BC Archives, Image A-00672.   89

11.5 Specimen of *Culex mitchellae*, courtesy of Walter Reed Biosystematics Unit (WRBU)   91

12.1 Cover of the *Journal of the New York Entomological Society* (March 1905) (*a*), courtesy of NYES; Henry Skinner (1890s) (*b*), courtesy of ANSP Archives Collection 457, Box#25, Folder#20; H.G. Dyar (c. 1905) (*c*), courtesy of OSIA.   96

12.2 Articles about Evelyn G. Mitchell's lawsuit against Dyar (*a.b*), courtesy of *Winston-Salem* and *Washington Herald* newspapers; Dyar's critique of Mitchell's *Mosquito Life* (*c*) and her rebuttal (*d*), both courtesy of ESC.   101

13.1 National Museum of Natural History (1912), courtesy of OSIA.   107

13.2 First issue of *Insecutor Inscitiae Menstruus* (*a*), courtesy of the author; Dyar's summary of species and genera from Mexico and Panama (*b*) (1913–1914), courtesy of Biodiversity Heritage Library, Smithsonian Institution (BHL).   107

13.3 Dyar's portrait (*a*), courtesy of AAS; Barnes Museum in Decatur, Illinois (*b*), courtesy of AES.   113

14.1 Wellesca Pollack Allen on her pilgrimage to Acca (*a*) and Bahá'í leader 'Abdu'l Bahá (*b*), both courtesy of National Bahá'í Archives (NBA).   118

14.2 Images of Washington apartments (*a,b*), courtesy of the *Washington Post*.   119

14.3 Dyar's Dupont Circle home (*a*), courtesy of the author; announcement of Bahá'í celebration (*b*) and envelope from 'Abdu'l Bahá (*c*), both courtesy of National Bahá'í Archives (NBA).   122

14.4 Labels on a limacodid moth, courtesy of the author.   124

15.1 Zella Dyar (*a*), courtesy of Handy Family; 804 B Street SW (*b*), courtesy of the *Washington Post*; 1910 U.S. Census of the Allen family (*c*), courtesy of NARA; map with 804 B Street SW across from the National Mall (*d*), courtesy of LOC; Roshan Allen (*e*), courtesy of NBA.   126

( xiv )  List of Figures

16.1  The beginning paragraphs of the separation agreement between Harrison and Zella Dyar from *Dyar v. Dyar*, courtesy of Washoe County Court, Reno Nevada.  136

17.1  Harrison G. Dyar (1913) (*a*), courtesy of LOC-LOH; *SS Kroonland* in the Panama Canal (*b*), courtesy of LOC; Mrs. Harriet M. Peabody (c. 1900) (*c*), courtesy of the Denver Public Library, Western History Collection, x-33018.  138

18.1  Zella's divorce bill in Washington, D.C., in November 1915 (*a*), courtesy of the *Washington Herald Tribune* (LOC); Dyar's 1916 divorce bill in Reno (*b*), courtesy of the *Reno Gazette*; Zella's attorneys' motions for Dyar to pay her legal fees in the Reno case (*c*), courtesy of the *New York Herald Tribune*.  145

18.2  Wilfred Allen as an imaginary spouse (*a*), courtesy of *Nevada State Journal* (Reno); halt of the Allen case when Mr. Allen could not be served with a summons (b), courtesy of *Reno Evening Gazette*; appeal to the Nevada State Supreme Court (*c*), courtesy of *Reno Evening Gazette*.  147

19.1  Wilfred P. and Wellesca P. Allen's marriage license, courtesy of Hustings Co. Courthouse, Richmond, Virginia.  155

19.2  Postcard of Hotel Mahackemo (*a*), courtesy of CardCow.com; and T.D.A. Cockerell (*b*), public domain.  158

19.3  Frederick Knab (*a*) and correspondence to H.G. Dyar (*b*), courtesy of OSIA  159

20.1  Portraits of August Busck (*a*) and William Schaus (*b*), courtesy of OSIA; and William Barnes (*c*), courtesy Rutgers University Press  166

20.2  Dyar (*a*), Schaus (*b*), John M. Aldrich (*c*), Mathilde M. Carpenter, and other entomology staff (1925) (*d*), courtesy of OSIA; illustrator and future aviator Mary Foley (Benson) (*e*), courtesy of Shorpy.  172

21.1  Portrait of H.G. Dyar (c. 1920) (*a*), courtesy OSIA; *Culex pipiens* mosquito (*b*), courtesy of Miroslav Deml  175

21.2  Dr. Clara S. Ludlow (*a*) and Army Medical Museum (*b*), courtesy of LOC.  177

21.3  Ketchikan, Alaska (*a*), courtesy of LOC; and *Culiseta alaskaensis* (*b*), courtesy of the author.  180

21.4  Passport photograph of Dr. and Mrs. Dyar (1923) (*a*), courtesy of NARA; Raymond C. Shannon (1925) (*b*), courtesy of OSIA.  182

22.1  The Dupont Circle home in 1990s (*a*), courtesy of the author; photograph of the tunnel (*b*), courtesy of the Shorpy; map (*c*), courtesy of LOC; satellite image (*d*) courtesy of Landsat (USGS).  185

| | | |
|---|---|---|
| 22.2 | B Street home (*a*) and tunnel (*b*), courtesy of the *Washington Post*; map (*c*), courtesy of LOC; satellite image (*d*), courtesy of Landsat (USGS). | 187 |
| 23.1 | Dorothy Dyar in yearbook (1918) (*a*), courtesy of University of the University of California, Berkeley; photographs of Otis Dyar (*b*,*c*), courtesy of the Handy Family. | 192 |
| 23.2 | Children from Dyar's second family (*a*), courtesy of OSIA; National Mall (1930s) (*b*), courtesy of the Historical Society of Washington, Kiplinger Research Library, Washington, D.C.; Wallace J. Dyar's high school yearbook photo (1931) (*c*), courtesy of Eastern High School. | 193 |
| 23.3 | Book of Wellesca's (Aseyeh's) lectures on Bahá'í matters (*a*) and cover page of *Reality* (*b*), courtesy of LOC. | 198 |
| 24.1 | William Barnes in his enormous collection (*a*), courtesy of AES; letter from his nephew (*b*), courtesy of NARA. | 202 |
| 24.2 | Foster Benjamin (*a*), courtesy of OSIA; Frank Morton Jones (*b*), courtesy of AES; W.T.M. Forbes (*c*), courtesy of *The Scarlet Letter*, Rutgers University. | 204 |
| 25.1 | Grinnell Glacier, courtesy of Glacier National Park Archives. | 208 |
| 25.2 | Dr. Harrison and Mrs. Zella Dyar (c. 1926) (*a*), courtesy of Archives of the Natural History Museum, London. Limacodid species *Sibine* (*Acharia*) *zellans* (*b*) and Dyar's handwritten title on the manuscript for "The Macrolepidoptera of the World" (*c*), courtesy of Department of Entomology, National Museum of Natural History, Smithsonian Institution. | 212 |
| 26.1 | The Chastleton (1921), courtesy of LOC. | 216 |
| 26.2 | H.G. Dyar (1928?) (*a*), courtesy of the Archives of the Natural History Museum, London, and job description (*b*,*c*) courtesy of NARA. | 219 |
| 26.3 | Last known photograph of H.G. Dyar with Schaus, Howard, Heinrich, Busck, and Francis Noyes (January 4, 1929), courtesy of NARA-LOH. | 223 |
| E.1 | Columbia University alumni form and card courtesy of Columbia University Archives. | 226 |
| E.2 | The final resting place of Harrison and Wellesca Dyar at Glenwood Cemetery, Washington, D.C. (*a*); grave of Charles Valentine Riley (*b*), courtesy of Terry L. Carpenter. | 228 |

The cover is: Harrison G. Dyar, Jr. with National collection of mosquitoes, courtesy of Visual Studies Workshop, Rochester, NY

(xvi)  List of Figures

## COLOR PLATES

Color plates are located between pages 168 and 169.

    1    Caterpillars of the Limacodidae. All caterpillar and moth images courtesy of Jane Ruffin except: red-cross slug, courtesy of David L. Wagner; hag moth, courtesy of Noble Proctor, and combined spiny oak and saddleback, courtesy of the author.

2 and 3    Dyar's field notebooks. Dyar's "Blue Books" and entries courtesy of OSIA and Department of Entomology, NMNH.

    4    Dyar's "Catalogue of Lepidoptera." Images courtesy of OSIA.

    5    Examples of owlet moth caterpillars reared by H. G. Dyar between 1882 and 1903. Caterpillars of the paddle moth, green marvel, 8-spotted forester, the herald, and *Cucullia laetifica* courtesy of David L. Wagner; adult paddle moth courtesy of Jason Dombroskie; caterpillars of the smeared dagger courtesy of Carol Wolf and the laugher courtesy of Herschel Raney.

    6    Some of the notodontid moth caterpillars and their adults reared by Dyar in the 1880s and 1890s. Lace-capped caterpillar, saddled prominent, Drexel's datana, and black-etched prominent all courtesy of David L. Wagner; adult of lace-capped prominent courtesy of Valerie Bugh; adult saddle prominent courtesy of Lynette Schimming; adults of Drexel's datana and black-etched prominent courtesy of Carol Wolf.

    7    Moths from assorted families reared by Dyar in the late nineteenth and early twen¬tieth centuries. Caterpillars of lappet moth and curved tooth geometer courtesy of D.L. Wagner; caterpillars of promethea and definite tussock courtesy of Stan Malcolm; caterpillar of laurel sphinx courtesy of Colin Gillette; caterpillar of faithful beauty courtesy of Carol Wolf; caterpillar of emerald moth courtesy of David Moskowitz; adult emerald moth and aspen leaf miner larva courtesy of Jason Dombroskie; adult of curved tooth geometer courtesy of Kimerly Fleming.

    8    Dyar's mentors and early correspondents. Augustus R. Grote and Charles V. Riley courtesy of OSIA; Joseph A. Lintner courtesy of the American Entomological Society; Charles H. Fernald courtesy Department of Special Collections and University Archives, W.E.B. Du Bois Library, University of Massachusetts Amherst; Herman Strecker (seated) courtesy of NYES; Strecker portrait courtesy of ANSP Archives Collection 457, Box#26, Folder#47; Alpheus S. Packard courtesy of *Popular Science Monthly* (public domain); George H. Hudson Courtesy Special Collections, Feinberg Library, SUNY College at Plattsburgh

*List of Figures* (xvii)

9   Entomologists of the U.S. National Museum in their younger days. All photos of August Busck and William Schaus courtesy of NARA LOH photo collection; Caudell photos courtesy of OSIA.

10  Dyar family photos along the northeastern seaboard in the early 1900s, courtesy of the Handy family.

11  Dyar with his children Dorothy and Otis courtesy of the Handy family. Harriet M. Peabody with the Navajo courtesy of The Denver Public Library, Western History Collection, x-33018.

12  Illustrations of immature stages of mosquitoes courtesy of the Carnegie Institution.

13  Entomological workers of the Division of Insects, United States National Museum (USNM) courtesy of OSIA

14  Hullabaloo surrounding the tunnel collapse in Dupont Circle. All images courtesy of the *Washington Post* except 14f and 14h, courtesy of American Antiquarian Society.

15  Diagrams of Dyar's home and labyrinth. Drawings based on illustration in *Modern Mechanix*; photographs of the house in the Washington Post and in the Washingtoniana room at the Martin Luther King. Library, Washington, D.C.; and descriptions from newspaper articles and an oral history interview with Wallace J. Dyar.

16  Dupont Circle tunnel collapse. Photo of alley collapse courtesy of Shorpy. Curd children in tunnel courtesy of Curd family and image of outside the blocked tunnel entrance courtesy of the author.

# ACKNOWLEDGMENTS

First, I acknowledge Pamela M. Henson, historian and director of the Institutional History Division of the Office of the Smithsonian Institution Archives (OSIA), for her collaboration over many years. Without her guidance, and knowledge of the Smithsonian's rich history and many other subjects, this biography would not have been possible.

For encouragement to write this book I'm abundantly grateful to the late William E. Miller, my friend and former Ph.D. advisor at the University of Minnesota. Bill nudged me in this direction because he was someone who clearly understood Dyar's significance to many areas of entomology. It was in 1991 when Paul Opler (Colorado State U.) suggested that Pam and I write the initial article for *American Entomologist*, for which he was the managing editor. Rather nonchalantly, after reading a draft of "Digging for Dyar," the late Howard E. Evans (Colorado State U.), who had been my undergraduate advisor in the 1970s, was the first to suggest that I write about the subject in book form. My wife, Joan, family, and friends all have helped me navigate the uncertainties I've faced with this project over the years. At the Plant Pest Diagnostic Center (PPDC), California Department of Food and Agriculture, here in Sacramento, California, my colleagues have been very supportive: in particular Obediah Sage, for his insight and moral support to keep up my spirits, Scott Kinnee who provided help with images and through his unique sense of humor, and Stephen Gaimari for his interest in entomological history and literature.

Wallace J. Dyar, the youngest of Dyar's second family and last survivor, provided Henson and me with irreplaceable anecdotes about growing up in the Dyar household. Wally became our good friend and played the key role in acquiring the Dyar papers, which included Dyar's short stories, being sold in New Market, Virginia. Because of the notoriety of H.G. Dyar's brother-in-law, George Freeman Pollock, who had some correspondence among the papers, there was concern that they end up scattered among those interested in the history of Shenandoah National Park. Mr. Dyar's wish to have the papers reside at the Smithsonian was testament to his trust and friendship. Fortunately Pam was able to have the Smithsonian Institution Archives purchase most of the papers in the nick of time.

Roberta Hill and Lowell Dyar, who knew of their granddad through family lore, provided Pam and me with a very thoughtful and informative oral history interview. Hill was the granddaughter of Dyar's first wife, Zella, and Lowell the grandson of Dyar's second wife, Wellesca Pollock. It is with great regret that Wally, Roberta, and Lowell were unable to see this book, but we were able to see their satisfaction in reading "Digging for Dyar: The Man and the Myth."

Complementing these oral histories were the vast genealogical resources at the Library of Congress and in searchable historic newspapers. This information dovetailed with Dyar's catalogue and notebooks in the Department of Entomology, Smithsonian Institution and the OSIA. These sources cryptically mention relatives, close friends, and neighbors that helped in his pursuit of Lepidoptera and their caterpillars. I also owe a major debt of gratitude to Terry L. Carpenter (Armed Forces Pest Management Board). Lt. Col. Carpenter, with the permission of his supervisor, Capt. Stanton Cope, did an incredible amount of sleuthing on a variety of topics including the mysterious marriage of H.G. Dyar and Wellesca P. Allen, the divorce of the latter and Wilfred P. Allen, and many more unexpected pieces of this story: I fondly refer to him as "my gumshoe," but he has given me with much deeper insight than this implies. Tammy Peters, Supervisory Archivist at OSIA, has provided me with essential support in rounding up images and other documents necessary for the completion of this project.

John Dodge (Arlington, VA), who volunteered for me for over a decade at the National Museum of Natural History (NMNH), performed the gargantuan task of double-sided copying the Shenandoah acquisition for the Dyar Papers for OSIA. In addition, he made multiple copies and placed them in folders and four boxes, enabling the archives, Henson, and me to have copies collated in the original order, which helped keep Dyar's stories intact. Martin Hauser (PPDC) gave me his valuable time at the Field Museum (Chicago) to photograph the Dyar letters among the Herman Strecker papers. Tim McCabe (Albany, NY) generously went through Dyar's letters to J. Lintner at the New York State Library's archives in Albany. William Leach (Columbia University) graciously provided copies of information from the university's alumni files and notes from early meetings of the New York Entomological Society. Illustrator Kandy V. Phillips (Gaithersburg, Maryland) provided help with various tasks relating to Dyar's work on Limacodidae. Kandace Muller (Shenandoah National Park) located images of the Pollocks.

I also wish to acknowledge my fellow lepidopterists, entomologists, and combined staff at the National Museum of Natural History, Smithsonian Institution, for fruitful discussion on H.G. Dyar that began in 1983. These include in alphabetical order: Nancy Adams, David Adamski, John W. Brown, John Burns, Jack Gates Clarke, Don Davis, Terry Erwin, Doug Ferguson, Ollie Flint, Dick Froeschner, David Furth, Patricia Gentili-Poole, Eric Grisell, Don Harvey, Tom Henry, Gary Hevel, Ron Hodges, Buck Lewis, Steve Lingafelter, Ron McGinley, Stuart McKamey, Vichai Malikul, Wayne Mathis, Scott Miller, Dave Nickle, Alan Norrbom, James Pecor, Mike Pogue, Robert Poole, Bob Robbins, Louise Russell, Curt Sabrosky, Floyd Shockley, Alma Solis, David Smith, Ted Spilman, Warren Steiner, Alan Stone, Chris Thompson, Adrienne Venables, and Norm Woodley. Sadly, Adams, Gates Clarke, Ferguson, Froeschner, Russell, Spilman, Stone, Sabrosky, and Venables were unable to see this book in print.

I wish to acknowledge the help of George Venable as my partner in cofounding the Department of Entomology and Illustration Archives at the NMNH and David Furth for insuring the growth of the archives that house Dyar's notes on rearing insects. Also, the availability of scanned copies of all of these notebooks at the NMNH through the help of the Smithsonian Field Book Project, made it possible for me to recover many aspects of Dyar's youth and early work. The project came to my attention from Scott E. Miller, and I am grateful for the help I received from Carolyn Sheffield and Kira M. Cherrix to insure the careful scanning and rebinding of the notebooks.

Steven Handy and his sister-in-law Dawn Handy (Pasadena, CA) provided of never-before-seen photographs of Dyar and his first family, as well as details of the personality of Dyar's son Otis Peabody Dyar. These came into the possession of Steven and his brother Douglas through their aunt Catherine Ross, wife of Otis P. Dyar ("Uncle Oat"), who was a pillar of strength for the Ross and Handy families.

I gratefully acknowledge Albert Rolls (New York, NY) for his assistance as my primary editor for the manuscript. His insights on establishing the priorities in deciding what to include in a biography have been extremely helpful. Jeremy Lewis of Oxford University Press made a number of helpful comments to get the final stages of the draft ready for review prior to my book contract and subsequently. Molly Morrison and a talented team of copyeditors from Newgen made a number of useful comments to get me through the final stages of book production. Editor Sheryl Englund (Ithaca, NY) provided insight on the arrangement of the book. William E. Miller, Art Shapiro (U.C. Davis), Obediah Sage, Joan Epstein (Sacramento, CA), John Calhoun (Palm Harbor, FL), Terry L. Carpenter, Annette Aiello (Smithsonian Tropical Research Institute), Daniel H. Janzen (U. Pennsylvania), Robert H. Stockman (Wilmette Institute), Elizabeth Foster (Davis, CA), Rosser Garrison (PPDC), Mark Schoen (Los Angeles, CA), Todd M. Gilligan (Colorado State U.), John W. Brown, Mike Collins (Nevada City, CA), William Blank (Sacramento, CA), Marilynn Price (Sacramento, CA), Steen Dupont (NHM, London), Jennifer Bundy (U. Arizona), and five anonymous reviewers all made comments to improve the manuscript. Deborah Smith Frost (Sacramento, CA) provided graphic advice and assistance with final preparation of images. Thanks also goes to my good friend and caterpillar co-worker Steven Passoa (National Identification Service, APHIS) for providing obscure bibliographic references on Dyar over the years.

I had discussions on the process of publishing the book with Pamela Henson, Barbara Mann (U. Toledo), Tom Zavortink (U.C. Davis), Dave Smith (NMNH), Scott Miller (NMNH), Todd M. Gilligan (Colorado State U.), David Wagner (U. Connecticut), John W. Brown, Larry Gall (Yale), Michael Pogue, Heather Ewing (New York, NY), Robert Price (PPDC), Joseph Scheer (Alfred University), Art Baum (Sacramento, CA), Mindy Conner (Winston Salem, NC), Joan F. Epstein (Sacramento, CA), Joseph N. Epstein (Washington, D.C.), Shirley Epstein (Denver, CO), Alan Epstein (Toledo, OH), Barbara and Steve Becker (Colorado Springs, CO), John Hessler (Library of Congress), William Leach, Carol Anelli (Washington State University), Jerry Powell (U.C. Berkeley), Tom Manos (Sacramento, CA), Lynn Kimsey and Jay Rosenheim (U.C. Davis), and Julian Donahue, Rudi Matoni, and Jeanine Oppewell (Los Angeles, CA). I'd also like to acknowledge the support of my friends and relatives in the Washington, D.C., area, including Matt and Catherine Kramer, Howard and Marji Epstein, Ellen Epstein and Will Guthrie, Dan Epstein,

Rivka Yerushami, and the Goldberger family. Likewise, in this bi-coastal project I had support from my friends in Sacramento, including Jerry Roth, Ben Glovinsky, and Miriam Steinberg. I would also like to thank my attorney Greg Victoroff (Los Angeles, CA) for his services relating to my book contract, Edgar James and David Dean (Washington, D.C.), and Daniel Weitzman and Mark Urban (Sacramento, CA) for book-related legal insights.

The late Alan Stone provided not only an anecdote about "dinner at the Dyars" that was passed to him through Raymond Shannon, but also one of his limericks about our 1992 publication on Dyar. Other than Wallace Dyar, the only two people I interviewed that had seen or met Dyar in person were F. Martin Brown and Louise Russell, now deceased, and I thank them for providing colorful anecdotes. Richard Brown (Mississippi State U.) kindly gave me the story he had heard about W.T.M. Forbes's alias: "Pink Whiskers."

I'd like to thank Rhinebeck Historian Nancy Kelly for her discussions about Linwood Hill and various aspects of Rhinebeck history. I owe a debt of gratitude to Susan D. Abele of the Newton Historical Society Archives for her sleuthing in the historical Boston directories and newspapers for information about Harriet M. Peabody. I appreciate the help of the reference staff at the British Library, London, for help in locating the patent application of H.G. Dyar, Sr., for tunneling and discussions about why it was never granted. Robert (Bob) Dyer is also acknowledged for sending me some additional useful information he found in his genealogical research of the Dyer/Dyar families.

In addition to images from OSIA, material was duplicated from the following collections: Archives at University of Colorado, Boulder (T.D.A. Cockerell Papers); Rutgers University (J.B. Smith Papers: Thomas Frusciano and Katie Carey); National Archives and Records Administration, College Park; Library of Congress (Dyar Genealogy Papers: Bruce Kirby); Department of Entomology Archives, NMNH (David G. Furth and Floyd Shockley); National Bahá'í Archives (Roger Dahl); University of Massachusetts, Amherst (C.H. Fernald papers); Field Museum, Chicago (H. Strecker Papers); Academy of Natural Sciences (American Entomological Society Papers and Photographs); Carnegie Museum (B. Preston Clark Papers: John E. Rawlins); National Library of Medicine (S. A. Knopf Papers); Cornell University (J.H. Comstock and W.T.M. Forbes, papers); Carson City, State of Nevada Courthouse and the Washoe County, Nev., Courthouse, Reno; SUNY Plattsburgh (G.H. Hudson Papers: Debra Kimok).

David Wagner (University of Connecticut)—gifted with singular talents of rearing numerous caterpillars simultaneously and can appreciate Dyar's earlier efforts—generously provided many of his excellent caterpillar photographs and helped me find others. Carolyn Rogers (Washington, D.C.) has braved the elements over the years to get me photographs of Dyar's former homes and other real estate and my friend Jane Ruffin (Philadelphia, PA) for her excellent limacodid caterpillar images. The many others who provided excellent nature photographs are acknowledged in the list of illustrations. I would also like to thank BugGuide.net, and MothPhotographersGroup.msstate.edu for helping me to locate these photographers and their unique images.

# INTRODUCTION

Friday September 26, 1924, was a normal day in the nation's capital, except for baseball. Washington Senators' pitcher Walter Johnson had his winning streak of thirteen games end at the hands of the Red Sox, threatening the team's goal of a pennant. The Yankees were now only one game back with three to play. But something rather peculiar happened the day before. In Dupont Circle a truck backed up in an alley and the ground collapsed, revealing an architectural marvel: a mysterious underground labyrinth. After a day of wild speculation in the newspapers about the origin, including it being a German spy tunnel, the trail led to a stooped and soft-spoken Smithsonian scientist: Harrison Gray Dyar, Jr. (1866–1929) (see frontispiece).

According to urban legend Dyar's tunneling had to do with moving from one household to another, and the secret was kept until the children met at school, discovering they each had a father who worked on butterflies at the Smithsonian. This myth lasted decades and produced at least one limerick, published in 1974 in a spoof of a scientific journal called *Frass: An Occasional Journal of Para-lepidopterology*, which recounts its more salacious element—that is, how Dyar was able to reach his mistress:

> A fabulous man was H. Dyar.
> As he aged he only got spryer
> He said to his sweet
> As he dug 'neath the street,
> "With each shovelful, darling, you're nigher."[1]

The true story is that there were tunnels at each house, but constructed mostly a decade apart and not interconnecting, being separated by miles in different parts of D.C. It is indeed true that Dyar was married to two wives at the same time, to the second under a fictitious identity, and that he did father children with each wife, though his two sets of offspring were of vastly different ages and did not meet at school. Although we can take Dyar at his word that digging tunnels was his form of exercise, perhaps it wasn't that simple.

※

Following the Civil War, at a time when the United States was becoming preeminent in science and industrial might, Harrison Gray Dyar, Jr., made significant contributions to our understanding the hidden world of insects. Dyar's interest in metamorphosis of Lepidoptera went back to a childhood when he referred to it as "transformation." Although this was not unusual terminology in the Gilded Age, it perhaps had another layer of significance in a home of believers in spiritual transformation. In fact, he called his rearing chamber a "house of transformations."

Dyar and other Gilded Age children, including future entomologists Leland O. Howard and William Schaus, came of age with a passion for Lepidoptera and other insects. Although they had books to guide them, at some point, the young naturalists pursued more knowledge. To fill this need would require the help of a local collector, perhaps a friend or a teacher. Fueling their passion was a continent teaming with new species awaiting discovery and a group of established scientists and collectors to mentor them.

The previous generation of scientist-collectors, dubbed by Conner Sorensen the "Brethren of the Net," was the first in North America with enough knowledge to make serious attempts to name swaths of the vast continent's insect fauna. Among these "brethren" were state entomologists, who published regular reports on insects from an economic perspective. The leader of this group was Charles Valentine Riley, who went from being State Entomologist for Missouri to the second federal entomologist in Washington in 1878, following Townend Glover. Others included Asa Fitch, Joseph Lintner, Benjamin Walsh, and Stephen Forbes. There also were those, such as John Henry Comstock of Cornell, who worked at land grant colleges, which were brought unto existence in 1862 by the Morrill Land Grant Act. The Hatch Act of 1887 established agricultural experiment stations at these colleges. Basic research was being conducted in places like Cornell, which had the first entomology degree program in the country. Here, Comstock applied Darwinian approaches to build a natural classification of insects based on anatomical parts and their ancestral origins: a major achievement of U.S. scientists.

Important to a new generation of lepidopterists was access to large private collections of knowledgeable amateurs, including those in homes of actor Henry Edwards and banker Berthold Neumoegen in New York City, and sculptor F. H. Herman Strecker in Reading, Pennsylvania. Eventually these holdings of specimens, each with their own strengths, would end up at larger museums. But in the meantime, these collections were critical to giving proper names to specimens for the growing scientific community.

At the age of twenty-four Dyar noted discrepancies in the literature in the number of molts (sheds) by particular species of caterpillars and developed what has become known as Dyar's Law or Rule of Geometric growth: a pragmatic method to standardize this information in making accurate larval descriptions and identifications, documenting variation, or predicting the size of other larval stages. Use of ratios in a similar manner, though not necessarily applied to physical growth, continues to be common practice today in disciplines even beyond the biological realm.

During his career as an entomologist, Dyar built natural classifications of two very different groups of insects: first Lepidoptera and then mosquitoes. His insect phylogenies (genealogies) were based on both larval and adult stages. His methods brought about major changes in our understanding of natural relationships, or at the very least a holistic approach to insect systematics. Along the way he named well over 3,000 species and genera of insects while establishing the Lepidoptera collection at the U.S. National Museum, Smithsonian Institution as one of the foremost in the world. Furthermore, his "List of North American Lepidoptera" was never equaled by his contemporaries nor was his role in advancing our knowledge of New World mosquitoes through monographs and numerous other publications.

The lack of a compendium of Dyar's important contributions to entomology is due in part to the fact that his influence spanned three unrelated orders of insects: the Lepidoptera (moths and butterflies), the Diptera (flies, including mosquitoes), and Hymenoptera (bees and wasps, including sawflies). There are bibliographies of Dyar's hundreds of mosquito publications and twenty-one on sawflies. A complete bibliography of his over 650 contributions to Lepidoptera will be published elsewhere.

But there is more to Dyar than can be found in his professional publications. His notebooks at the Smithsonian Institution and his scientific correspondence in many archival collections reveal that significant pieces of Dyar's and his contemporaries' stories remain untold. Taking advantage of such material, this book presents a much clearer picture of Dyar's life, including his early career and his relationships with his boss Leland O. Howard and other contemporaries at United States National Museum (USNM) of the Smithsonian Institution

Another previously unknown aspect of Dyar's career discussed in the following pages is his extensive travels, particularly in North America. These include two extended transcontinental trips by rail, the first a "collecting honeymoon" in 1889–1890 after his marriage to Zella Peabody, followed soon after with a second trip with her for fifteen months in 1891–1892. That these two early trips were taken with his wife, and the field assistance she provided, is revealing of the way in which Dyar's personal and professional life were intertwined.

Just as fascinating is the world Dyar constructed in his fiction. Dyar's two published novellas and the several stories that he wrote for *Reality,* the magazine affiliated with Bahá'í, a nascent religion that Dyar became involved with in the teens and 1920s, represent only a small fraction of his literary output. He wrote around 200 unpublished, previously undiscovered stories, the interpretation of which will surely challenge future researchers. Dyar's story topics are a blur of fact and fiction from his marriages, his childhood among his spiritualist parents, and other miscellaneous interests such as eugenics. In a number of cases the stories were written on the backside of personal correspondence, real estate dealings and other investments, and scientific correspondence, including the largest collection of letters from Frederick Knab, Dyar's collaborator and friend. Other personal notes submitted as exhibits in his divorce and the testimony of Dyar and many other witnesses provide many new insights into Dyar's personal drama.

Dyar's legendary productivity and interactions with his peers, his unusual childhood, complex personal life, and strange pastimes make the examination of his life so compelling to this biographer. Unfortunately, the legacy of his personality has obscured his significant scientific contributions. Even within his own field of entomology he is best known in relation to feuds with colleagues and harsh critiques of their work. In this biography I hope to tip the balance a bit more toward his bequest to science, while chronicling his illustrious personal life and complexities. Although perhaps clichéd to say this, I believe you will find, as I did, that Dyar's true story is indeed stranger than fiction, and from the pages that follow you will gain better insight into his "Eccentric Life."

# MAPS

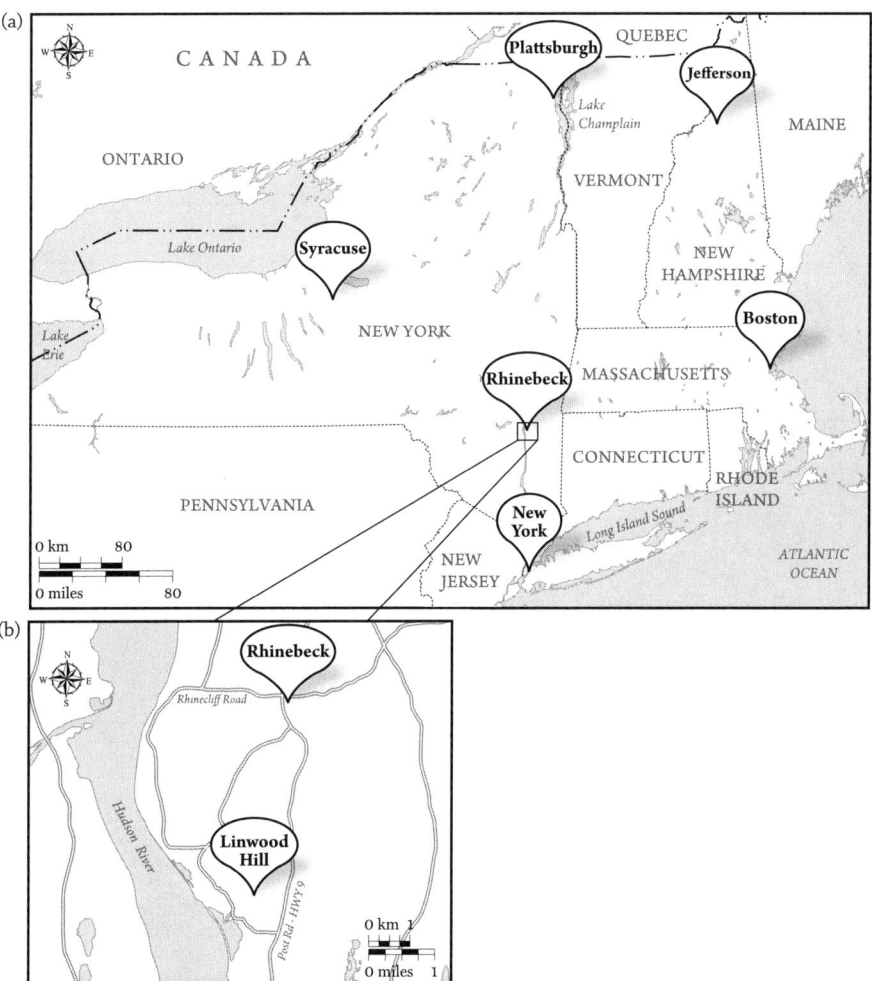

**Map 1** Known locations of Harrison G. Dyar, Jr., from 1866 to 1886: Early homes and collecting grounds (*a*); detail of where Dyar's home in Linwood Hill is located in relation to Rhinebeck, where he went to school (*b*).

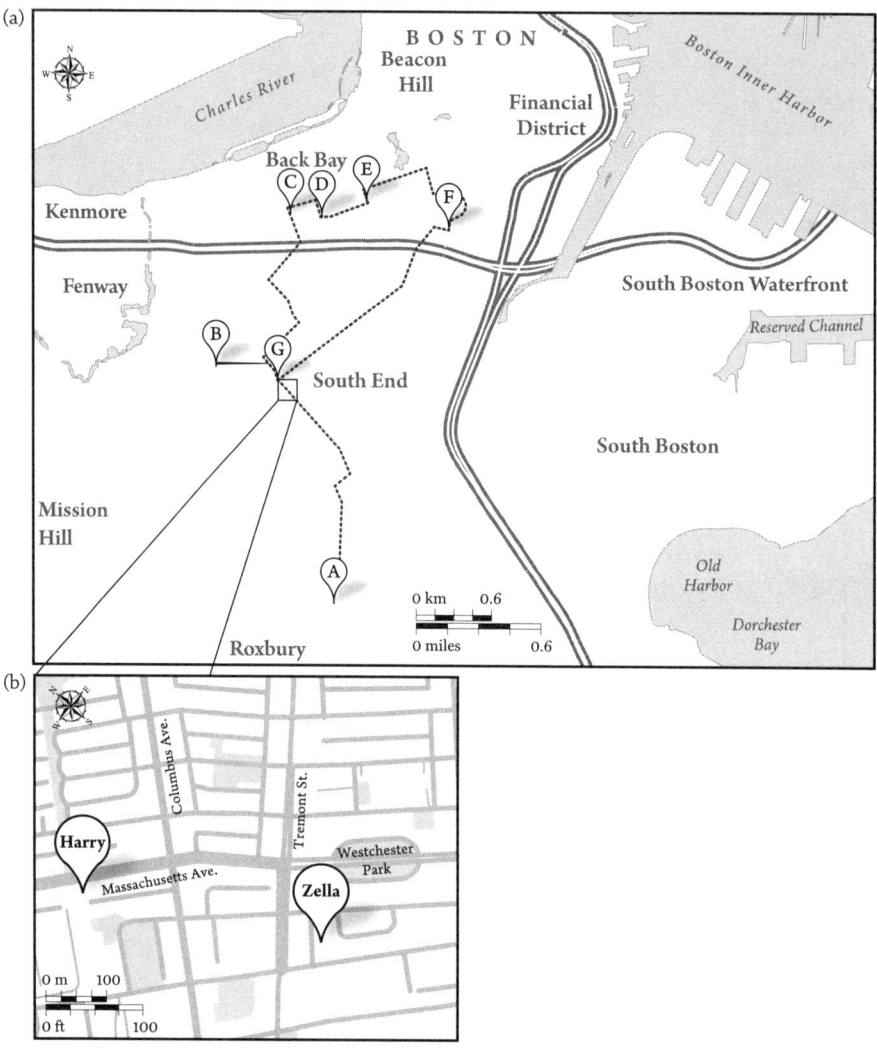

**Map 2** Locations and walking tour (dotted line) of homes, schools, societies, and places of worship for Dyar, Jr., Zella Peabody, and their mothers in Boston (1882–1889): (A) first residence (123 Mount Pleasant) in Boston near Roxbury Latin School (old location); (B) second residence at 170 Westchester Park Road (near Massachusetts and Columbus Avenues today); (C) First Spiritual Temple of Boston (Exeter and Newbury Streets); (D) Massachusetts Institute of Technology (old location in the Back Bay with "C & D"); (E) Boston Society of Natural History (Boylston and Berkeley Streets); (F) Children's Mission (277 Tremont Street, Matron: Mrs. Harriet M. Peabody); (G) Everett Girl's School (Northampton and Tremont Streets) (*a*); location of the Dyars' flat on Westchester Park Road (1883–1889) and Zella's school, Everett, where she graduated (1884) (*b*).

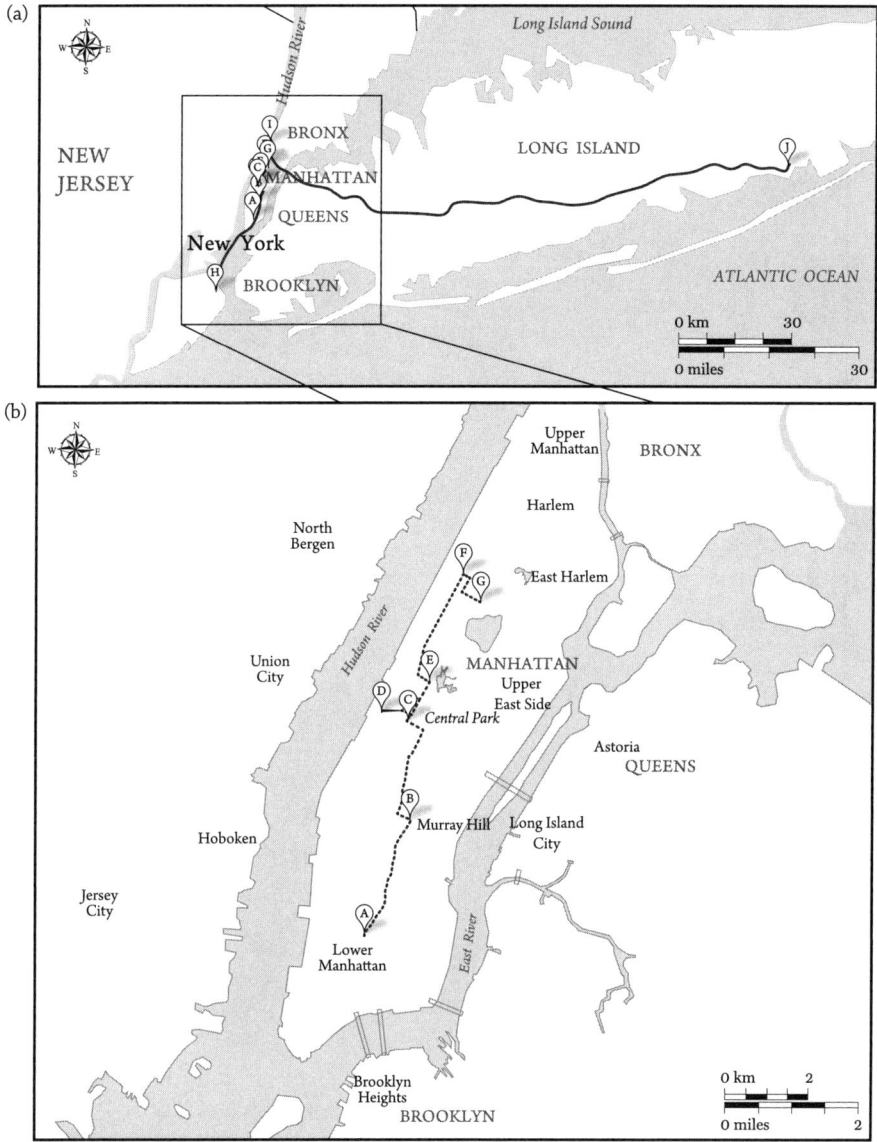

**Map 3** Dyar's life in New York City and environs (1866–1897): overall map, including Staten Island (H), where his uncle Josiah lived, and Bellport, Long Island (J), where Dyar vacationed with his family (*a*); walking tour (dotted line) of (A) Probable location of Dyar's collection until 1896 at 599 Broadway; (B) Early property of Dyar, Sr., and the first residence of Dyar, Jr., at 331 Fifth Avenue; (C) flat where Dyar and Zella lived in 1890 at 400 West Fifty-seventh Street; (D) Department of Pathology, Columbia University, where Dyar did his Ph.D. at 437 West Fifty-ninth Street; (E) flat where the Dyars lived while he attended Columbia (1893–1895) at 76 West Sixty-ninth Street; (F) Dyar flat at the time his daughter was born (1896–1897) at 243 West Ninety-ninth Street; (G) brownstone owned by Dyar and later exclusively by his sister Perle Knopf at 16 West Ninety-ninth Street (*b*).

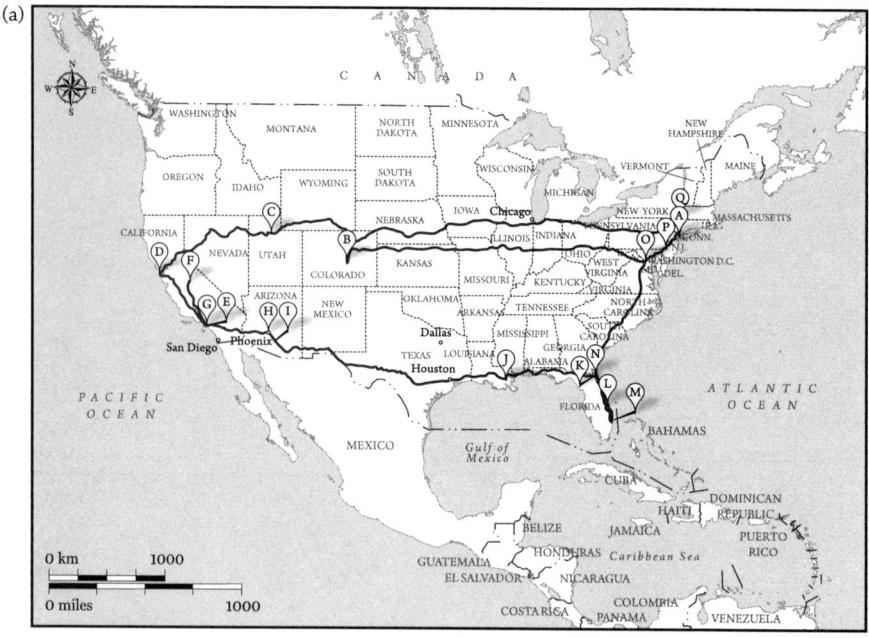

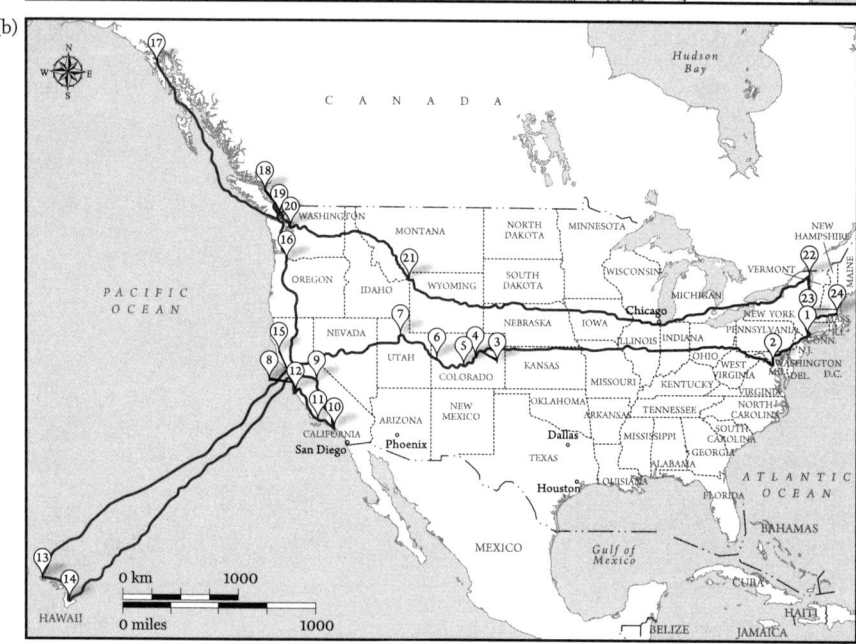

Map 4  Dyar's two longest trips (rail and ship routes only approximate): the 8-month collecting/honeymoon trip in 1889–90: (A) New York, NY; (B) Denver, CO; (C) Salt Lake City, UT; (D) San Francisco, CA; (E, G) Los Angeles Basin, CA; (F) Yosemite, CA; (H) Prescott Junction, AZ; (I) Phoenix, AZ; (J) New Orleans, LA; (K, N) Jacksonville, FL; (L) Melbourne, FL; (M) Palm Beach, FL; (O) Washington, D.C.; (P) Reading, PA; (Q) Rhinebeck, NY (*a*); the 15-month trip in 1891–92: (1) New York, NY; (2) Washington, D.C.; (3) Denver, CO; (4) Colorado Springs, CO; (5) Salida, CO; (6) Grand Junction, CO; (7) Great Salt Lake, UT; (8, 12) San Francisco, CA; (9) Yosemite, CA; (10) Los Angeles, CA; (11) Santa Barbara, CA; (13) Honolulu; (14) Big Island of Hawaii; (15) Watsonville, CA; (16) Portland, OR; (17) Sitka, AK; (18) Nanaimo, BC; (19) Victoria, BC; (20) Seattle, WA; (21) Yellowstone Park, MT; (22) Plattsburgh, NY; (23) Rhinebeck, NY; (24) Boston, MA (*b*).

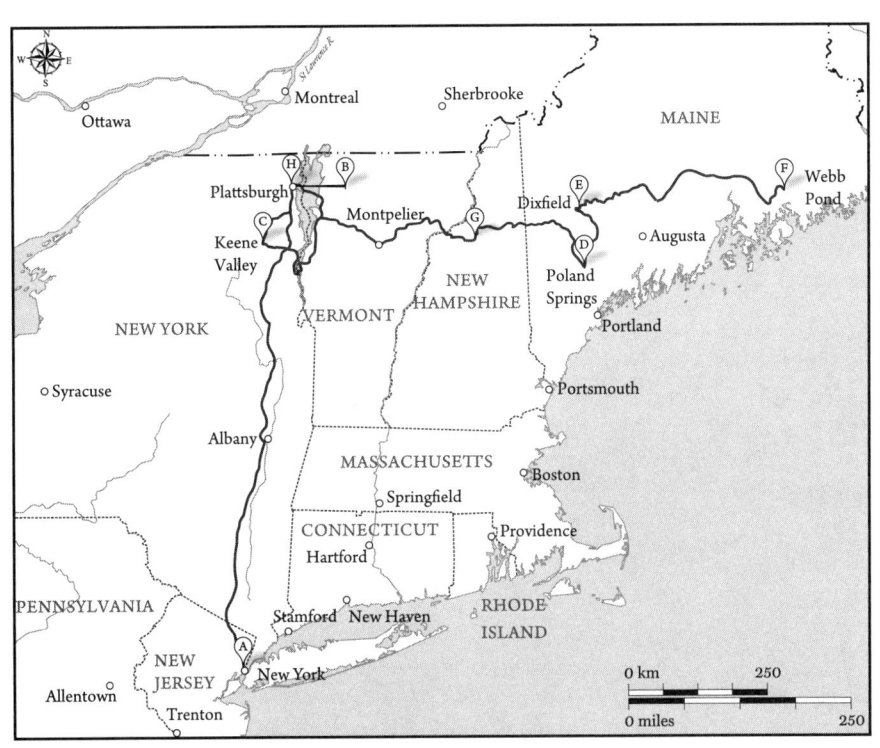

**Map 5** Dyar's trip to New England in 1894 from June to September, including two months of fieldwork in Keene Valley (C), New York: New York, NY (A); Plattsburgh, NY (B,H); Poland Springs, ME (D); Dixfield, ME (E); Webb Pond, ME (F); Faybyan, NH (G).

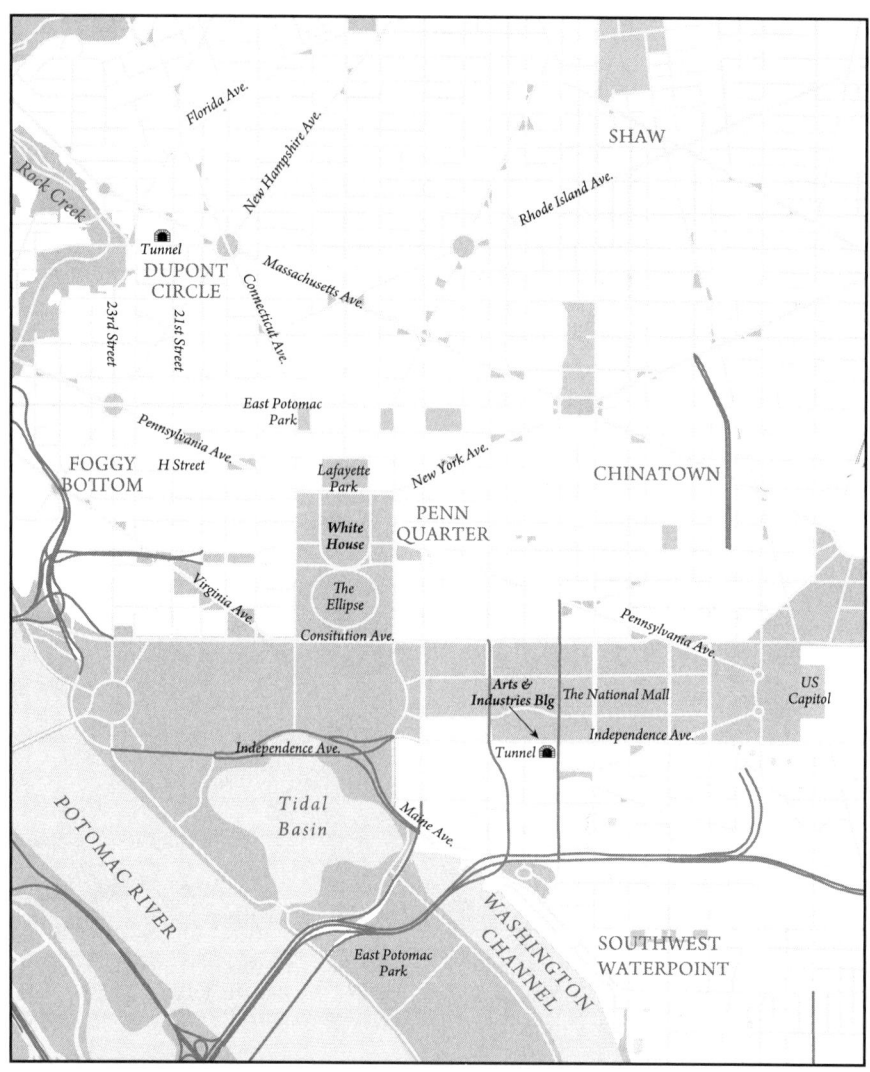

**Map 6** Location of Dyar's Dupont Circle tunnel (1512 Twenty-first Street NW) and B Street (Independence Ave.) tunnel (804 B Street SW). Soil maps of the District of Columbia show that Dyar had a maximum of 50 feet to reach bedrock in Dupont Circle and up to 200 feet near the National Mall: This, along with his property boundaries, helps explain why the Dupont Circle tunnels were horizontal and the B Street tunnels three deep.

# PART I

## Preparatory Stages

## CHAPTER 1

## The Dyars and the Hannums: 1805–1875

*I intended to send strokes of electricity in such a manner, as that the diverse distances of time, separating the diverse sparks, should represent the different letters of the alphabet and stops between the words H and C.*[1]

Harrison G. Dyar

Born to Jeremiah and Susanna Wild Dyar in Harvard, Worcester, Massachusetts, on March 1, 1805, Harrison Gray Dyar, Sr., a middle child of twelve siblings, grew up with "hard, dark Calvinism," the polar opposite of the modern spiritualism that he adopted later in life.[2] Schoolmates and contemporaries considered him a "real genius." He lived up to this characterization through engaging, by age twelve, in fundamental chemistry experiments and mastering, as an adult, "most of the principles of science" known at the time.

As a young teen Harrison lived in Concord, Massachusetts, with his brother Joseph, who was ten years his senior and who worked for the watchmaker Lemuel Curtis. Eventually, Harrison became the watchmaker's apprentice, and the brothers gained a reputation for being sharp dressers, "bright and intelligent," and "skillful in their trade and of faultless character." Young Dyar also pursued his interest in science, sometimes with dangerous results: He caused a chemical explosion in Curtis's shop and could have been struck by lightning, "if he had been holding the kite," when he recreated Ben Franklin's kite experiment.[3]

The brothers moved to Middlebury, Vermont, in 1822. Joseph opened a shop. Harrison attended Middlebury College and invented a new type of clock with a pendulum that moved "in a cycloidal arch, and perform[ed] long and short vibrations in equal times."[4] His skills as a clockmaker and his knowledge of chemistry enabled him to invent a telegraph in 1826. "I intended to send strokes of electricity in such a manner, as that the diverse distances of time, separating the diverse sparks, should represent the different letters of the alphabet and stops between the words H and C," Dyar explained. The letters were indicated by red spots on moist, blue

( 4 )  *Preparatory Stages*

Figure 1.1  Harrison Gray Dyar, Sr.

litmus paper formed of nitric acid created by each spark. Dyar used a pendulum, as he did with the clock he invented, to regulate the electricity.[5]

Dyar proposed a field test with a wire along the "Causeway" (today Lowell Road) in Concord, Massachusetts. When he brought his plans to potential investors, he was met with "laughter and ridicule." He thus conducted the experiment at a Long Island racetrack. The results were positive, and Dyar decided to attempt to transmit a message through a wire from New York to Philadelphia. The experiment was never completed because of lawsuits and fears of the new technology by the legislature, nor apparently were others Dyar wrote about in 1828. Disheartened, Dyar left the country in early 1831,[6] going to England, perhaps to Middlesex. This left Samuel F. B. Morse, who apparently learned of Dyar's invention from his brother-in-law and Dyar's legal counsel, Charles Walker, to develop the "electric telegraph" with Dyar's and others' research.[7]

The year he arrived in England Dyar began to apply for his first patent, a machine described in the *London and Edinburgh Philosophical Magazine and Journal of Science*, as being "for an improvement in tunneling, or method of executing subterraneous excavations."[8] Seven years later Dyar was still in England and applied with J. Hemming for a second patent, one for a process to improve "the manufacture of carbonate of soda" that was granted. Then, in 1839, Dyar, along with Charles Button, was granted a U.S. patent on the manufacture of white lead.[9]

Dyar moved to Paris by 1841, and during the 1840s, he amassed wealth from dye patents through an award of $300,000 from one of the Royal Societies of France.[10] He returned to live in the United States, probably near the end of the 1840s and became president of the New York and New Haven Railroad. He held this position on and off for the first half of the 1850s, though he was apparently unhappy, expressing, in May 1855, "his determination to retire from the unpleasant position" (see Figure 1.1 for a portrait of Dyar, Sr., from the next decade).[11]

**Figure 1.2** Parthenia Hannum (c. 1860) (*a*) and (*b*) Hannum with Nettie Colburn and President Lincoln.

Dyar continued to register patents in the United States, including one for the "time division telegraph" in 1857.[12] He also developed a steam-powered precursor to the automobile, building, after 1859, steam carriages contemporaneously with William T. James, Joseph Dixon, and Rufus Porter.[13]

In 1860 Dyar purchased the Manhattan property at 331 Fifth Avenue, across the street from the present day site of the Empire State Building (Map 3). He also owned other buildings on the same corner.[14] That same year, Eleonora Rosella Hannum was living with her parents and a couple of siblings in South Adams, Massachusetts. The future Mrs. Dyar's father was a paper mill operator. Her sister, Parthenia "Parnie" Robinson Hannum (1832–1922) (Figure 1.2a), was born on the same date as Eleonora but a decade earlier. She was a divorcee who had been making a name for herself since 1858 as the assistant of the medium Nettie Colburn Maynard (1841–1892), who rose in notoriety as modern spiritualism, which took off in the 1840s, grew in popularity.[15]

## THE HANNUMS AND PRESIDENT LINCOLN

Nettie Colburn would play a significant role in the Hannum family, expanding the horizons of the young Hannums' world. Indeed, according to Dyar family lore, a relative participated in séances with President Lincoln (Figure 1.2b).[16] That relative was Parnie. Colburn went to Washington to obtain a furlough for her injured brother in Alexandria, Virginia, and Parnie went with her. Mary Todd Lincoln heard about Colburn and Hannum and insisted both remain in the Washington area,[17] helping them get jobs at the Department of the Interior at Seventh and F Streets, sewing together little sacks containing seed corn or beans.[18] Thereafter they were involved in séances that Mrs. Lincoln attended.

Before one such séance with Mrs. Lincoln, which took place in Georgetown in February 1863 and in which Parnie participated, Colburn's spirit guide revealed, it is said, that the "long brave . . . Mr. Lincoln, would also be there!" Mrs. Lincoln had not mentioned that he would be attending, so "[t]he girls were surprised. . . . And so was Lincoln when he arrived, being told he was expected!"[19] Another visit with the Lincolns took place at the White House, purportedly during the Battle of Chancellorsville in May 1863. Parnie said Colburn, through her spirit guide Pinkie, reported the devastating losses on the front.[20]

Josiah Cushman Hannum (1832–1911) was in New York City around the time his sister Parnie was in Washington, D.C. He was a first lieutenant in 1862 and, in June 1863, was promoted to captain of the 28th Battery Light Artillery of the New York Volunteer Army, the battery that confiscated a stock of arms hidden by a group of rioters opposed to the draft into the Union Army: The New York Draft Riots.[21]

## MARRIAGE TO ELEONORA HANNUM, BIRTH OF DYAR, JR., AND DEATH OF DYAR, SR.

Little is known about how Dyar, Sr., met Eleonora. However, the interest he developed in spiritualism late in life—an interest that led him to conduct experiments in an "endeavor to find . . . material foundation" of the spirit world—is said to have brought him and Eleonora together.[22] She may have gone to New York, if she was in that city when they met, to visit her brother. Eleonora appears to have been close with her brother, as he, along with Parnie, was later involved in Dyar, Jr.'s life. Whatever the case may be, at the time of their marriage, Miss Hannum was a medium and she would later gain prominence using the name "ER Dyar," or "ER Dyar Clough." Despite the great disparity in age between Dyar, who was sixty, and Eleonora, who was twenty-two, the two were married on May 9, 1865, in New York City, only two months after Eleonora's mother died. Harrison G. Dyar, Jr., was born on Valentine's Day 1866 in New York City, precisely forty weeks and a day after the couple exchanged vows. His sister Nora Perle (later known as Perle Nora) was born two years later, also in New York.[23]

In 1869, the Dyars moved to Linwood Hill in the environs of Rhinebeck, New York, acquiring 178 acres on the Hudson River and Landsman's Creek near Rhinecliff, southwest and close to Rhinebeck (see Map 1b).[24]

Dyar, Sr., may have wanted to retire. In 1870, after all, he was listed as having no occupation, meaning that he was wealthy and needed no income. The value of his real estate, presumably at Linwood Hill, was $25,000. Another estate was valued at $10,000, perhaps in the town of Rhinebeck itself, as Dyar, Jr., would attend school there.[25]

Dyar, Sr., died on January 31, 1875, after a lengthy illness, two weeks before his son's ninth birthday. At the funeral Rev. Charles S. Harrower suggested that Dyar was not well known in Rhinebeck. Harrower, who got to know Dyar toward the end of his life, spoke about their discussions of science, spiritualism, and even some intelligent design, on which Dyar was purported to have said:

> It is incredible that the universe, which so answers to intelligence, should have come from any source less than an intelligent author. Intelligent and moral he must be; but what there may be, more and above these qualities, some mind besides man's, must be left to discover or to understand.[26]

## CHAPTER 2

# Dyar, Jr.

## Early Growth Stages and Development: 1875–1889

*Got 2nd good Green Sphinx at Mrs Asher's. Butterfly net made. Caught* Papilio turnus *and male* Melitaea pharos.[1]

Harrison G. Dyar, Jr., from his Blue Book, July 27, 1882

For young Harrison Dyar, Jr. Linwood Hill was more than just a Victorian home overlooking the Hudson River (see Figures 2.1, 2.2b, Map 1b), and situated at woods edge. He and his sister, Perle, roamed the forests, encountering such creatures as the "cats," their name for caterpillars, which came in all sizes, shapes, and textures, including woolly, brightly colored, or camouflaged (Plates 1, 5–7). Some even had hairy appendages on their backs (Plate 1g) or head ends that look like the faces of clowns.

At the time of his father's death in 1875, Harry, as he was known to family, lived in a stimulating home environment. Along with his mother, Eleonora (see Figure 2.4a, b), and Perle, there was homeopath Lucy Ann Hudson (1813–1898) as well as a domestic, a dollhouse maker, farm help, and a friendly group of neighbors.[2]

Eleonora was known in Rhinebeck as a wealthy widow[3] and was active in Rhinebeck affairs. This was illustrated by her role as the generous, gift-giving Mrs. Santa at a church Christmas festival in 1879 and, the following month, as an entertainer one evening at the Rhinebeck Free Academy, when she read Elizabeth Barrett Browning's "The Lay of the Brown Rosary."[4]

The widow Dyar, in 1876, began to use a small, blue, cloth-bound diary, known today as the "Blue Book," as an accounting book of expenditures: "Cash on hand" on January 1 was the only entry before she switched to larger, leather-bound books in 1877 (Plate 2a,b). Her son would write over her records during the next decades,

# EARLY GROWTH STAGES AND DEVELOPMENT: 1875–1889 (9)

**Figure 2.1** Dyar's childhood home at Linwood Hill above the Hudson River near Rhinebeck, New York, designed for Federal Vanderburgh by Alexander Jackson Davis.

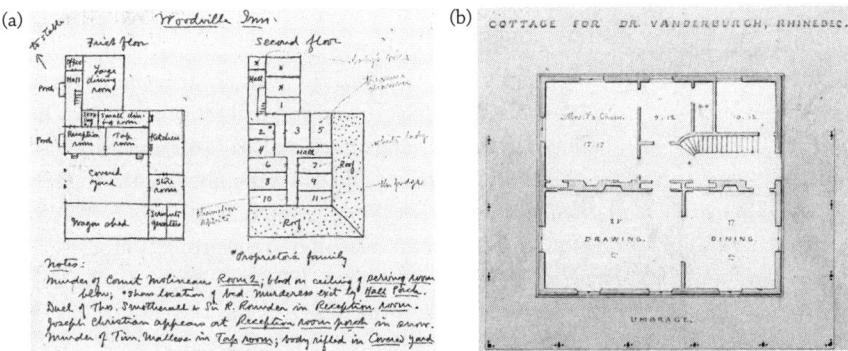

**Figure 2.2** Floor plan for Dyar's fictitious "Woodville Inn" (a) and (b) floor plan of the Linwood Hill cottage.

using the "Blue Book," the earliest surviving document from Dyar's childhood, and leather-bound books to record his scientific observations (see Plate 2b).[5]

To say that the Dyar children had an unusual childhood is an understatement. One key element in their upbringing appears to have been the involvement in the spiritualist movement by Eleonora and her sisters, Parnie Colburn (see Figure 1.2a) and Ruth Marble, who apparently made regular visits from White Plains, New York. They were very fond of Dyar, as is evident from their correspondence and moth finds they bestowed upon him (see Plate 4c,d). They formed, along with his mother, the doting "trio of femininity," as he referred to mother and two aunts of the protagonist, Reverend Mott, in Dyar's unpublished story written in 1915.[6] Mrs. Dyar likely held séances in Linwood Hill, Rhinebeck, and other places with her children in the 1870s, perhaps to contact their departed father.

Suggestive of Dyar's upbringing among spiritualists is a 1901 letter to his sister in which he mentions a story, or anecdote, as an "event." In another instance he wrote: "Mr. Banks' dreams came later."[7] The Dyar siblings continued to exchange anecdotes and ghost stories between 1901 and 1913, apparently recollecting relatives' or other members of the spiritualist community's stories, dreams, and séances or trances.[8] The stories from 1901 include "Ida Talks," "Various Things," "July 8th," "Mr. Banks Dreams," "The Three Green Sisters," "Robert Goes Exploring," and "Christmas of year 29."[9] There was some overlap among their characters. "Mr. Banks Dreams" includes music notation for "Sleep Baby Sleep" written with the story. It begins,

> Mr. Banks had a dream. He saw Ida standing by a heart-shaped lake in a pretty dell in woods.... The dream came again and again and this is what she sang. Mr. Banks now remembered well both words and tune. "Sleep, Lugula, Sleep. Queen Lethe the earth toad keep. While the darkened moon of the southern sky, Dream thee an endless lullaby, Sleep, Lugula, sleep."

In "The Three Green Sisters" there is a character named Sir Wilfred: The name Wilfred would become an important part of Dyar's life.[10]

Embedded with the ghost stories he sent to Perle in 1913, Dyar included instructions: "Read the enclosed lot and at your leisure and you will know all about a bully bunch of ghosts—all I believe that our friends have met so far. I don't say it's all, for that old inn is fairly stuffed with them, but all I can remember that we have met to date. This letter is marked KEEP and not 'Burn immediately.' . . ." The "old inn" referred to in the letter is the "Woodville Inn" featured in the group of stories. Dyar included a floor plan of the fictitious inn (see Figure 2.2a), noting on it where each ghost story was to have taken place. Perhaps the fictitious inn was loosely based on the Linwood Hill cottage (see floor plan in Figure 2.2b).[11]

"The Banshee" was also among the ghost stories from 1913 and begins with "Old man Cruikshank" who,

> runs down the road to take sanctuary in the little brick church, but Dame Cruikshank, who would follow him in nothing . . . runs the other way, and whether from a bad conscience or by accident she falls off the bridge into Mad River and was drowned. The Banshee keeps a-squealin' on the gable, so no one durst go near the place for weeks and weeks . . . 'tis a very necessary place for travelers, ye see, and I've not heard that many have been disturbed by it of late; but then 'tis not the right season of the year.[12]

During the same period, Dyar wrote to Perle about her reading more "anecdotes . . . [perhaps taking place in his Woodville Inn] before I come up [to New York for a visit], otherwise you wont be fully cognizant of all the gristly apparitions. . . . If they materialize all these people they'd have the place well peopled."[13] The last sentence suggests accounts of spirits materializing during séances much like his mother participated in as a medium in the 1880s.

∽

Along with Dyar's own relatives, Lucy Hudson, who was living with the Dyar family by 1875 and remained at least five years after, and her family appear to have had a significant influence on the early development of Dyar's interests. The Dyars knew her as "Aunt Lucy," although she was probably not a relation. That she was a homeopath suggests she was brought in to help the ailing Dyar, Sr. at the end of his life.[14] Hudson's relatives lived on the shores of Lake Champlain in Plattsburgh, New York (Map 1a). Rev. Henry James Hudson (1821–1901), a Unitarian minister in Chelsea, Massachusetts, in the 1850s and also a musician, adopted spiritualism, as had Dyar, Sr. He and his abolitionist wife, Hannah Elizabeth Blake, encouraged their two children, George Henry Hudson (1855–1934) (Plate 8b) and Charles, to pursue interests in natural history and music. George—who was eleven years Dyar Jr.'s senior and who, despite beginning his professional life as a musician, became a prominent paleontologist and professor[15]—pursued his interest in natural history with Dyar. This is evidenced by his being named in the initial entry in Dyar's catalogue in the mid-1880s, the recording of an unidentified moth or butterfly from August 1885 collected in Plattsburgh. Hudson, however, likely began encouraging Dyar to become a naturalist in the previous decade. Dyar's Lepidoptera collection, in any case, would be filled with hundreds of moth specimens taken by Hudson from electric lights.[16] Dyar, who was to become an accomplished classical pianist, perhaps received his musical education through George or other members of the Hudson family.[17]

It is unclear when Dyar actually began attending school. However, he graduated during the spring of 1882 from the DeGarmo Institute in Rhinebeck, founded in 1860 by Dr. James M. DeGarmo, a naturalist with a large butterfly collection and a skilled microscopist. From Dyar's perspective, Dr. DeGarmo's presence was perhaps the most important element of the school.[18]

## LIFE IN BOSTON: SCHOOL, MOTHER, AND FUTURE WIFE

Dyar lived with his mother in Boston in the 1880s. They moved to a house at 123 Mt. Pleasant Avenue in 1882 when first attending nearby Roxbury Latin School (see Figure 2.3a, Map 2a).

For the widowed Eleonora, Boston became a permanent residence. During the summer break, Dyar returned to Linwood Hill to collect and rear Lepidoptera. Perle seems to have migrated to Boston by late 1886. Notes in her brother's catalogue suggest she visited Linwood Hill, perhaps adding the latest specimens he raised in Boston to his growing Lepidoptera collection, which was housed there.[19]

During his first year at Roxbury, Dyar was ranked fourth of nineteen students. He received a 100% score in algebra, along with two other students. His poorest showing was in Latin and Greek. During his second year, Dyar rose in rank to second of twenty-one students, but still did poorly in Latin, scoring in the bottom

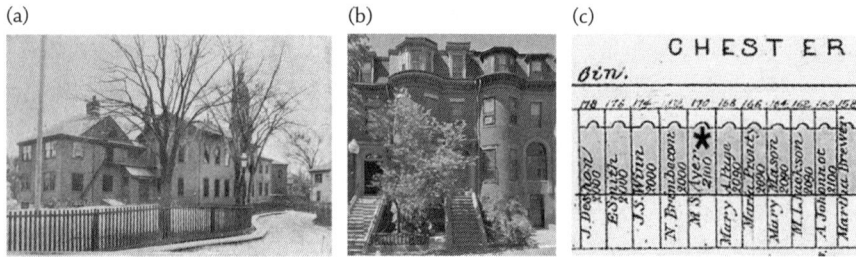

**Figure 2.3** Roxbury Latin School (a); West Chester Park brownstone (b); location of brownstone listed under E.G. Ayer at 170 West Chester Park Road marked on map (c).

third.[20] Ironically Dyar, who was less capable in Latin, used Latin to create new names for over 3,000 species and genera of Lepidoptera, 48 genera and 628 species of flies, and 28 species of wasps![21]

In his final year Dyar dropped to a rank of seventh[22] but passed the Harvard entrance exam along with all of his fellow graduating class, receiving college credit in Prescribed Mathematics, Prescribed Physics, and French, but he would attend the Massachusetts Institute of Technology (M.I.T.) instead.[23] While at M.I.T., Dyar was a sergeant in the corps of cadets during his freshman year (1885–1886) and lived on the second floor of the West Chester Park brownstone (see Figure 2.3b, c, Map 2b) (present day Massachusetts Avenue) rented by his mother, who lived below.[24]

In spring 1889 Dyar completed a bachelor's degree in chemistry, following in his father's footsteps. During his senior year, he wrote a thesis called "Investigation of a Proposed Synthesis of Tartaric Acid from Butyric Acid" and took a course relating to carbon chemistry from Thomas M. Drown (perhaps his thesis advisor). Another one of his professors was geologist William Otis Crosby, whom Dyar and other students assisted in an 1889 publication on felsite and conglomerate in the Boston Basin.[25]

✧

Dyar's mother may have moved to Boston to keep close watch over him but she also served as a medium (see Figure 2.4a, b). She had formed connections with the Working Union of Progressive Spiritualists (WUPS) and its founder Marcellus S. Ayer (Figure 2.4c), who had once lived with younger brother Eugene G. Ayer at the West Chester Park brownstone she rented (Figure 2.3b,c). Ayer's wealth enabled him to finance the construction of the First Spiritual Temple of Boston, completed in 1885, at the corner of Exeter and Newbury Streets in the Back Bay.[26] By that time "Mrs. E.R. Dyar," as she was known in spiritualist circles, was a trance and physical medium for at least a year within the WUPS, which met on Tremont Street, near Ayer's home in West Chester Park.[27]

People with Tremont Street connections would be significant in the lives of Mrs. Dyar and Harrison. A few blocks away on Tremont Street, between their West Chester Park home and the WUPS meeting place, was the residence of Dr. Jacob N.M. Clough (1828–1907). Eleonora married the doctor on May 22, 1887. She

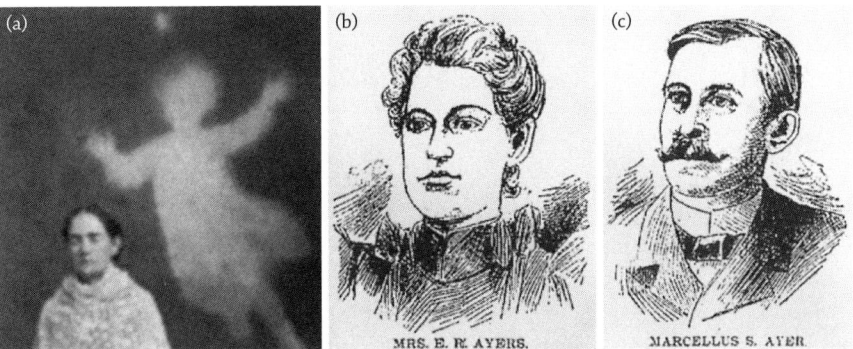

**Figure 2.4** Mrs. Eleonora Rosella Dyar with her spirit guide Christal (*a*); Mrs. Dyar (note "Ayers" misspelling) (*b*); Mr. Marcellus S. Ayer, founder of First Spiritual Temple in Boston (*c*).

continued her involvement with the First Spiritual Temple and the newlyweds lived at the West Chester Park address.[28] Several months later Mrs. E.R. Dyar Clough was referred to as "the priestess" of the group, now named Spiritual Fraternity, in a newspaper article about a temple service celebrating the eighty-fifth birthday of recently departed Allen Putnam. The involvement of Eleonora as a medium would continue only a short time longer: On May 18, 1888, she died at age forty-five. Mrs. Dyar's death may have been expected, as according to the death record she suffered from "valvular disease of the heart."[29]

After a funeral service at the West Chester Park home on May 22, Mrs. Dyar was buried in Rhinebeck next to her first husband.[30] A six-sided memorial star dedicated to her at the First Spiritual Temple read: "Mrs. E.R. Dyar Clough.... Her work was done well," along with her birth and death dates. Five years later Dyar Clough is reported to have materialized, along with President Lincoln, during an 1891 séance given by Nettie Colburn Maynard, attended by Eleonora's sisters Parnie and Ruth.[31]

During Mrs. Dyar's trance discourse in 1885, she spoke of death being a celebration of the spirit. Perhaps this explains why Dyar, Jr. attended to his usual activities between his mother's death and funeral. He documented emergences of 70 moths and butterflies he had found around the Hudson Valley during the previous summer and had spent the winter as cocoons or chrysalids. Among the moths were several limacodids including hag moths (*Phobetron pithecium*) (see Plate 1g). Their caterpillars were Dyar favorites because of the special furry, armlike structures on their backs. We often think of "hag" in rather derogatory terms; however, an alternate meaning may have seemed appropriate to Dyar now that his mother had passed on: "a sorceress, enchantress, or wizard." Following his mother's burial in Rhinebeck, the orphaned son made a quick return to Boston and transported his cocoons and chrysalids back to Linwood Hill.[32]

As a young adult, the blue-eyed and thin-framed Dyar reached 5'10." He had fine features and dark hair, and, by the age of twenty-seven, had a mustache and patch of hair below the lip, along with a receding hairline (see Figures 5.1b, 5.3a).[33]

Zella M. Peabody (1869–1938) was the only child of Philo Peabody (1839–1879), who worked as a farmer, straw worker, carpenter, and joiner, and Harriet M. Holland (1842–1922), a homemaker from Canton, Maine.[34] How Zella, a thin adult, perhaps as tall as 5'10," with dark hair[35](Plate 10), met Harrison G. Dyar remains obscure; they most likely met in Boston. Not only did their lives in the city overlap for five years, but the First Spiritual Temple, M.I.T., and the Boston Society of Natural History in the Back Bay (see Map 2a) were all very close to Dyar's West Chester Park address and the Everett School attended by Zella (see Map 2b). Further, her mother had "a wide experience in things spiritual, and she was fond of investigating each new idea ... as it appeared,"[36] so it is possible she became involved with the Boston spiritualist community.

In 1887 Zella and her mother moved to southern California. Perhaps coincidentally, the following year a girl named Susie began sending Dyar specimens of Lepidoptera from the Los Angeles Basin (Plate 4c,d). These specimens were recorded in Dyar's catalogue along with various eastern butterflies from Arthur Hudson, perhaps a cousin of the Plattsburgh family. February 1889 brought Arthur and Susie together to collect at electric lights in Pasadena.[37] It is plausible both knew Zella because in July 1888 she also sent specimens to Dyar from Pasadena, including those she reared. Zella may have also been there with Perle Dyar, perhaps meeting her earlier in Boston. Perle's fiancée, Dr. Siegmund (later "Sigard") Adolphus Knopf (1857–1940), was also living in Los Angeles at the time.[38]

Whether or not Dyar proposed marriage to Zella Peabody in 1888 remains an open question, but Zella's summer collecting suggests she was prodding him. Zella raised a total of 39 butterflies and moths from caterpillars that she found near Pasadena over July and August and sent to Dyar. Clearly Zella knew the way to her future husband's heart. One of her first captures was the day-flying California

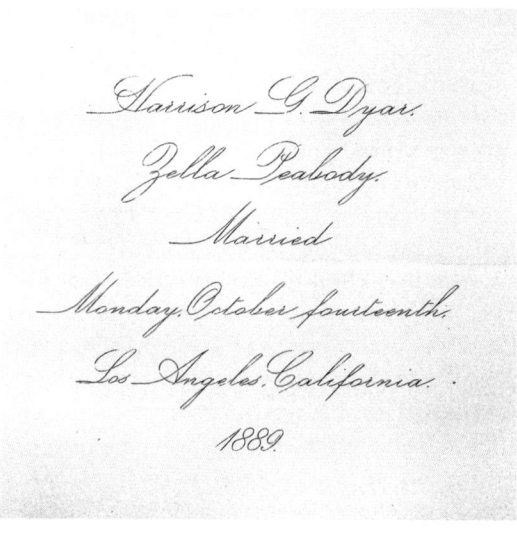

**Figure 2.5** Wedding announcement for Harrison G. Dyar and Zella Peabody.

oak moth, *Phryganidia californica* (Dioptinae) at Eagle Rock in the South Pasadena area.[39] The genus "*Phryganidia*," which sounds like "Frick'n Idiot," conjures up folks cursing these caterpillars. This would be appropriate because when abundant they severely denude the live oak trees in coastal parts of the Golden State. However, the name came from Alpheus Packard, who saw a superficial resemblance between the moth and the caddisfly genus *Phryganea*.[40]

⁂

Zella M. Peabody married Harrison G. Dyar on October 14, 1889, in Los Angeles, roughly a year after her collecting on his behalf (Figure 2.5). Rev. Conger of the First Universalist Church officiated. According to the *Pasadena Daily News*: "Mr. Dyar is a wealthy young scientist who has been traveling for several months on this coast. The bride came from Boston with her mother about two years [earlier] and made her home in Pasadena for some time, where she is known as an accomplished lady and a fine pianist." Perle married Dr. S. Adolphus Knopf in "an impressive ceremony" five days later on October 19 in Los Angeles.[41]

The Dyars stayed in Los Angeles at least through the end of October. All the while Harrison checked electric lights for moths before they went off on their extended honeymoon: a bicoastal collecting trip.[42]

CHAPTER 3

## Collecting and Rearing of Lepidoptera, and Dyar's Law: 1882–1891

The genius of H.G. Dyar, Jr., is perhaps best reflected in his work on the larval stage of insects, the hallmark of his career. The skills he honed rearing caterpillars and writing detailed descriptions of each larval instar (from 4 to 10) as a young naturalist would enable him to develop his "Law of Geometric Growth" at age twenty-four. Dyar's earliest known rearing and collecting records were written in pencil in the Blue Book (Plate 2a,b) when he was sixteen. In the middle of the book he wrote the title, "Diary of Lepidoptera (June 17–October 6) 1882" (Figure 3.1). The level of expertise displayed in his first records from 1882 (Plate 3), as well as his reference in early July to caterpillars as "described in other book," suggest his entomological explorations began much earlier.[1]

Dyar's penmanship and use of such shorthand as "babe chrysalis" for a newly formed pupa, "cats" for caterpillars, and "out" for adult moth and butterfly emergences from cocoons or chrysalids—the last two, notation forms that were used in later notebooks—show his youth. However, his use of Latin names for many of Lepidoptera show sophistication. Dominant among the first entries were notes on the metamorphosis of the red admiral butterfly, for example, "*Cynthia atalanta* made chrysalis" (Figure 3.1). Similar notes followed for common butterflies and moths, among them mourning cloaks, monarchs, black swallowtails, hornworms, and such giant silk moths as io and luna. Caterpillars of slug caterpillars (limacodids) (Plate 3d) and prominents (notodontids) followed later that summer.[2]

Dyar pursued fieldwork during the next few summers, mostly near Linwood Hill or Rhinebeck, though he mentions Syracuse (Map 1a) as early as July 1882.[3] Family members, neighbors, and friends, among them his sister and Norman Asher and Grace Woods, accompanied him on these excursions.[4] Indeed, Dyar's collecting seems to have been a communal affair throughout his youth.

Mrs. Asher gave him, in late July, a "2nd good Green Sphinx," the day his "Butterfly net [was] made" and with which he captured "*Papilio turnus* and male

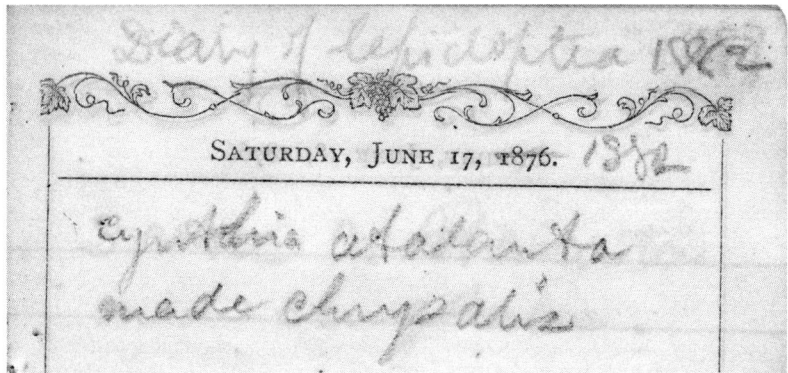

**Figure 3.1** Earliest entry in Dyar's "Blue Book" for the caterpillar of the red admiral butterfly, *Cynthia* [now *Vanessa*] *atalanta*, which just became a chrysalis on June 17, 1882 (note: the heading "Diary of lepidoptera 1882" above and how the year 1876 has been crossed out).

Melitaea tharos" (now *Papilio glaucus* and *Phyciodes tharos*, the eastern tiger swallowtail and pearl crescent)" (see Plate 3a).[5] Dyar would also get caterpillars and moths from his sister, his aunts, Lucy Ann Hudson, and from his mother, who found, on August 3, 1882, "a caterpillar of a lappet moth ... Gastropacha Americana," the only specimen she is known to have given him (see Plates 3b, 7a). That same day, he arose at 6 AM, released several black swallowtails with damaged wings, and went collecting on the other side of the Hudson River.[6]

Even relatives who lived relatively far away became involved. On August 13, 1882, his "Uncle Joe" (Josiah Cushman Hannum) came from Staten Island. The two went out collecting on Post Road near Rhinebeck, a site Dyar would refer to as "Cresphontes Hollow." Uncle Joe captured a giant swallowtail butterfly (*Papilio cresphontes*), and Dyar located io moth and regal moth caterpillars.[7] Hannum probably accompanied his nephew for a visit to Boston the next day, where Dyar found "an awful lot of cocoons & batches of eggs of O. [*Orgyia*] *Leucostigma*" (Plate 3c).[8]

After returning home, Dyar continued to collect, capturing regal fritillaries (*Speyeria idalia*) in Ulster County (now rare in the Northeastern United States) and reporting a "Limacode," known today as the smaller Parasa (*Parasa chloris*) (see Plates 1e and 2e,f), on wild cherry, his first slug caterpillar on record (see Plate 3d). The index of the first two Blue Book volumes listed the caterpillars of 157 species found between 1882 and 1896. Seven out of the first twelve were slug caterpillars, which had become his "cherished treasures."[9]

## COLLECTING AND THE CATALOGUE

Over the coming decades Dyar went on to collect thousands of moth specimens at sugar baits and electric lights, which became commonplace in the United States after the development of commercially viable bulbs around 1880.[10]

Among the first moths in his catalogue, which he began in 1885 (Plate 4a), was a cutworm moth (*Hadena devastatrix*) from an August day in Plattsburgh, New York, the home of the Hudson family (Plate 4b). His early records also show the communal nature of his efforts, as Louisa Hoff (1868–1945), a domestic and friend of the Dyar family, contributed an underwing moth (*Catocala ultronia*).[11]

In 1885 Dyar was gathering live specimens from electric lights at night or dead ones trapped inside their covers by day. Concurrently he collected and reared in the area surrounding his Linwood Hill home at, for instance, Esopus Creek, Hussey Mt. Spring, and Linwood Cove. In September his collecting grounds included Livingstons, a stretch of land between Rhinebeck and Linwood Hill and Red Hook (Fraleigh's Hook) in Dutchess Co., and Ulster Co. across the Hudson River.[12] After he finished his fieldwork, Dyar brought cocoons from caterpillars raised during the previous summer to Boston and recorded adult emergences in May 1886.[13]

From mid-June until late August 1886, Dyar visited the White Mountains of New Hampshire, collecting at Campton Village, North Woodstock, Bethlehem, and Jefferson (Highlands) (Map 1a). At the end of the visit Dyar wrote a brief, but unpublished, life-history description of a pink-striped oakworm (*Anisota virginiensis*) from Jefferson.[14] Similar accounts by Dyar would soon appear in scientific journals.

The next summer the now twenty-one-year-old stayed closer to home, collecting in Linwood Hill, Rhinebeck and other localities nearby. He received a major acquisition for his collection: George Hudson (Plate 8b) gave him hundreds of moth specimens collected at lights or their globe fixtures from around Plattsburgh, New York. Dyar also received several butterflies from Hudson collected in Switzerland. It appears Hudson had initially been more interested in beetles and shells, but as a professor he switched to Lepidoptera and had perhaps the largest collection of species (800+) from the local region in the 1880s.[15]

Harrison wasn't the only family member pursuing Lepidoptera. During the winter of 1888, Perle collected in Florida[16] prior to the death of their mother. Among the first of these Florida butterflies recorded in Dyar's catalogue was a butterfly that "Perle got" at Altamont Spring on February 25 and two others from Rockledge on April 5 (*Ascia monuste*). There were other butterflies (and some moths) nabbed in late February from Altamont and through April 25 from Merritt's Island, Rockledge, and St. Augustine.[17] In the end of May and June, Dyar was at Saratoga Springs, Plattsburgh, Raven Pass, and Elizabethtown, New York. After collecting in Ulster Co. at the end of July, Dyar frequented the electric light globes in Poughkeepsie, emptying them during the day for his bounty of dead moth specimens.[18]

Although there are scant reports of Lepidoptera found by Dyar in Boston in the 1880s, he did record findings at Dartmouth Street, Columbus Avenue, Franklin Park, and the Arnold Arboretum near his home at West Chester Park at the time of his college graduation in May 1889. On his summer return to Linwood Hill, Dyar collected at his usual haunts. Perle also found a moth for her brother in the library window and the visiting Aunt Parnie captured a tiger moth *Arctia* (now *Grammia*) *virgo* (Plate 4c,d).[19] That summer Dyar went to Poughkeepsie once again to gather moths from the light globes.[20]

## EARLY MENTORS AND PUBLICATIONS

As an upperclassman at M.I.T., Dyar was in contact with lepidopterists whose reputations extended beyond the Hudson Valley. Among them were Ferdinand Heinrich Herman Strecker (1836–1901) (see Plate 8g,h), whom Dyar wrote for information contained in one of Strecker's books missing from the Boston Society of Natural History Library, Joseph Albert Lintner (1822–1898) (see Plate 8c), and Charles H. Fernald (1838–1921) (see Plate 8e).

Strecker, who sold and traded Lepidoptera, had among the largest collections of Lepidoptera in North America at his home in Reading, Pennsylvania. His book, "Lepidoptera, Rhopaloceres [butterflies] and Heteroceres [moths], Indigenous and Exotic ..." was the sort that would capture the imagination of a serious student such as Dyar.[21] Profusely illustrated by the author, it included hand-colored plates of butterflies and moths. Dyar was able to make personal visits to Lintner, the New York State Entomologist who curated the state's insect collection in Albany, New York, a convenient distance from Rhinebeck. Both Lintner and Strecker provided the neophyte taxonomist with access to scientific literature and assisted him in putting names on the various moths he found or reared from caterpillars; however, both would soon be the beneficiaries of Dyar's talents.[22] Fernald, a professor at University of Massachusetts, identified "Microlepidoptera" for Dyar.[23]

In his second letter to Strecker, Dyar gave an account of his Lepidoptera interests, showing he had field experience with a notodontid, or prominent, caterpillar of *Datana robusta*, found in Strecker's 1872. His collection, Dyar noted, was "not over 200 species," but he hoped to "enlarge it," observing he could prepare specimens "as well as most if I take the pains." Living in Boston with his collection in Rhinebeck, Dyar wrote, "I am away from it now, and cant [sic] tell just what I have, but am afraid there is nothing you want, though I wish there were."[24] Dyar also drew attention to his inexperience, noting that he could identify caterpillars by appearance and the food plants where he found them, but admitting he had some problems separating caterpillars disassociated from their habitats in his rearing box.

Dyar's increased acumen was soon evident. In November 1888 he challenged Strecker's contention that several pairs of *Datana* were the same species: "*Ministra* [have] narrow yellow stripes and black head, *Major* (at maturity) [have] white or yellow spots and dark red head. I know of no two *Datana* larvae that differ more." Dyar also noted biological differences between the caterpillars: "*contractas*, especially when young, spin a sort of slight web over the leaves and place their heads around among each other's in a very peculiar way, such as I have never observed in *integerrima*." Dyar's observation of physical and behavioral traits of *Datana* suggested, correctly, that there were four species based on the caterpillars, even though Strecker suggested two from the moth stage (see another, Drexel's datana, in Plate 6e,f).[25]

Given his level of knowledge, it is not surprising Dyar would soon publish. His first scientific paper in December 1888 was titled "Partial Preparatory Stages of *Dryopteryx* [*Oreta*] *rosea* Wlk" in *Entomologica Americana*, the journal of the Brooklyn Entomological Society. The article described an odd moth larva with a long tail commonly known today as the rose hooktip (Drepaninae). Dyar produced one more in the same journal, a notodontid (*Oligocentra lignicolor*) (see Plate 6a-b), and four other brief moth publications in *The Canadian Entomologist* in 1889, all of them descriptions of complete larval stages of different species with very distinct caterpillars. The first was a notodontid (*Datana major*), followed by a crown slug caterpillar (*Isa textula*) (see Plate 1b), a noctuid (*Euplexia lucipara*), and a thyatirine (*Thyatira pudens*).[26]

Dyar's contacts with entomologists grew to include two of the principal federal entomologists of the day: Charles Valentine Riley (1843–1895) (Plate 8a), the Chief Entomologist of the United States Department of Agriculture, and Riley's assistant, John Bernhardt Smith (1858–1912) (see Figure 10.1b). In February 1889 Dyar wrote Riley, also the editor of *Insect Life*, that he had found the imported elm leaf beetle abundant in Poughkeepsie, injuring the city's elm trees, but less attracted to "electric light globes" than other beetle species, including over 6,000 of one in particular he was unable to identify. Riley published Dyar's letter and his reply in *Insect Life*, indicating the abundant species was a ground beetle (*Harpalus pensylvanicus*).[27]

The previous month Dyar began corresponding with John B. Smith, who would soon publish his "List of Lepidoptera of Boreal [North] America." This began a relationship that developed into one of the more intriguing rivalries among entomologists in the early twentieth century. Smith, in a reply discussing the identification of the enigmatic *Nola* (Nolidae) moths, noted, "At the present day it is difficult to name a species satisfactorily, and . . . in our own collection the species are mostly unnamed. . . . [I]t would require more time and material to do so than I have at command."[28] Dyar would go on to describe over sixty new species and three genera in Nolidae (1898–1923).

A burgeoning amount of taxonomic work and life-history descriptions of moths followed Dyar's return from his first lengthy trip to the West in 1890, particularly with datanas (Notodontidae). Dyar wrote Strecker about *Datana floridana* and *D. palmii*. His letter suggests that a closer bond had developed between the two correspondents: "I enclose a stamp that I think your son has not got in his collection and would like to have perhaps."[29] Dyar also remained in touch with Lintner, identifying, by June 1890, datanas for him.[30]

To complete a work on datanas, Dyar included a species Riley found in California and used, with his permission, Riley's manuscript name "*Datana californica*." In May, Riley suggested Dyar publish an article on these moths and others in *Insect Life*, writing, "Why not . . . send me your notes on *Datana* . . .? I should like to have something from you from time to time."[31] Dyar, however, had committed the work to *Entomologica Americana*, edited by J.B. Smith, who first suggested Dyar write the treatise. The article included descriptions of the adult stage of all known *Datana* in North America, including a key to aid identification. Ironically, *D. californica*, discovered by Riley, became the first of approximately 3,000 species that Dyar would name.[32]

In the postscript of his May letter, Riley also wrote Dyar about two articles "now in printers' hands," enclosing "duplicate type-written copies" to be returned "with corrections [and] I will see that they are made in the proof." These *Insect Life* articles were life histories: *Syntomeida epilais* Walker on oleander and *Nola sexmaculata* Grote on witch hazel.[33] Dyar continued corresponding with Riley and his editorial assistant, Leland Ossian Howard (1857–1950), about a lost manuscript,[34] which Dyar soon rewrote. It included descriptions of caterpillars and their food plants from his visits to Florida and Arizona.[35] Howard had replaced Smith as Riley's assistant the previous year and would play an important part in Dyar's life.

In spite of Riley's encouragement, Dyar published only five brief papers in *Insect Life*. Among the longest was about moths in Poughkeepsie's electric lamps in 1890.[36] Dyar may have refrained from submitting articles to Riley's journal because it was more for lay audiences, or he may have lost patience with Riley due to errors in the Poughkeepsie paper: Riley were never published the *errata* sent by Dyar. However, it is plausible that Riley can't be faulted for the initial transcription errors of moth names, as evidenced by Dyar's handwritten manuscript submission on the larval stages of *Nola* and other moths, which are difficult to decipher.[37]

While entomology journals were accepting Dyar's life-history papers, he desired to learn more basic external anatomy and went to Lintner for instruction. Dyar requested Lintner to critique his knowledge: "I find that I am a little confused as to the names of the parts of the head of a larva and I would like to ask you to set me straight. I make here a rough drawing of a head and would like you to tell me if I have the names right. . . . I shall be very glad to know if I have these right and if not to be corrected." Dyar confused the frons with the clypeus, the clypeus with the labrum, and did not know what to call the maxillary palpi. He referred to them as "a pair of jointed appendages like the antennae just below the mouth which I suppose to be the palpi."[38]

In the same letter, Dyar noted, "I should like to go to Albany some time next week to see you and consult some of your books." He went on to ask, "Shall I bring you some larvae of *Datana perspicua* to raise and if so would you like them young or nearly mature? They are quite common here now and I believe you said they were rare in Albany. I suppose the food plant—sumach—grows where you could get it."[39]

His travels to the West, beginning in 1889, familiarized Dyar with that regions moth fauna, enabling him to differentiate adult moths of eastern and western faunas as well as such caterpillars as the eastern (*cinerea*) and western (*cinereoides*) forms of the puss moth *Cerura* (now *Furcula*) *cinerea* (Notodontidae). Its long, spindly "tail" or caudal filament is actually a modified proleg on the end of the abdomen (see Plate 6g for a related species with a "tail"). Dyar observed in an 1891 article for *Psyche*: "The larvae of the two forms are much alike . . . greater prominence of crimson in *cinereoides* and its somewhat shorter caudal filaments."[40]

Dyar also depended on large private Lepidoptera collections and their owners to put accurate names on his reared specimens. One such collector was Henry Edwards (1827–1891)—an actor, the editor of the Lepidoptera journal *Papilio*, and the author of an 1889 bibliographic catalogue of reared North American Lepidoptera[41]—who was the de facto expert of the western fauna. Edwards had collected in California for a number of years before moving to New York in the late 1870s.[42] Thus, between his two big field trips to the West in 1889 and 1891–92, Dyar sought Edwards out, writing to him. Edwards answered Dyar in early December 1890:

> My dear Sir: I delayed replying . . . because I wished very much to be able to say when I could . . . myself [have] the pleasure of seeing you. . . . But I want to see you very much. How would Saturday morning suit you? I must leave home at ½ past 12—but I am free up to that time from 10 o'clock . . . as 2 hours at least. Let me know if you can bring *Ellida gelida* [a notodont] with you. I am greatly interested in this species. This is about the only one or almost the only one of the Bombycidae that I do not know.[43]

Edwards died seven months later, and Dyar wrote Lintner: "I was sorry to hear of the death of Mr. Henry Edwards in New York on the 9th inst. [June 9, 1891]. Have you heard what disposition is to be made of his collection? If it is to be sold I should be glad to know of it." To Dyar the fate of the Edwards collection was of practical concern. Left untended for months or years, insect specimens can turn to dust from museum pests such as carpet beetles. Large private collections often went to museums, sometimes located in other cities. The Edwards collection eventually went to the American Museum of Natural History in New York, though Dyar's letter suggests he may have been interested in purchasing it for his own.[44]

## DYAR'S LAW OF GEOMETRIC GROWTH

"The Number of Molts of Lepidopterous Larvae" is among the best known of Dyar's scientific works. It was published in *Psyche*, the journal of the Cambridge Entomological Society, in late 1890,[45] early in his career. The article presents what we now know as Dyar's Law (or Rule) of Geometric Growth, which assumes that geometric growth occurs between successive stages (instars). In practical terms, the size ratio of successive instars (a constant), in this case derived from measuring the width of the head, can be used to predict the number of instars, or the head width of earlier or later instars. The work that would lead to the formulation of this law began in December 1889, when Dyar was rearing caterpillars during his honeymoon with Zella in Palm Beach, Florida. He continued working on the project in Dutchess and Ulster Counties, New York through the measurement of a fifth instar of *Heterocampa unicolor* on September 21, 1890.[46]

Dyar was resolved to take the guesswork out of determining how many instars or molts there were (note: it takes four molts for a caterpillar to have five instars). In his words, what led to the development of this technique came from "considerable confusion" over the "number of molts of certain species" of Lepidoptera. For example, Dyar noted that in earlier *Psyche* articles, the number of molts presented by William Henry Edwards for the promethea moth (*Callosamia promethea*) (Plate 7b) did not agree with those of Lintner (3 vs. 4 molts) and Anna Katherina Dimmock showed variable results in tussock and giant silk moths.[47] Dyar's background in chemistry and math seems to have led him to seek the level of precision of those fields in the realm of biology. He chose the width of the head to calculate growth ratios because it was "not subject to growth during the stage, and ... [was] the most convenient measurement to take." In other words, to make meaningful growth measurements for each stage between molts, the hard, chitinized head is fixed whereas the body of a caterpillar expands and contracts too much during feeding and fasting. Dyar tested his method on twenty-eight species of Lepidoptera—among them the definite tussock moth (*Orgyia definita*) (Plate 7e), the famous woolly bear (*Pyrrharctia isabella*), the giant swallowtail butterfly (*Papilio cresphontes*), and the fall webworm (*Hyphantria cunea*; see Figure 3.2).[48] He arranged the species in tabular form by the number of instars they had, from 4 to 10. "I give ... calculated widths of head under each species, with the ratio, followed by those ... actually found," Dyar wrote. "All measurements ... may be considered accurate to within .1 mm.... [I]f two sets of observations show a different number of stages for the same insect but each follows its own progression ... [we] may conclude that this variation is actual; but if either set shows a lack of regular progression that one we must regard with suspicion."[49]

Dyar's approach was not completely original. William Keith Brooks (1848–1908) developed a similar system a few years earlier, but used body-length ratios that matched to identify crustacean species. Length works for finding crustacean growth factors because their exoskeleton is more fixed than in insect larvae. It has been suggested that this be termed the Brooks-Dyar Rule. However, each rule can remain separate because the goal and approach, and organisms are different. For example Dyar's head-width ratios determine the number of instars rather than species identity. In 1895 Dyar attempted a mixed approach in limacodids to predict instar length by using growth factors derived from head widths. These results

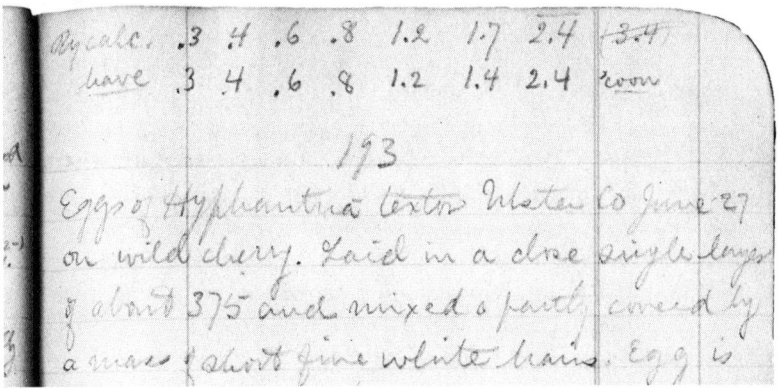

**Figure 3.2** Dyar's measurements of the head widths of fall webworm caterpillars (*Hyphantria cunea* ["textor" above] from June 1890 was one of the trials he used to develop his Law of Geometric Growth. The estimates ("By calc") are on the top row and the actual measurements ("have") are directly below.

were not published perhaps because limacodid instar lengths were too variable to be predicted.[50]

Dyar's results in 1890 showed the caterpillars grew in a geometric progression by a factor between 1.3 and 1.7 to the next stage, or decrease by .6 to .8 to the prior stage. He did not explicitly state how he calculated the growth ratio, his "calculated widths," but sought "corroborative observations" before expressing confidence in the "number of molts of any species." Thus, he suggested "all who hereafter describe larval stages give the width of the head for each stage."[51]

A number of problems remained in the measurements Dyar reported, as indicated in his footnotes. Some did not fit the geometric progression, whereas others were missing stages or were distorted in shape. Other inconsistencies, ones he did not note, can be found. For example, how the ratios were derived. Researchers have assumed these ratios resulted from dividing the head width of the penultimate instar over the ultimate, or last instar, but Dyar did not state that. In fact, only three calculations of the twenty-eight replicate the ratio of the last two stages rounded to .01. How, then, did Dyar calculate the other twenty-five? In at least ten instances, he appears to have determined the growth factor by averaging the last two or three ratios. In three other instances Dyar's growth ratio for a caterpillar matches those averaging all of the ratios, from first to last, and in one case they match the last four or five ratios. In eight instances the numbers do not match Dyar's calculation, perhaps due to rounding error: They were off by .01 or .02. Alternatively, particularly for problematic examples such as those missing data, Dyar may have used averages of ratios from estimated values for the head widths. For example, Dyar's published growth ratio of .64 for *Heterocampa unicolor* is arrived at by averaging ratios for calculated head widths of .63, .62, .65 and .65, rather than the actual ratios of .55, .65, .65 and .68 (average = .60) (see *Heterocampa guttivitta* larva, Plate 6c).

Dyar likely excluded the ratio of the first over the second instar in many instances because the growth was less dramatic than between the remaining instars. In other words, these did not fit his model for a geometric progression. However, in a few instances, Dyar appears to have averaged all of the ratios, particularly when the initial growth ratio matched the next or the average of all ratios. He used this approach when ratios in middle or late instars had a much larger ratio (less growth) than the others. For example, the sixth instars of the io moth have an average ratio of .69 with the first over the second being .67, whereas the fourth over the fifth is .76. Thus, it appears Dyar attempted to use data in a way that fit a consistent (geometric) growth factor to get his results, whether it meant excluding the ratios of the earliest instars, selecting the final few, or averaging all of them.

Not only was Dyar vague about where or when his samples were collected or how he derived his growth factors, he was also unclear about how he made the measurements. He surely had a microscope with a micrometer to measure the head widths and often used multiple individuals preserved in alcohol from the same egg masses, particularly for the tiny first instars, which would have been impossible to keep alive to accurately measure.[52] Dyar presents more than one set of measurements, indicating more than one set of caterpillars found elsewhere in space or time, when some stages are missing. The determination of growth factors and calculations may have been done in late summer or fall of 1890, following his work rearing them, as Dyar prepared his manuscript. This is suggested by the fact the measurements were copied from the original descriptions, often in margins before or after the descriptions along with the predicted measurements (see Figure 3.2).[53]

Dyar continued to publish head-width measurements in his life-history descriptions of caterpillars but soon quit the practice of including the calculated widths from growth ratios. In fact, a paper in 1892 on *Nadata* (Notodontidae) was one of the few times he published calculated widths after his seminal work of 1890.[54] Therein Dyar used the calculations to spotlight what he considered to be normal (six) versus abnormal (five) numbers of stages. However, he continued to make calculations in the Blue Books, perhaps as a check to see if he included the actual number of instars or whether there were any growth anomalies in head-width size or number of stages. Certainly research after Dyar's shows there are exceptions to his model of geometric growth for head widths. However, 80% of examples in the entomological literature concerning Lepidoptera and other insects between 1980 and 2007 fit the model.[55]

## CHAPTER 4

# Long Collecting Trips and Sawflies

## 1889–1897

*Stoneman House, Yosemite, CA*
  *I have found an Orgyia here on the oak but am at a loss to know what it is. . . . Can you, when you get time, turn to* [the journal] *Papilio vol 1 p. 60 and note for me the characters of the larva of O. vetusta? I have left the book behind.*[1]

H.G. Dyar to J.A. Lintner, May 18, 1891

Dyar's fieldwork, like that of many avid naturalists, was more constrained by season and climate than by personal events, hence it is unsurprising that his nuptials and honeymoon took place in the midst of his first lengthy trip: eight-months (Map 4a). Indeed, Dyar took advantage of his wedding to expand his collection, gathering specimens while en route to California in Denver, Colorado, at the lights near the Albany Hotel. He also collected at Union Station (see Figure 4.2a), less than a mile away, which he described as "one of the best collecting grounds at night,"[2] where he "collected 150 specimens."[3] Dyar would also collect in Salt Lake City, Utah, and Humboldt, Nevada. Once in California, he continued collecting in, among other places, San Francisco's Golden Gate Park, the Los Angeles Basin, Yosemite (Figure 4.1), and the Big Trees Mariposa Grove. Starting in September Dyar remained close to Los Angeles, collecting at Eagle Rock Valley in Pasadena, as Zella had done before his arrival, and, in October, sampling moths at electric lights and sugar baits in the city. On the day he got his marriage license, he also caught a large yellow butterfly, the cloudless sulphur (*Phoebus sennae*).[4]

At the onset of his honeymoon in November 1889 Dyar collected in alfalfa fields and along the arroyos of the Salt River around Phoenix, Arizona, where he found, in

**Figure 4.1** The Stoneman House, Yosemite Valley, California (c. 1890).

irrigation ditches, drowned western poplar sphinx moth caterpillars (*Pachysphinx imperator*) "in considerable numbers."[5] After collecting in New Orleans, the Dyars went to Florida, staying through February 1890 and collecting in Jacksonville, Palm Beach, Lake Worth, and Jupiter.[6] Dyar brought live caterpillars from the West, including those of tiger moths (*Notarctia proxima*) along for the ride: "I carried some from California to the East and fed them on the native plants by the way. . . . During all the larval stages they feed only at night, and are very lively in their attempts to run and hide if disturbed."[7]

The lengthy honeymoon ended in Washington and Philadelphia in March 1890. Dyar wrote Herman Strecker from Philadelphia, telling him, I am . . . on my way home from my trip west and would like to see you and your collection if you are willing. So if there is no objection I will come out to Reading and make you a call."[8] Soon afterward Dyar reported 65 moths and butterflies from Strecker, including some from the western United States, China, India, Assam, and Sikkim.[9]

Once in New York, Mr. and Mrs. Dyar went to Rhinebeck, apparently the last time Dyar visited his childhood home.[10] That summer (1890), in nearby Poughkeepsie, Dyar spent many hours gathering moths.[11] The couple moved to New York City in late September and lived in a flat at the Windermere at 400 West Fifty-seventh Street.[12]

## THE FIFTEEN-MONTH TRIP

Harrison and Zella Dyar began their second transcontinental trip in late April 1891 (Map 4b), stopping for a day in Washington, D.C., to visit the Department of Agriculture to verify the identity of a the white-headed prominent moth (*Symmerista albifrons*) Dyar had reared in New York[13] and then heading to California. Dyar again collected at the lights of Denver's Union Station (Figure 4.2a), and other lights while visiting the Colorado Springs area, and at train stops in Salida and Grand Junction, Colorado, as well as on the shore of the Great Salt Lake.[14]

The first intensive collecting of the fifteen-month trip began in May 1891 at Yosemite. While in the park for four months, Dyar carried out the most intensive survey of Lepidoptera in Yosemite to date.[15] Butterflies were especially abundant during the early part of the visit, including the western and Becker's whites (*Pontia sisymbrii* and *P. beckerii*) at Cloud's Rest, a 10,000 foot. peak overlooking the Yosemite Valley. Dyar also encountered eggs of a monarch, oddly found on wild gooseberry, which he regarded as accidental on the part of the butterfly because "no milkweed was near." When they hatched, the larvae "refused gooseberry, and would have certainly . . . died had I not supplied them with their ancestral food," the milkweed.[16]

Other places they visited in the park included Anderson's Trail and Tenieya Canyon in Yosemite Valley, as well as Inspiration Point. Dyar found few moths and had difficulties finding caterpillars in the typically dry summer. He noted that butterflies and moths such as the golden hairstreak (*Habrodais grunus*) and the ceanothus silk moth (*Hyalophora euryalis*) preferred newly emergent leaves.[17] In order to attract moths at Tenieya Canyon, as he had in Los Angeles several years earlier, Dyar used the technique known as sugaring, a method in which a sweet and semi-fermented liquid is painted on surfaces as bait. In nature liquids such as these are used to fuel the moths in flight and for nutrition needed to produce eggs.

(a)  (b)

**Figure 4.2** Denver's Union Station (between 1881 and 1894) (*a*); John Lembert, collector, and the Dyars' guide in Yosemite (*b*).

Dyar wrote to butterfly expert Henry Skinner (1861–1926) (see Figure 12.1b), of the Academy of Natural Sciences in Philadelphia, during a week of collecting at Stoneman House (Figure 4.1) near the end of their stay, revealing he had found 48 species of butterflies at Yosemite.[18] To help him find more, Dyar hired as a guide the fifty-one-year-old John B. Lembert (Figure 4.2b), who had been living like a hermit at his property at the headwaters of the Tuolumne River at 9,000 feet and would become Dyar's protégée for rearing Lepidoptera.[19]

In mid-September, after nearly all of Dyar's caterpillars had stopped feeding, he left them "tied out in bags on their food-plants," and Dyar, along with his wife and Lembert, went on horseback to climb the 13,000 foot Mt. Lyell to find the green Sierra sulphur butterfly (*Colias behrii*), as Dyar wrote:

> On September 20th we made the ascent of the peak and when about a quarter of the way up on a spur of the mountain overlooking the end of the glacier, a specimen of *Colias behrii* was started up which I succeeded in capturing. No more were seen that day as the weather soon became threatening, and by the time we reached the top of the glacier, the clouds had begun to float in over the peaks. On another day, September 22d, we went to some mountain meadows, 10,000 feet high, where Mr. Lembert had formerly seen some of the 'little green butterflies' (*C. behrii*), but met with no success, and were obliged to return almost immediately in a dense snow storm.[20]

On their return Dyar found that the caterpillars he had left in bags "had been cut off and destroyed by a crazy Indian, called 'Loco' by the children of the valley," making "a rather disastrous ending to an otherwise quite successful collecting season."[21] Lembert made up for the loss by sending Dyar several hundred butterfly and moth specimens over the next several years. Some that Dyar identified were returned to Lembert, exchanged for specimens in the Dyar or Strecker collections, and others were given to J. B. Smith to name.[22] Among them was a new species of ghost moth, *Hepialus lembertii*, named for the collector by Dyar in 1894.

Then, in April 1896, "a passing Indian found the body of Mr. Lembert lying dead in his cabin, with a large bullet-hole in his head, over the right temple. He had evidently been murdered ... [by] some Indian whom he had offended."[23] Lembert's chance encounter with Dyar in 1891 had changed his life. Dyar taught Lembert the fundamentals of entomology, including how to collect and observe insects and his contributions to the field were reported in *The Canadian Entomologist* and other entomology journals.[24]

❧

Following four months in Yosemite, the couple took a stagecoach out of the national park[25] and headed to Santa Barbara, California, where they stayed at the Arlington Hotel through mid-December 1891. Dyar collected at electric lights, placing the specimens in paper triangles and stamping the date and location on them with dark blue ink. Because it is often inconvenient to spread Lepidoptera specimens in the field when collected, they are stored in envelops until they can be "relaxed" in a moist chamber and spread on mounting boards. As he had at other stops on the trip, Dyar tested the rubber stamp destined for the envelopes on pages of his notebooks.[26]

Dyar wrote several letters in November 1891 to Henry Skinner, both as editor of *Entomological News* and an authority, to identify a hairstreak butterfly and publish his larval description of it. Dyar thought the species might be "*Thecla auretorum* or *T. sylvinus* of Boisduval," but if it were new, he suggested Skinner name it for the caterpillar's food plant, the willow (*Salix*), or indigenous tribal Miwok names in Yosemite, such as "Pobano (for Bridal Veil Falls) or Toya (for the Sentinel), etc."[27] Skinner stated that the butterfly was *T. sylvinus* in 1892 but delayed publishing the paper. Dyar wrote, asking if his manuscript had "been consigned to oblivion?" By the time the life history was published in December 1894, the butterfly had been determined to be yet another species: the California hairstreak, *Thecla* (now *Satyrium*) *californica*.[28] Whether Skinner or Dyar made the final identification is unknown, but clearly identification of western butterflies had its difficulties.

At the beginning of the fifteen-month trip Dyar had brought lepidopteran livestock along, for example, eggs of the definite tussock moth (*Orgyia definita*). These he sent back to Lintner once they had hatched, requesting that Lintner keep track of the precise number of stages (instars) before pupation.[29] Periodically during the trip Dyar also sent his specimens home to New York City, to the care of his Uncle Josiah Hannum, who was keeping watch of his nephew's growing collection, now apparently situated at Dyar's 599 Broadway address. Hannum and his wife Elizabeth lived in Staten Island but perhaps stayed at the Broadway flat while Dyar was away. Sending his bounty back home made perfect sense, as carrying fragile specimens and accumulating the baggage on the long trip would have been difficult. Further, Hannum not only could receive but also forward specimens, dead or alive, to Dyar's contacts.[30]

The Dyars sailed to the Pacific Kingdom and future State of Hawaii from San Francisco in January 1892. They were in Honolulu by January 26 and visited the Big Island of Hawaii in March.[31] Dyar, of course, collected butterflies, including the endemic Hawaiian blue (*Udara blackburni*), and reared the pea blue (*Lampides boeticus*) at Nuuanu Valley on Crotalaria flower. He also reared two owlet moths, *Spodoptera* and *Chrysodeixis*, an injurious crambid on coconut palm (*Omiodes blackburni*), and a sphinx (*Agrius convolvuli*).[32] Adult moths from Dyar's livestock from all over the continental U.S., including those from Rhinebeck and Yosemite, emerged from the pupal stage while in Hawaii, and to our knowledge, none escaped captivity. At the end of the stay on March 28, Dyar collected moths at electric lights in Honolulu.[33]

Upon landing on the mainland Dyar received a letter from Lintner, who thought he had returned to "the Atlantic Seaboard." Dyar updated Lintner on western tussock moth caterpillars (*Orgyia cana*) that he had placed under the care of famous lepidopterist Hans Hermann Behr in San Francisco. In the same letter, Dyar expressed his desire to go collecting with Lintner in the Adirondacks the coming summer. Lintner, who had been ill over the winter, replied: "I hope to be in the mountains in July and August and if my locomotive power were as good as yours will be delighted to have you for a companion."[34] They were not to meet that summer, as Lintner once again became ill.

After a stay in San Francisco, Dyar collected moths from electric lights for one night in Watsonville, California, in mid April 1892 and then for a month in the vicinity of Portland and Albina, Oregon. He also took a day trip to Mt. Tabor, where he took two male anise swallowtails (*Papilio zelicaon*).[35] The remainder of the trip included a stop in Sitka, Alaska, collecting only a few "evening deltoid" moths (owlets) in late June.[36] On the return they visited Nanaimo and Victoria on Vancouver Island, and Seattle, Washington and stayed several days at their final stop at Yellowstone Park, where Dyar collected a *Parnassius* butterfly at Mammoth Hot Springs.[37]

## SAWFLIES: A NINE-YEAR INTERLUDE

One of the lesser-known facts of Dyar's scientific resume was his major focus on sawflies, begun in 1893. These primitive wasps (Hymenoptera) would capture his fancy for a nine-year period. Superficially Dyar's segue into studying wasps appears as incongruous as the name "sawfly" is for a cousin of wasps, bees, and ants (Figure 4.3a), but the transition to sawflies was seamless. After all, sawfly larvae are caterpillar-like and feed on the same plants and often on leaf surfaces as did Lepidoptera caterpillars (Figure 4.3b). Further, the process of rearing the two insects is identical. Sawflies, like Lepidoptera, also form plant galls and mine leaves, so Dyar did not need to alter his rearing protocol: He fed the caterpillars the plant he found them on, kept their cocoons outside over winter, and waited for adults to emerge the following spring to identify or name them. We could say that Dyar, the Darwinian, was preadapted to rear sawflies. Indeed, the accounts from his Lepidoptera and sawfly notebooks show he was rearing both groups simultaneously, often from the same places and plants.[38]

Dyar's new passion became evident in his notes soon after he found a sawfly caterpillar on gooseberry near his home in Roxbury on May 31, 1893. By the time Dyar arrived at Woods Hole Biological Station for a summer course, he also sported

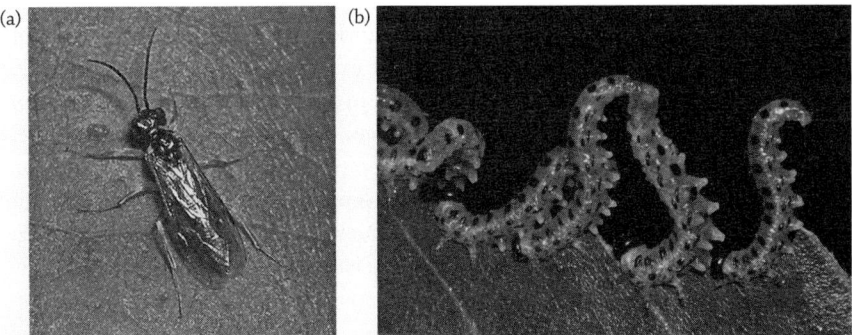

**Figure 4.3** Sawflies reared by Dyar: adult of *Ameteastegia pallipes* (*a*); caterpillars of *Craesus latitarsus* (*b*).

sawfly records from Franklin Park and Forest Hills, Massachusetts. Seventeen food plant records for the sawflies at Woods Hole followed from early July to early August, including: white birch, sugarplum, alder, willow, larch, and black and white oak. He continued seeking larval sawflies in Plattsburgh in late August and early September.[39] During the summer Dyar also reared moths from various stages and collected moths at electric lights at Woods Hole and Plattsburgh.

In April 1894 Dyar found the adult stages of sawflies in Plattsburgh, New York. The larval search continued in Central Park, followed by a trek back up North from June to September, including two months in Keene Valley, while he sought bacterial cultures for his Ph.D. research and Lepidoptera caterpillars (Map 5).

The voracious sawfly caterpillars were eating leaves of poplar, linden, and many of the same trees as the previous year. In September Dyar found moth or sawfly caterpillars in Maine, including Webb Pond (near Bangor) and in Dixfield, where Zella's maternal relatives lived. The Dyars then traveled inland to Fabyan, New Hampshire, before a longer stay in Plattsburgh (Map 5).[40] Running out of space in his latest Blue Book, Dyar used the back of a Poland mineral water receipt to write a description of a moth caterpillar found in the area.[41]

During that summer Dyar had several breaks from fieldwork. He and Zella attended a mock trial suit against Judge David McAdam in the vicinity of Lake Placid and Saranac along with 500 others in late July. Dyar also attended a meeting in Brooklyn of the American Association for the Advancement of Science (AAAS) in mid August. At this time his interest in meeting other scientists, including entomologists, was in full throttle. After the meeting he joined a group of 210 AAAS "learned persons" for a social gathering on a steamer on the Hudson River from "the battery Pier at 9:30 a.m."[42]

Zella remained in Keene Valley when Dyar was at the AAAS meeting and continued to assist him with his cultures while he was away. She wrote several descriptions of moth caterpillars on hotel stationary from "The Adirondack House." When he returned, Dyar corrected and annotated Zella's descriptions, including one of a beautiful owlet larva known as the paddle caterpillar (*Acronicta funeralis*) (Plate 5a). Although Zella reared caterpillars for Dyar prior to their marriage, this description and one of another caterpillar are the only known examples she had written.[43]

Dyar began his third season of intense sawfly rearing in May 1895, along with his habitual work with Lepidoptera, in New York City by finding sawfly caterpillars on yellow dock plants "in the vicinity of unoccupied building lots" near his residence (243 West Ninety-ninth Street), and in Van Cortlandt Park, Plattsburgh, and Rouse's Point.[44] He and Zella went to New Hampshire in mid-June, where they spent most of the next two months at Jefferson Highlands. Dyar took along a letter of introduction from Rodrigues Ottolengui, a member of the New York Entomological Society who had collected specimens there, to the owner or manager of the Waumbek Hotel in Jefferson so that he would be able to collect specimens without suspicion.[45]

Dyar found sawfly caterpillars on a number of plants he had not previously reported, including strawberry, serviceberry, viburnum, and elderberry.[46] There were two side trips to Mt. Washington from Jefferson. On the summit in early

July, Dyar found the butterfly, *Chionobus semidea* (Say) (now *Oeneis melissa semidea*), "stuck in fresh paint."[47] On a return trip in late August, Zella picked up a sawfly larva on sedge near the summit. After returning home in early September, Dyar found more sawflies in Fort Lee, New Jersey, Van Cortlandt Park, and New York City.[48] The couple also took some personal time, and their first child was conceived on the trip.

Field seasons in 1896 and 1897 were from May to September, entirely in New York and New Jersey. Dyar found various sawfly species on oak, hickory, and chestnut trees. In mid-May 1897 he found sawfly larvae on rock oak at Pelham Manor, New York; later that evening he gave a talk on sawflies at the New York Entomology meeting in Manhattan: He noted that sawfly caterpillars have large thoracic legs and small abdominal prolegs, the opposite of Lepidoptera and their larval setae. Perhaps the most interesting of his observations was that although parasitic wasps and flies attack sawfly larvae as Lepidoptera caterpillars, they have their own unique species of parasites.[49]

༺ঞ༻

During his sawfly work Dyar needed assistance to identify species. To this end he corresponded with Charles Lester Marlatt (1863–1954) of the Division of Entomology, U.S. Department of Agriculture, and William H. Ashmead.[50] In July 1897, while in Bellport, Long Island, Dyar wrote Marlatt a reminder: "I think you must have forgotten about the sawfly I sent you which you said you would look up on your return from Boston." Marlatt clearly understood the necessity of Dyar's requests, writing in an article: "Mr. Dyar is anxious to publish the descriptions of the larvae, and the technical descriptions of the species presented herewith are made to enable him to assign his larvae to described species and avoid the difficulties which would arise from the description of larvae before the adult insects have been characterized."[51] In other words, descriptions of unnamed larvae and associated information on their food plants have limited scientific value.

Dyar's contributions to the knowledge on sawflies drew praise in 1896 from Marlatt, who wrote in *The Canadian Entomologist*, while describing new species of these primitive wasps, "The valuable work which Mr. Dyar is doing in rearing larvae is resulting in the clearing up of some puzzles in the classification of insects, and has no more interesting outcome than the fact that many of his rearings, at least in the line of sawflies, prove to be of species hitherto undescribed, showing how little we really know of the insects of this group in this country."[52] Marlatt, himself, was no slouch. He made major contributions to entomology in early efforts to control the San Jose scale and developing the system of naming the broods of the periodical cicada. He would become one of Dyar's most important advocates.

Among the new species of sawflies mentioned by Marlatt that were discovered from Dyar's rearing, Dyar chose to describe twenty-eight himself over the next decade or so. To this day, Dyar's body of work on the life stages of North American sawflies stands alone as the most significant ever done.

CHAPTER 5

## Postgraduate Education on Classification of Moths and Bacteria at Columbia: 1892–1895

On his return home in August 1892 from his transcontinental and Pacific travels, Dyar wrote New York State Entomologist Joseph Lintner that he "did not get all the Natural History subjects" as a chemistry student at M.I.T.[1] Consequently, he re-enrolled at M.I.T. for the 1892–1893 school year and was under the supervision of professor William Thompson Sedgwick (1855–1921), "America's foremost sanitary bacteriologist." Although Sedgwick may have steered him toward doctoral research on bacteriology a few years later, Dyar's decision to pursue "Natural History subjects" was in line with his passion to be a lepidopterist rather than a chemist.[2]

Harrison and Zella settled in Roxbury, Massachusetts during fall 1892 at Forty-nine Winthrop Street. They collected moths nearby, including an adult tussock moth found by Zella.[3] While in the Boston area Dyar attended monthly meetings of the Cambridge Entomological Club, whose journal *Psyche* published his papers. At Dyar's first meeting, Samuel H. Scudder read an address by the outgoing president William Jacob Holland (1848–1932) entitled "Communal cocoons and the moths that weave them." Scudder mentioned the habit described by Westwood of the Mexican madrone butterfly caterpillar (*Eucheira socialis*: Pieridae). Dyar interjected that some tent caterpillar individuals "usually remain in the nest and undergo their transformations." Dyar presented new records of moths found at electric lights in Poughkeepsie in February, and at the next month's meeting he was chosen as secretary, Dyar's first position as an officer in a scientific society (four years later he would serve as president).[4]

Dyar's next step in adding to his biology curriculum was his participation in a six-week course beginning in July 1893 at Woods Hole Biological Station. He took an advanced course in embryology (Figure 5.1a, b), the first at the station and taught

**Figure 5.1** H.G. Dyar (*back row, fourth from right*) with Professor Charles Otis Whitman and classmates of the first embryology course taught at Woods Hole, 1893 (*a*); closeup of Dyar (*b*).

by the director Dr. Charles Otis Whitman,[5] and developed his interest, apparently independent of course work, in sawflies. That fall, Dyar began graduate school as a master's student at Columbia. He came under the guidance of Henry Fairfield Osborn. Two courses Osborn taught during this time included "Evolution of the Vertebrates" and "Mammals, Recent and Extinct." An early proponent of eugenics, Osborn was to become the director of the American Museum of Natural History and president of the International Congress of Eugenics when it was held at the Museum in 1915.[6] The professor likely influenced Dyar in his adoption of eugenic philosophy, one shared with many of his contemporaries.

The new graduate student showed a clear passion to explore Lepidoptera classification, especially silk moths and their relatives (known then as Bombyces), for his master's research. While fulfilling his degree requirements Dyar, along with the avid collector Berthold Neumoegen (see Figure 5.2a), revised the taxonomy of the Bombyces in North America and worked alone on higher classification of the same. For his studies Dyar requested, from his contacts, specimens that he and Neumoegen lacked. They were sent to his home address, now 76 West Sixty-ninth Street.[7] Soon after his return to New York, Dyar also attempted to gather specimens by setting female moths outside to attract their mates, or "send," via their own sex pheromones. This was a useful technique to obtain specimens or for mating to keep a species in culture, even though the chemical makeup of the pheromone attractant wasn't known at the time. In this instance Dyar's attempt failed. "I have had the same ill luck with my last [female] *Orgyia definita* that befell you. I exposed it for several nights here in Central Park but attracted no [male]," he wrote Lintner.[8]

Showing an interest in moths other than Bombyces, Dyar wrote Lintner in the same letter: "Do you recognize the work of the curious leaf miner enclosed?" He found these particular caterpillars, among many that tunnel between leaf surfaces, on sour gum (*Nyssa sylvatica*). In the final stages the caterpillar spins silk between the layers inside of the leaf and then cuts the leaf around itself in order to fall to the ground where it spins a cocoon. Adults emerged were tiny fairy moths (Heliozelidae), which have long fringes on their hind wings. The following week Dyar wrote again to Lintner about an owlet moth caterpillar's behavior: His childhood friend George Hudson had "observed the larva of *Harrisimemna trisignata* performing its curious trick of boring a hole in a host for pupation."[9]

After his fieldwork in New York, Dyar made a visit to the United States National Museum in Washington, D.C., to examine Bombyces in the Lepidoptera collection but found them split between two separate collections, one at the National Museum, today the Arts and Industries Building (see Figure 9.1a), and the other, blocks west at the U.S. Department of Agriculture. Dyar only visited the Museum and was disappointed the Bombycidae were at the Agriculture building.

Museum aid Martin Linell informed C.V. Riley, the Curator and Chief Entomologist, of Dyar's difficulties and the snafu resulted in the following reproachful letter from Riley:

> Dear Sir: I inquired this morning of Mr. Linell, my aid ... in reference to what transpired ... he informed you that the Bombycidae were ... not arranged in the Museum but were ... still over at the Department of Agriculture, and that you could see them by speaking to me about it. This plainly indicates that if you did not see the Bombycidae that were not in the Museum rooms, it was your own fault.[10]

Undeterred, Dyar requested specimens of "South American Bombyces" from Strecker to include as "good representation of as many genera as possible."[11] Dyar's interest was in obtaining adult moths for dissections, to corroborate expanding larval classification in line with Darwinian principles suggested by A. Weismann and J.H. Comstock.

## NEUMOEGEN AND THE MYTH OF THE NAME "DYARIA"

Perhaps the best-known story about American entomologists doing battle in the literature is between Dyar and John Bernhardt Smith (see Figure 10.1b). Dyar was to have named a species "corpulentis" or "smithiformis" after his rotund colleague Smith, while Smith was to have retaliated with the name "dyaria." The story is apocryphal, although the two taxonomists would become fierce rivals during decades following the naming of the actual moth: *Dyaria*.

Rather than Smith, it was Berthold Neumoegen (1845–1895) (Figure 5.2a), a Jewish immigrant and Wall Street banker and well-respected member of the entomological community in New York City who named *Dyaria*. This is a genus rather than a species and a name Neumoegen did not intend as derogatory (Figure 5.2b).

Mr. Neumoegen had a large private collection of Lepidoptera of around 100,000 specimens. The collection was elegantly housed on the top floor of his home in cork-lined black walnut drawers.[12] Dyar, according to legend, had few friends.

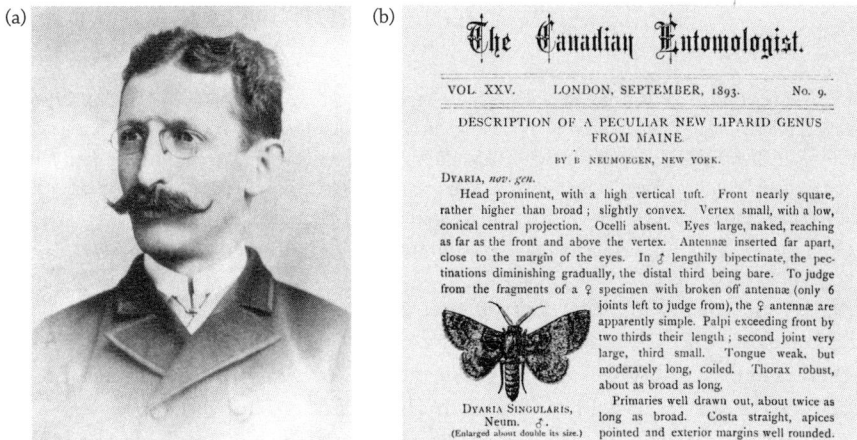

**Figure 5.2** Berthold Neumoegen (a) and his 1893 publication in *The Canadian Entomologist* describing *Dyaria singularis* (b).

However, he was on good terms with Neumoegen, Herman Strecker, Joseph Lintner, and Augustus Grote. Neumoegen, a friend of Strecker's, met Dyar in New York City in 1891 after Neumoegen wrote him two letters requesting he visit his collection.[13] Their shared interest in Bombyces brought them into a partnership, which eventually led to a monograph on these moths.

Neumoegen and Dyar described a few new species and varieties of Bombyces in the inaugural issue of the *Journal of the New York Entomological Society* (*JNYES*) in March 1893. This was in advance of a monograph titled "A Preliminary Revision of the Bombyces of America North of Mexico," which ran through the last part of volume 1 and all of volume 2 of the *JNYES*. The contribution represented a total of 160 pages of short descriptions, citations, and identification tables (now called keys). Neumoegen, whose collection made the work possible, was the senior author.[14]

The story of the naming of *Dyaria* began when Carl Braun in Bangor, Maine, collected a moth specimen at electric light and gave it to Neumoegen. Neumoegen wrote Dyar, informing him of his plans to dedicate the new genus to him, and sent the specimen along with several tiger moths: "I send you to-day, by express prepaid a box containing: . . . Bombyx, nov. gen. & nov. spec. I call it *"Dyaria singularis*, Neum. from Maine. I . . . want to describe it over my own name, in your honor."[15]

Neumoegen's article with the Dyar patronym titled "Description of a peculiar liparid genus from Maine" appeared in the September 1893 issue of *The Canadian Entomologist*. He named the genus *Dyaria* and it was "Dedicated to my faithful colabourer and friend Mr. H. G. Dyar." Neumoegen named the species *Dyaria singularis* because he thought it to be an odd member of Liparidae (now Lymantriinae, the gypsy moths), ". . . one of the most *singular* [italics mine] . . . of our fauna."[16] Neumoegen was right that the moth was odd, but only because it was misplaced as a gypsy moth rather than its proper placement as a snout moth (Epipaschiinae).[17]

Less than two years after the paper's appearance Neumoegen died of tuberculosis. The completion of his joint publications with Dyar on "Bombyces" turned out to be his finest hour, entomologically speaking.

## RECLASSIFICATION OF LEPIDOPTERA

*Which of the two inherited classes of characters have been the most distinctly and completely preserved, and which of these, through its form-relationship admits of the most distinct recognition of blood-relationship, or inversely, which has diverged the most widely from the ancestral form? [The answer] will be arrived at in most cases as soon as the ontogeny of the larvae, and therewith a portion of the phylogeny of this stage, can be accurately ascertained.*[18]

August Weismann, "Studies in the Theory of Descent," 1882

Dyar's exposure to Darwinian principles at Columbia combined with his passion for documenting the early stages of moths motivated him to overturn older, artificial schemes of classification even though he and Neumoegen made few changes above the genus or species level in the taxonomic hierarchy in their work.

August Weismann's (1834–1914) philosophy, expounded in "Studies in the Theory of Descent" (1882), was significant to Dyar, especially because most insect taxonomists simply ignored the larval stage as part of their toolbox to name, classify, or study insect evolution. Characteristics of the wings of the adult, for example, were heavily emphasized by most. Weismann wondered whether a classification based on caterpillars would produce one congruent with the adult stage. His suggestion was to determine which inherited characters of each "have been the most distinctly and completely preserved," thus most closely illuminating the ancestral relationships, or in reverse, "which has diverged the most widely from the ancestral form?" He suggested the answer was to be found in Haeckel's biogenetic law, that is, "ontogeny recapitulates phylogeny." In other words, through knowledge of larval ontogeny (the developmental stages, or instars) one could witness "a portion of the phylogeny of this stage."[19]

Cornell's John Henry Comstock (1849–1931), who presented his Darwinian views in his essay on the evolution of insect wings as applied to Lepidoptera, was also an influence on Dyar. In a glowing review of Comstock's approach, Dyar wrote, "[A]ll scientific entomologists will be gratified at the appearance of this paper, which is an attempt to base a classification of the Lepidoptera upon the ground of evolution." Showing his distaste of the new Lamarckian philosophy, or "Neolamarckian School" practiced by Alpheus Spring Packard (1839–1905) (Plate 8d), he added that Comstock looked at "evolution by natural selection, not befogged by the questionable action of so-called 'acquired characters.'"[20]

Comstock divided Lepidoptera into two suborders, "the Jugatae and Frenatae," each with a different structure latching the wings in flight. The primitive Jugatae have a lobe on the forewing base (a jugum), which connects it with the hindwing, whereas the more derived Frenatae have a hook at the hindwing base (a frenulum), which hooks into a forewing latch. Comstock considered "jugate" moths more primitive because their more intricate pattern of wing veins are similar to more primitive insects outside of Lepidoptera.[21]

Dyar's hope was to develop a natural classification of Lepidoptera, one that represented characteristics of the organisms and their evolutionary history. In his seminal paper on Lepidoptera classification in the *Annals of the New York Academy of Sciences*, published in May 1894 (Figure 5.3a, b), Dyar wrote,

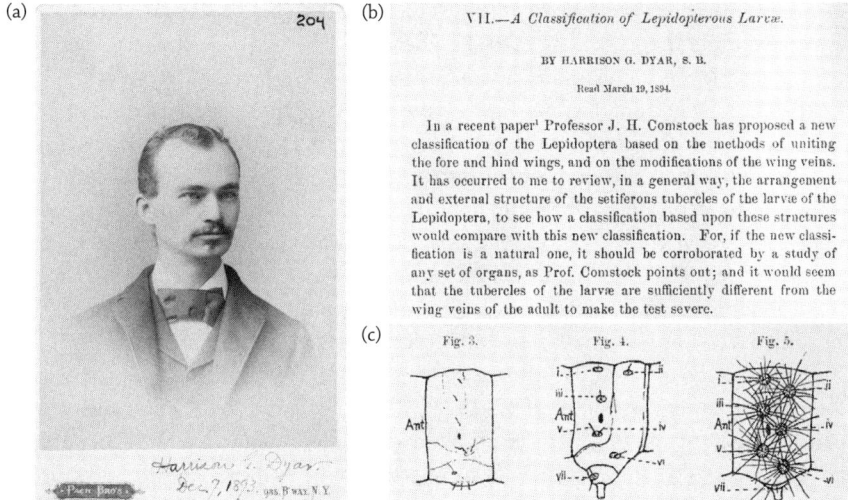

**Figure 5.3.** Dyar portrait in December 1893 while attending Columbia (*a*); part of Dyar's thesis on the classification of Lepidoptera based on larval setae (*b*); Dyar's comparative figures on how larval setae evolved from simple to hairy or spiny in various Lepidoptera families (*c*).

It has occurred to me to review . . . the arrangement and external structure of the setiferous tubercles [setae, or hairs, on warts or pinacula] of the larvae of the Lepidoptera, to see how a classification based upon these structures would compare with . . . [Comstock's] classification. For if . . . [his] new classification is a natural one, it should be corroborated by a study of any set of organs, . . . and it would seem that the tubercles of the larvae are sufficiently different from the wing veins of the adult to make the test severe.[22]

He used the setal arrangement, or map, found in sawfly larvae as the ancestral type to formulate his hypothesis, and preferred to use the derived conditions found in the larval stage upon hatching, the first instar.

In a series of papers over the next several years, Dyar included original figures of setal maps on two segments, one each for thorax and abdomen. The ability to compare position, fusion, or reduction determined whether particular setae were homologous. Dyar's comparative figures (Figure 5.3c) would lead to advances in our knowledge of Lepidoptera evolution and classification. He constructed genealogical trees by placing the ancestral condition at the base and defined each branch point by derived setal characteristics.[23]

In order to complete the work, Dyar needed critical specimens among the primitive Lepidoptera suborder Jugatae. Of prime importance was obtaining *Micropteryx*, a living fossil in the most primitive family Micropterygidae. Adults of these moths have mandibles: a primitive condition for adult Lepidoptera. They use the mandibles to feed on pollen grains rather than imbibe liquids for nourishment or hydration with a proboscis, as is typical. To obtain specimens Dyar wrote Riley: "Will you kindly inform me whether there are specimens of the larvae of any species of

*Micropteryx* in the . . . collection. If there are, I would highly appreciate the opportunity to examine them, as I desire to see whether the larv[al] structure corroborates Prf. Comstock's recent classification of moths."[24]

Riley was away from Washington, unable to answer the query. His absence, however, led to the first correspondence between Dyar and Leland Ossian Howard, second in command at the Division of Entomology. Although his answer on March 5, 1894, was inauspicious, as he could not fill the request because there were no specimens, Howard was destined to be the most influential person in Dyar's career.[25] Less than three months later and in poor health, Riley resigned as chief of the Division of Entomology and was replaced by Howard, who would become the diplomatic buffer to Dyar's acidic personality. In 1895 Riley died from a bicycle mishap in D.C.

Dyar finally obtained *Micropteryx* caterpillars from Scottish Entomologist Thomas Algernon Chapman (1842–1921), who considered their larvae allied with Limacodidae because of a retractile head, body shape, and unusual numbers of ventral appendages on the abdomen. Dyar rejected this view, which never found broad acceptance, and concluded that setae of *Micropteryx*, along with the ghost moths (Hepialidae), were more similar to each other than to Comstock's Frenatae. Dyar had corroborated Comstock's classification, which placed both in Jugatae based on the presence of the jugum on the hind wing.[26] For another related study in 1894, Dyar needed larval specimens of the tiny delicate plume moths (Pterophoridae), and his sister Perle assisted: She was now living in Paris, and made sketches of specimens found in the city's Natural History Museum.[27]

Other specimens came to Dyar as a result of his own fieldwork. For example, he found larvae of a tiny scythrid moth (*Butalis basilaris*) in 1894 in Keene Valley, New York. Dyar was able to identify the moth only through the help of Professor C.H. Fernald of the University of Massachusetts. After Fernald sent back the identification a tiff resulted. The professor scolded Dyar for sending him so poor a specimen "in such shape that I was obliged to soften, pin and spread, before it could be determined." Fernald continued, "One of the first lessons you should learn before doing anything with small moths, is to put them in proper shape . . . for those who can aid you are very busy men whose time is worth much more than your own."[28] When Dyar sent the specimen he suggested to Fernald that he would send it elsewhere if Fernald could not identify it, to which the professor responded: "Who can you get to name N.A. [North American] Micros for you: I know of no competent person who will undertake it for they are usually sent to me to be named finally, though others sometimes take the credit of it after I have named them."

The following day Dyar replied to Fernald:

> I am much obliged to you for the identification of the moth. I must, however, resent the remarks with which it is accompanied as overbearing and rude. While you may prefer material mounted, others would perhaps prefer it uninjured and fresh and you could have easily exposed the upper surface by removing one pair of wings without any loss of time. I do not think the circumstances warrant your language and I anticipate an apology from you. Had you been unable to determine the

moth, I should have tried Dr. Riley and Lord Walsingham. Are these the gentlemen to whom you [refer?][29]

As Fernald suggested, he was the foremost taxonomist of North America "micros", a most difficult group to identify and from whom Dyar would continue to need assistance. The tiff bothered Dyar enough to write Grote, who was a friend of Fernald's. Grote's response in December 1895 suggests that Dyar's critical nature had created a backlash. "As to Fernald I don't understand his conduct to you [or] want to excuse it.... He writes me as if he had nothing against you but says—'I have been told strange things about' you. Now I told him you were OK. But I would like to know who told Fernald."[30] Fernald was later reluctant to work with Dyar on his "List of North American Lepidoptera."

As Dyar worked on Lepidoptera classification, so too did A.S. Packard at Brown University. His goal was similar to Dyar's, but he included pupal along with larval characteristics and stuck with the older classification, one inconceivable to Dyar as it did not reflect his view of what were ancestral and derived larval characteristics. Packard considered Limacodidae to be a specialized member of the silk moths (Bombycidae) rather than in the "Generalized Frenatae" of Comstock and Dyar.[31] Although Dyar rejected Packard's view of evolution and classification, he helped the professor obtain specimens for study and allowed him to quote extensively from his original life-history descriptions. Packard changed his classification of silk moths to one more in line with Dyar's when T.A. Chapman found similar classifications by looking at pupae, and from his own examination of pupal skins of various moths.[32]

## TALKS ON LEPIDOPTERA CLASSIFICATION

Dyar gave a number of talks and published many papers related to his caterpillar-based classification of Lepidoptera after completing his Ph.D. in 1895 and while holding a teaching position at Columbia in bacteriology. His talks at entomology societies often coincided with his publications in their journals.

One of Dyar's more significant findings on Lepidoptera classification was presented at an April 1895 meeting of the New York Entomological Society and in a short paper that soon followed in *Psyche*. In the paper Dyar presented a series of larval setal maps—each of one abdominal segment—that represented seven moth families ordered in a line of descent (see Figure 5.3c). Included was the tiny scythrid moth identified earlier by Fernald. Dyar was intrigued by the close proximity of two dorsal setae, which formed a link (evolutionary bridge) between "Microlepidoptera" and the burnets (zygaenids), limacodids and others that were thought to be "Macrolepidoptera" on which the setiferous tubercles were fused.[33] This was compelling evidence to overturn old views of what constituted the superfamily Bombyces. A natural classification would require component families of Bombyces be moved to form natural groupings. Dyar's work also showed that groups referred to as "micros" or "Microlepidoptera" were also artificial constructs. For example, limacodids would now be considered "micros" rather than "macros." Dyar, in short, demonstrated that "micros" and "macros" were artificial groups, for characteristics other than adult size. For example anatomical features (morphology) of many macro-sized moths, it turns out, belong in groups with tiny "micros", and vice versa.

In June 1895 Dyar's scheduled talk at the Brooklyn Entomological Society on "Classification of Lepidoptera based on tubercles of the larvae" never took place. The punctual Dyar left before the tardy members showed up. He received two letters of apology, one from Archibald C. Weeks, an attorney and the Corresponding Secretary of the Society, and the other from Rev. George Duryea Hulst (1846–1900) (see Figure 10.2a), an active member and well-known expert on inchworm moths. According to Weeks, John B. Smith, then the New Jersey State Entomologist, was usually on time for the meetings but did not attend that evening. Other members were normally "half to three quarters of an hour late." Hulst wrote, in November 1895, "I was extremely sorry I was not able to get to the meeting at which you were to read a paper, and extremely mortified at the last meeting to learn that you had come over and had found no meeting. Several members said they came somewhat late but did not find you. I my self had to officiate at a funeral and could not get away."[34]

Dyar continued to speak tirelessly on Lepidoptera classification based on larval characteristics, including at the February 1896 meeting of the New York Entomological Society, when he gave examples of the slight variation in the position of head setae found among caterpillars in a number of Lepidoptera families. Dyar illustrated how the flattened setae of a sack-bearer moth larva had been mistakenly described as antennae based on the flat setae appearing in the same positions as normal setae (homologous). He also gave another example of flat setae in the last larval stage of an owlet moth (*Acronicta funeralis*) (Plate 5a).[35]

Dyar published a genealogical tree of owlet moths and their relatives in another important paper in the June 1896 issue of the New York Society's journal. Among those included were pericopines, noctuids, lymantriines, lasiocampids, arctiines, and dioptines. The tree is nearly obscured by Dyar's illustrated life history of the faithful beauty (*Composia fidelissima*) (Plate 7d), a stunning pericopine moth found in Florida and the Caribbean, in the same article. Dyar considered alternative hypotheses about the pericopid origin, including derivation from dioptines based on larval setae and wing vein branching of the adults. Although the tree would not be accepted today, it was an innovative and significant approach to look at both larval and adult lines of evidence, and one Dyar would adopt for the rest of his career.[36]

Another paper in 1896 (in *Proceedings of the Boston Society of Natural History*) was on the inter-relatedness of these same owlet moth families based solely on larval setae. Dyar again preached the importance of larval characteristics in classification, a view not shared by others, including Grote and Skinner. A bit off course, he used the name Bombycides instead of Noctuina for the owlet moths and relatives previously mentioned because he included the silk moth genus *Bombyx* in this group rather than with the giant silk moths ("Saturnians").[37] For his next publication, "Notes on the phylogeny of Saturnians," Dyar discussed differences in phylogenies based on his larval data versus those found by Grote on the wing veins of the adults.

Dyar believed mapping the adult characters of these moths on larval trees provided a sounder explanation of the evolution of the adult wing veins and antenna than these characteristics by themselves.[38]

## DOCTORATE IN BACTERIOLOGY

Given a dearth of professional positions in the field of insect taxonomy, Dyar, with the encouragement of his brother-in-law, S. Adolphus Knopf, and his M.I.T. professors William Thompson Sedgwick and Thomas M. Drown pursued a Ph.D. in bacteriology related to public health at Columbia College in the "University Faculty of Pure Science." He was the first graduate student under Dr. Theophil Mitchell Prudden (1849–1924), who was increasingly involved in public sanitation and bacteriology, and encouraged Dyar to do research in bacteriology.[39]

Prudden's "interest in dust-borne spread of tuberculosis, and for bacteriologists to determine the connection bacteria had to the long-standing notions that infectious disease was spread by bad air" led to Dyar's dissertation, "On Certain Bacteria from the Air of New York City." Ironically, Dyar did his dissertation reading (Feb. 18, 1895) only several weeks after Neumoegen's death from tuberculosis. Dyar mentioned the difficulties determining "the identity of the bacteria commonly occurring in the air of New York" because of the lack of monographs on bacteria, and believed that this "could be overcome by the facilities possessed by the bacterial laboratory in the College of Physicians and Surgeons of Columbia College, with its considerable collection of living species, which might be directly compared with those obtained from the air."[40]

Dyar found a few reliably identified specimens in the Columbia collection, however, including some from Cheesman. He looked at 50 species from Kral's bacterial laboratory in Prague and some additional specimens from Sternberg. He, therefore, relied primarily on published descriptions to identify his cultures, and while hesitant to describe new species he saw "no other ... satisfactory way of treating this without the application of a scientific name."[41] In naming nearly 50 new species or varieties of bacteria, Dyar believed he took "the more conservative course" by naming all "differing forms" species until he obtained a better understanding of their inter-relatedness. Part of the difficulty was that his identifications were made by biological differences rather than difficult to detect morphological differences. One problem with this approach was that the manner in which the bacteria reacted to different media was highly variable.[42]

As he was with insects, Dyar was interested in bacteria as organisms, not for their medical or economic importance. His dissertation research was related to his desire to visit his favorite haunts. By "planting" cultures in the countryside, however, he was less likely to find bacteria harmful to humans. Dyar's sites for cultures during the summer of 1894 included Plattsburgh and Keene Valley, New York, the same places he did fieldwork on Lepidoptera and sawflies (Map 5). Dyar also obtained samples from his flat on Sixty-ninth Street and near the lab in his building at Columbia on Fifty-ninth Street (Map 3).[43]

Dyar named some of his more interesting bacteria species—found in Plattsburgh—after carnivorous plants such as *Bacillus vacuolatus* from a trap on the greater bladderwort (*Utricularia vulgaris*) and *Bacillus sarracenicolus* from a fresh leaf of a pitcher plant (*Sarracenia purpurea*). George Hudson would also figure into Dyar's work on bacteria and sawflies during the summer of 1894, as Dyar named one species of each—a bacteria cultured by Hudson (*Bacillus hudsonii*) and a large sawfly (*Nematus hudsonii magnus*)—after him. Other interesting bacteria were found in Lepidoptera caterpillars including the owlet moth *Scoliopteryx libatrix* from Keene Valley.[44]

Experience gained from the taxonomy of Lepidoptera during his masters helped Dyar deal with the taxonomy of bacteria, which were poorly described in the scientific literature, and his criticism of bacteriologists was similar to his criticism of lepidopterists. For example, in his dissertation he wrote, "Frankland's synoptic table, which, for example, contains ninety species under a single heading (bacilli which liquefy gelatin . . .), these species only to be distinguished by laboriously reading through the several descriptions many of which present no tangible points of difference."

In 1896, after earning his Ph.D., Dyar continued working on bacteria based on recommendations in the literature by J.G. Adami, who considered species of bacteria to be new to science only if they were kept unmodified in continuous culture for a year on standard media. Dyar cultured roughly a third of the 125 species from his dissertation, including those named for Prudden and Hudson. Over half he maintained were either constant or more vigorous, while nine died and six were less vigorous.[45]

Late in life Dyar confided that his work did not meet with his or his advisor's expectations, writing: "at the time I remember that Dr. T. Mitchell Prudden and myself were disappointed, and he expressed himself that I 'had not done myself proud.'"[46] This was perhaps a reflection on the obstacles he faced in making advances in a new field. However, Philip Hadley noted in his study on the variability of two species of bacteria during his dissertation work in 1927: "Dyar . . . [found in] 125 subcultures made from isolated colonies of the 'wrinkled mutant' . . . all but three remained true to the wrinkled type. . . . The wholly wrinkled type, however, bred true for many generations, and in the hands of a less careful observer than Dyar might easily have been reported as involving a permanent mutation." Hadley was impressed by Dyar's forethought and concluded: "After more than thirty years, we are just beginning to appreciate the truth of this view."[47]

Dyar advanced the thesis that bacterial variation was often a reflection of life stages and environmental factors and thus was not an indicator of separate species. The notions of variation and bacterial life stages was rejected by Dyar's colleagues at the time and by most scientists since, although now regaining some traction by the pleomorphists.[48] Dyar was indeed correct that biological features were inadequate to distinguish species of bacteria, but none of his new species were considered valid for the 1980 approved list because they lacked an adequate published description or "if cultivable" they had no type specimens or reference strains available.[49]

## CHAPTER 6

# Genealogies of the Limacodidae and the Dyars

CONSTRUCTION OF THE GENEALOGICAL TREE ... *The larvae divide at once into two groups, the "smooth" and the "spined," separated not only by the differences between the tendency to atrophy of the warts on the one side and hypertrophy on the other, but by the peculiar structure on joint 5 in the spined group.*[1]

H.G. Dyar, Jr., "Life-Histories of the N.Y. Slug Caterpillars," *Journal of the New York Entomological Society*, 1899

For much of Dyar's life his pet group was Limacodidae, the moth family known as the slug caterpillars. They were referred to by Dyar and his peers as 'Codes (pronounced codeez), short for *Limacodes*, the name of the genus from which the family name Limacodidae was derived.[2] The term "pet group" meant more than simply a favorite to Dyar because 'Codes required special care to raise in captivity, taking six to eight weeks to grow from egg to cocoon, nearly twice the time for larger moths or butterflies. Dyar carefully moved the newly hatched caterpillars, which are smaller than letters on this page, to fresh leaves with a tiny camel hair brush.[3]

The name "slug caterpillar" or "slug moth" (Plate 1, Figure 6.1a) comes from the caterpillars in the group that are smooth on the backs (gelatines), although all have sticky underbelly, which has abdominal suckers instead of the typical fleshy pods (prolegs). Dyar referred to the suckers when crawling as "creeping disks." Many of these caterpillars are called "nettles" because they have stinging spines on the back (Figure 6.1a), also often with bright colored bands or spots (Plate 1c,d,f). Less common among slug caterpillars are those called "hairys," which have long, spidery tubercles on the back that pull off (Plate 1g) and the aforementioned "gelatines" with smooth or granulated surfaces that look like leaves, leaf buds or galls, and often have bright-colored ovals or blotch patterns resembling partially eaten or dead parts of leaves (Plate 1a,i).[4]

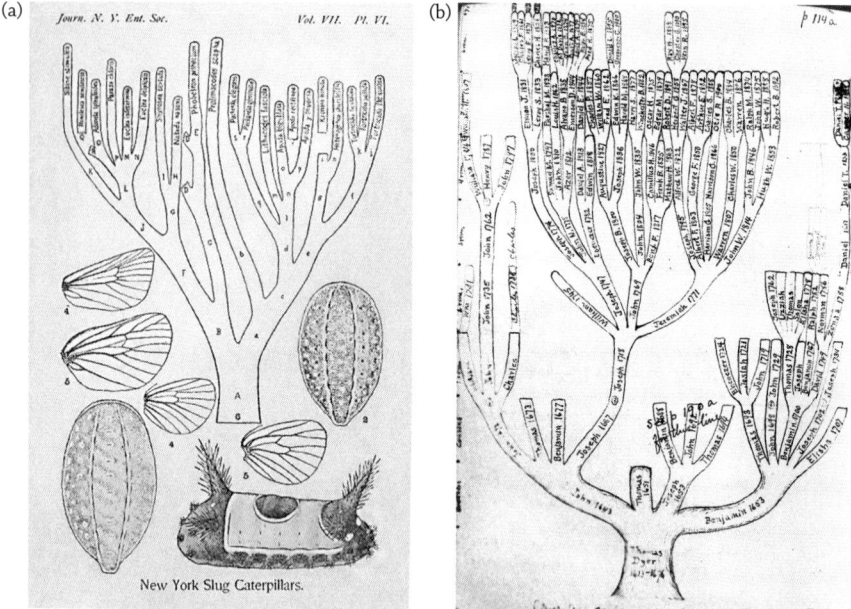

**Figure 6.1** Dyar's genealogies of similar styles at roughly the same time: "The New York Slug Caterpillars" (1899) (*a*) and his own Dyar family's (unpublished) (*b*).

Dyar was not only captivated by the odd specializations or major developmental transformations that limacodid caterpillars go through from the hatchling first instars through many molts (sheds) until they spin a special oval cocoon. He was also interested in the adult moths (Plate 1f), which though often small, have intricate wing patterns with beautiful shades of greens or browns and spidery legs, and are able to contort their bodies.

## THE NEW YORK SLUG CATERPILLARS

Dyar wrote many treatises on moths and mosquitoes of local, state, country, or hemispheric proportions. That he chose the slug caterpillars for his first geographically delimited project, however, is no coincidence: There were only twenty species, so focusing on them presented a rare opportunity to describe all life histories of a particular moth group. His multipart series on the moths, "The Life-Histories of the New York Slug Caterpillars," begun in 1895, had descriptions of each larval stage (eight or more), what they ate, and their other habits. These descriptions were challenging because changes in appearance between molts can be extreme.[5]

Dyar, although well trained for the task, needed the help of Miss Emily L. Morton (1841–1920) (Figure 6.2a) of New Windsor, New York "on the Hudson"—one of among several women actively describing Lepidoptera life histories in U.S. entomological circles—to acquire, rear, and illustrate the life stages.[6] Dyar and Morton's collaboration seems to have begun because they placed mutual ads for exchanges of

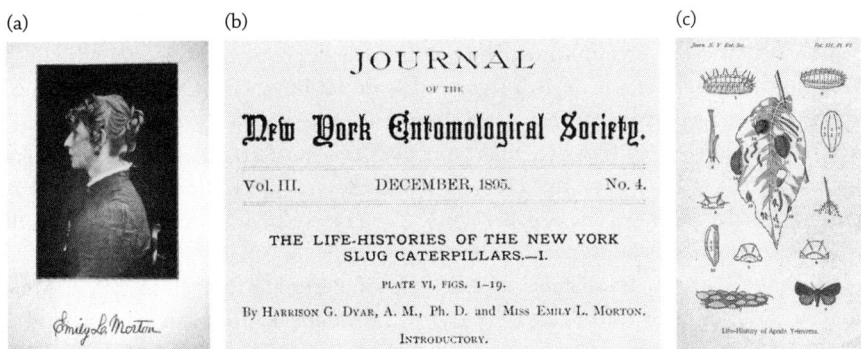

**Figure 6.2** Miss Emily L. Morton, coauthor, illustrator, and caterpillar rearer (*a*); the first page (*b*) and hand-colored reproduced in B&W plate (*c*) of *Apoda y-inversa* from the first of "The New York Slug Caterpillars" series (1895).

moths, including limacodids, in the March and April 1893 issues of *Entomological News*. She, consenting to collaborate with Dyar, sent him specimens in April 1893, writing, "The moths you requested were sent by mail this afternoon. The dates on labels refer to their emergence from cocoon.... I am surprised you want my notes after your criticism of my paper on *Isa textula*" (italics mine). Morton was referring to Dyar's March 1892 critique in *Entomological News* pointing out that she–with the help of Prof. Packard–had misidentified *Isa textula* in her recent paper in the same journal. It was actually another limacodid, *Tortricidia pallida* (Plate 1h).[7]

Adults from her limacodid cocoons and those Morton received from correspondents such as J.B. Angelman of Newark, New Jersey, emerged the following spring. They were mated to produce the eggs and caterpillars Dyar needed to write his descriptions.[8] The two divided their responsibilities to produce the first two of the resulting articles in the following way: "The accounts of the habits are by Miss Morton, and the labor of obtaining fertile eggs has been performed by her. Dr. Dyar has prepared the technical descriptions." Dyar also did a number of the original drawings in his "Blue Books." The first installment of "NY Slugs"—the purpose of which was "to present jointly a full account of the life-history of each species" found in the state—was published in December 1895 issue of the *Journal of the New York Entomological Society* (Figure 6.2b).[9]

Each "NY Slugs" paper included descriptions of the caterpillars and eggs, along with drawings of the adults. The level of detail and accuracy of the drawings, particularly of the tiny first stages and eggs were clearly done by Dyar with the aid of a compound microscope, although Morton also drew and colored a limited number of them (Figure 6.2c, shown in B&W). The description of each species was accompanied by an extensive list of literature citations of adult and the larval stages, followed by "Special Structural Characters" (unique features), "Affinities, Habits, Etc." (relationships with other species, number of broods of adults, food plants), "Criticism of Previous Descriptions (including those by Packard, Dyar, and others), and the "Description of the Several Stages in Detail" (included eggs, up to ten instars, and cocoons).

Morton abandoned the project after the appearance of the second paper in March 1896 for reasons unknown. She, nonetheless, continued to contribute livestock and remained in touch with Dyar and seems to have valued their connection, something suggested by the last letter she is known to have written to him: "I cannot tell you how sorry I was to miss your visit last autumn—the only two weeks I was away the whole season from May to September. . . . [I]t would have been a great pleasure to me, to show you my collection. . . . I so rarely have an opportunity to talk to people who know about insects."[10]

Without Morton, Dyar solicited the help of Perle and her close friend Miss Louisa (Lou) Hoff, who collected for Dyar in Rhinebeck. They helped him find a newly hatched red-cross button slug (*Tortricidia flexuosa*) (Plate 2f) on July 2, 1897, in Bronx Park after a two-day search. Both this caterpillar and a sawfly larva were found on American chestnut.[11] The innovative "concluding paper" on the "NY Slugs" was published in December 1899 after Dyar had moved to Washington, D.C. (He published two more "NY Slugs" papers in the twentieth century.) The 1899 paper included a revised identification table to reflect knowledge of the caterpillars discovered after Dyar began publishing the series, as well as a primer on how to find and rear the caterpillars, gleaned from his work with Morton. Tips on rearing the caterpillars included wintering the cocoons in backyard flowerpots, something Dyar also noted in his catalogue, and using glass jelly tumblers for female moths to lay eggs in and for him to raise the emergent, slow growing caterpillars on leaves placed inside.[12] Dyar made his own quince jelly in the fall, providing plenty of tumblers for preserves or caterpillars.[13]

Among Dyar's most impressive contributions to the field in the concluding paper was his "Genealogical tree of New York Slug Caterpillars" (Figure 6.1a). This tree of relatedness was based solely on characteristics of the caterpillars uniquely derived from a generalized caterpillar. This ancestral form was to have had three rows of tubercles (warts) and fed upon hatching from the egg on smooth leaves. On the tree Dyar subdivided the Limacodidae into four different sections based on the general appearance of the upper surface texture and a geographic component: Tropic hairy, Tropic spiny, Tropic smooth, and Palearctic smooth. Within these broad categories he broke them into a total of seven types of caterpillars. Other branches from this main stem had various modifications or reductions of warts. One major stem, The Tropic spiny, lost the ability to feed upon hatching from the egg.[14]

The use of structural and biological features of the hatching stage, the first instar, was very innovative for the classification of insects and may hold the key to understanding the phylogeny (genealogy) of this group to this day. As he had done in previous work around this time, Dyar went on to place features of the adults onto his larval tree, assuming these caterpillars more accurately reflected the genealogy. He also broke down the number of times an adult feature, primarily the branching vein pattern on the forewing or the length of labial palpi, the fingerlike appendages surrounding the proboscis on the head, was gained through convergent or parallel evolution.[15] To conclude, Dyar—revisiting a topic that he spoke about at the April 1897 meeting of the New York Society[16]—offered a preliminary explanation of the geographic distribution of the various groups of limacodids. It included six maps, developed with the help of Columbia University professors Henry F. Osborn and Gilbert Van Ingen that showed, complete with land bridges, "the probable distribution of land and water in the Present, Eocene, Upper Cretaceous, Lower Cretaceous, Jurassic and Triassic periods."

"NY Slugs" and other serial papers, including those with Neumoegen earlier, made Dyar the dominant author in the *Journal of the New York Entomological Society* over its first seven years. Eighteen articles on the "NY Slugs" were published through 1899; two more would follow over the next fifteen years. Concurrently in the 1890s Dyar wrote several other life history accounts of limacodids for the journal, including species from Florida and Europe.[17]

∽

While writing "NY Slugs" and other life histories of limacodids, Dyar took to giving lectures on his pet group. One of his talks at the New York Society meeting in October 1896 was on the first-instar larvae, or hatchlings, of ten species. In it Dyar suggested the ancestral limacodid larva was "more like *Lagoa* (Megalopygidae) than any other known larva, a conclusion entirely in harmony with . . . [my] previous results."[18] The following February, after his winter trip to Florida, Dyar again gave a talk on a Florida limacodid at the New York Society: *Calybia* (now *Alarodia*) *slossoniae*.[19] In a *mea culpa*, Dyar admitted that he had been wrong by previously considering it to be a flannel moth (Megalopygidae), in disagreement with Packard who believed it was a Limacodidae based on the adult stage. Dyar's discovery of the caterpillar, which fed on mangroves in the vicinity of Miami, led him to concede Packard was correct. In a paper published in the autumn, Dyar described the moth's complete life history, including illustrations of developing embryonic stages inside the translucent eggs—perhaps the first time all these stages of a moth had been shown in natural position while still inside the egg. The form of the larva was close to the hag moth (*Phobetron pithecium*) (Plate 1g), which Dyar considered to be most primitive of the limacodids.[20]

The published version also included Dyar's observations on wasps, which, during their larval stages, attacked and killed some of his mangrove limacodids. Each wasp was identified and described as a new species by William H. Ashmead at the request of Howard of the U.S. Department of Agriculture. One of the wasps was named for Dyar (*Crypturus dyari*) and according to its namesake, the larval stage of the wasp fed inside the limacodid caterpillar but did not destroy it until after it built a cocoon. The adult wasp cut its own hole in the cocoon rather than "emerging by the lid," as moths do.[21]

Dyar's next talks at the New York Society focused on the form and function of a caterpillar's fleshy abdominal prolegs (not to be confused with the six legs on the thorax). His subjects were flannel moths (megalopygids), which are related to limacodids, and sawfly caterpillars: each have an inordinate number of prolegs—seven or more pairs (five or less is normal in Lepidoptera). In April Dyar illustrated that two characteristics of prolegs in flannel moth caterpillars have different functions. "First the ordinary [pro]legs with hooks on segments 3 to 6 and 10 used for prehension and second a series of paired soft pads on segments 2 to 7 used as sucking disks for adhering to smooth surfaces." He went on to describe how the odd undersurface could evolve to sucker disks in limacodid caterpillars, "where the prehensile legs have disappeared and the disk is formed by an extension of these soft pads."[22]

Dyar's last talk to the New York Society was in October 1897 and described the larval stages of three species of burnet moths in New York. Burnet moths, Pyromorphidae (now Procridinae), are related to both slug and flannel moths. The talk's subject included the well-known grape leaf skeletonizer (*Harrisina americana*), which feeds in large aggregations causing economic damage to grape crops.[23]

## DYAR'S GENEALOGY AND CONNECTION WITH THE HANNUMS

Dyar was a genealogist in perhaps the broadest sense imaginable. As he was assembling the genealogy, or phylogeny, of the New York slug caterpillars and the same for other Lepidoptera in the late 1890s, he also actively documented the genealogy of his own family (see Figure 6.1b), the Dyars, published in 1903.[24] Both Dyar and limacodid family trees were strikingly similar in style (see Figure 6.1a)—having volumetric branches with the names inside—in contrast to most of his familial trees of Lepidoptera, which followed the traditional stick-figure model.

In gathering genealogical data Dyar asked for information from relatives, including Daniel Everett Dyar and Charles Warren Dyar, to help him construct a Dyar-family tree much in the way he requested moth specimens with a similar tree-building objective.

In scientific precision, Dyar tabulated the number of male and female Dyars born between 1740 and 1900—109 males and 103 females—and calculated the proportion as being 21 males to 20 females, although this was not included in the published genealogy. He also tallied birth dates for the same Dyars by date and month (July had the least at 14 and October the most at 29). Dyar's February 14 birthday was unique among those tallied.[25]

Although Dyar published only the genealogy of his paternal family, extant notes show his extensive work on the maternal one as well. Dyar's maternal aunts and uncle were clearly important to him, particularly because he lost his father just shy of his ninth birthday. Furthermore, all of his father's siblings preceded his death in 1875 except one, John Wild Dyar of Romeo, Michigan (1814–1889). Six of Eleonora Dyar's siblings, all preceding her in birth, survived her into the 1900s.[26]

Among those who remained close to Dyar were Josiah, or uncle Joe—who sent him the Hannum family bible along with his civil war artillery sabre—and Ruth Banister Marble and Parnie (see Figure 1.2). All stayed in touch with their nephew Harry and his wife Zella, writing affectionate letters to him, Ruth up to a few months before she died. The sisters, living in White Plains, also kept connected with Eleonora Dyar's second husband Dr. Clough and with Nettie Colburn's husband William P. Maynard.[27]

CHAPTER 7

ᴄᐯᴐ

# Last Days in New York, L.O. Howard, and a Move to the U.S. National Museum

In late 1895 Dyar rented a flat a mile from the Columbia University Campus at 243 West Ninety-ninth Street (Map 3), less than a mile from Perle and her husband Dr. Knopf's home at 16 West Ninety-fifth Street (see Figures 7.1a, b and Map 3), about a half block from Central Park West.[1] Having a doctor in the family, and particularly one close by, had its benefits, especially as the Dyars' first child, Dorothy, was on the way. She was born on April 10, 1896; soon after, her father Dyar named a new species of moth for her: *Macrurocampa dorothea* (Notodontidae).[2]

Another factor in Dyar's family relocation to West Ninety-ninth Street may have been related to housing Dyar's growing collection. Likely kept previously at 599 Broadway, a property leased to the Shakers in January 1896, the size of Dyar's collection of Lepidoptera in early 1895, at the time of Neumoegen's death, was over 11,000 specimens. His collection grew partially as a result of gifts, exchanges, and purchases from dealers and auctions.[3] The auctions benefited entomology societies of the day, including the New York Society, which used the funds raised to publish its journal.[4]

The immense wealth Dyar and Perle received from their father was invested largely in New York City real estate. The most notable property was at 331 Fifth Avenue, which was estimated to be worth over $250,000 in 1907. They also owned equal shares of the "four-story high stoop stone dwelling" at 16 West Ninety-fifth Street (Figure 7.1). Dyar sold his share of the brownstone to the Knopfs in 1899.[5]

Dyar siblings' leasing of both the 331 Fifth Avenue and 599 Broadway buildings in 1896 was perhaps related to Perle's return to the city from France that year. The Fifth Avenue building went to Albert H. Davenport of Boston for "12 years . . . total rental amount of over $100,000."[6] The 599 Broadway building went to Robert Valentine, as trustee of United Society of Shakers of New-Lebanon for sixty-three years, at a rate of $16,000/year. When the lease on Fifth Avenue expired in 1909, the following update appeared in the New York City newspapers: "Plans have . . .

**Figure 7.1** Sixteen West Ninety-fifth Street, the last of the New York City properties owned by Harrison and Perle Dyar that remain today (*a*); along with residence next door at 13 West Ninety-fifth Street, apparently where Dyar and Zella were later photographed on the stairs in the 1920s (*b*) (see also, Figure 25.2a).

been filed for remodeling the five-story basement store at the southwest corner of Fifth Avenue and 33rd street . . . owned by Harrison G. Dyar and Dr. P. N. Knopf [*sic*]." It was to become "a studio building, with elevator service, and shops on the ground floor."[7] The Dyar siblings sold the Broadway property in 1899 to Frederick Ayer (no relation to spiritualist Marcellus Ayer of Boston) for $300,000 ($8.3 million, today).[8] Income from rent or sale of these properties gave Dyar assets to purchase other real estate or later to cover living expenses in Washington, D.C. This was to become an important cornerstone to aspects of the Dyar legend: both above and underground.

## L.O. HOWARD AND BRINGING DYAR TO WASHINGTON

*I do not like very well the idea of only 30 days vacation in Summer. I am accustomed to go to different regions for the purpose of working out the life histories of the Leps there and this length of time would prove insufficient. I do not want to have to give up these researches, so would ask you if it is possible to make a different arrangement in this respect. . . . 243 W 99th.*[9]

H.G. Dyar to L.O. Howard, March 5, 1897

Leland Ossian Howard was the primary force behind Dyar's move to Washington (Figures 7.2b, c). In November 1895, Howard filled Dyar's loan request for giant skipper caterpillars (*Megathymus*), which bores into the roots of yuccas and agaves.[10] The frequency of the correspondence between Howard and Dyar increased the following month. The U.S. Department of Agriculture (USDA) needed to identify tent caterpillars, potentially serious forest pests. Evincing his appreciation of

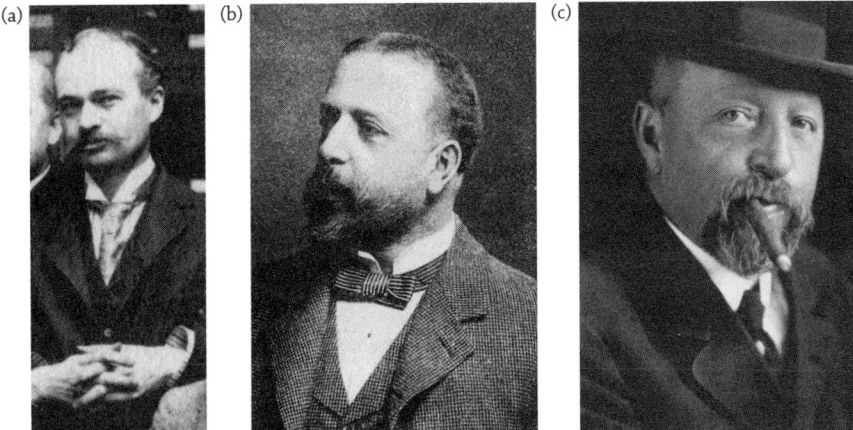

**Figure 7.2** Dyar, Jr. (c. 1905) (*a*), who was courted by L.O. Howard (c. 1890s and 1920s) (*b–c*) to come to the U.S. National Museum.

Dyar's skills, Howard sent Dyar a package from Tucson, Arizona, containing the tent caterpillar, which spun a social web on emory oak (*Quercus emoryi*) writing:

> "[W]e have reared them ... thought at first the species was *Gloveria arizonensis*, but I see it is very different. ... I send you two moths and a vial of larvae. ... Should the species ... prove to be new, I will send you another pair, to be used in drawing up descriptions and afterwards returned here, so that we may have the co-types, at least, in the National Museum collection."[11]

Dyar named the species after Howard, who had been correct in his assertion that it was a tent caterpillar moth (Lasiocampidae) new to science. "I propose to call them," Dyar wrote Howard in January 1896, "*Dendrolimus* [now *Gloveria*] *Howardi* with your permission. ... The earliest stage sent (II) shows the arrangement of the tubercles in the typical Lasiocampid fashion with considerable clearness." Thanking Dyar for his proposition, Howard suggested the collector Prof. James W. Toumey of the University of Arizona be the recipient of the honor, but Dyar insisted, and the species remained *howardi*. "If you wish to suggest a name for the new *Dendrolimus* I will adopt it with pleasure," he explained. "Otherwise I would prefer the one I have proposed."[12]

The two continued corresponding about tent caterpillars. Howard soon asked Dyar if he was familiar with the western tent caterpillar of *Clisiocampa* (now *Malacosoma*) *californica*, which can cause severe defoliation in trees and shrubs. At least that was what Dyar believed, based on the description by Henry Edwards, and was "pleased to get any information about the California species."[13] Per Dyar's request Howard had Theordor Pergande write up a summary of all the tent caterpillar specimens, including all life stages, in the USDA and U.S. National Museum (USNM) collections. The depth of knowledge Dyar displayed on the economically important tent caterpillars was impressive enough for Howard to invite him to examine more specimens in Washington: "You will see that we have considerable biological material ... of interest to you.... Can you not arrange to come down some time [sic] soon and look it over?"[14] At some point during 1896, probably before he began his fieldwork on sawflies in May, Dyar accepted the invitation, and Howard discussed with him the idea of his taking up a position as the lepidopterist at the USNM.

Howard demonstrated his growing confidence in Dyar, whimsically asking him to recommend a preparator for the National Museum: "[Do] you know any young man, more or less expert in mounting and preparing insects, who would take a position in this office at $60 per month. One of my clerks has just left me, and I should like to fill the place with such *a male specimen, if one is to be captured.*"[15] Dyar suggested C.L. Brownell, a botanist and ornithologist with some insect collecting experience,[16] but Howard wanted a true entomologist: "I should much prefer some one [sic] with a better knowledge of insects and of entomological technique."[17] That Howard didn't follow Dyar's suggestion did not harm the growing friendship that would serve as the foundation of their future scientific partnership. At the end of the year Howard complained of a wasted opportunity to meet Dyar, writing: "I just missed you at the Onteora Club in the Catskills last summer. I dropped in at the Bear and Fox Inn on Tuesday, and found that you had registered there the previous Sunday. Why did you not make a call at my cottage?"[18]

Howard felt his mission as curator to expand the national collection, move it to a more spacious building, and hire select scientists was best served by choosing Dyar as the lepidopterist. An added benefit was that Dyar's knowledge of caterpillars was sorely needed to make accurate identifications of agriculture and forestry pests for the USDA. Howard was also likely impressed by Dyar's attempts at making more scientific classifications of moths using larval and adult stages, and Dyar's visit to Washington gave Howard the opportunity to test Dyar's abilities as a taxonomist. Indeed, from late December 1896 through January 1897 he forwarded moths to Dyar in New York for his identifications, and although Howard must have been aware of Dyar's growing reputation for being highly critical of lepidopterists, he saw that the benefits to hiring him outweighed the risks.

Following up on Howard's interest in getting him a position at the USNM, Dyar wrote in February 1897 for clarification, noting that Howard had mentioned "the departure of Mr. [Martin] Linell" was necessary for a position to open and musing that Linell "would be needed for the Coleoptera." Linell, Dyar seems to have thought, was going to be dismissed. During the spring Dyar told Howard that he would soon know if he would be reappointed at Columbia University. If he were to go to Washington instead, he would need to "have a flat rented and family to arrange about." He also wondered about "the duties of the position and how much time can one have to go off collecting?"[19] Howard intended to personally discuss these matters with Dyar during the summer, writing, "I shall be passing through New York several times and should like to see you." In his response, Dyar noted, "I have a cottage at Bellport Long Island (2 hours from New York ...) where I shall be most of the time in June—September. Before June I shall be here and shall be pleased to see you if you come up that early."[20]

Regardless of his success getting a position at USNM, Dyar was thinking about going to Washington, writing to Joseph Lintner that he would "probably be moving" but did not know his future mailing address in D.C.[21] Two weeks later, on May 3, Martin Linell unexpectedly died of a heart attack,[22] and Dyar indicated to Howard that he had not yet accepted the reappointment he had been offered by Columbia, noting the university might "want me sooner than . . . expected." While hinting he was inclined to a move, Dyar did not make a firm commitment.

Still waffling, Dyar wrote to Howard on June 5, 1897:

> I would not have bothered you with writing again on the subject [of moving to Washington] except that I naturally supposed that as you suggested my taking Mr. Linell's place in case he were discharged last year, that now his death could certainly make the place open. . . . I hope I shall be able to come across you during the Summer. If not, I shall see you in September as I have about decided to move to Washington in the Fall, where I can use the library facilities to finish a work on Bacteria that I have started and also be on hand if you are able to carry out your plans about the museum.[23]

Clearly pleased, on June 7, 1897, Howard pledged to give Dyar "more money when . . . [he came], and earnestly hope that the plan which I outlined in my letter to you from Boston can be realized not later than July 1st, 1898." Howard also mentioned he did not expect Dyar "could afford to come to Washington for Linell's salary" and was "delighted to hear that . . . [he would] come . . . in the fall."[24] Dyar does not seem to have been worried about money, writing Howard in June that that he expected the cost of living in D.C. to be half of what it was in New York City, which "costs me here $5000 [$143,000 in today's dollars] a year" and that it was not an "insufferable obstacle" to start work in Washington without a salary. The bigger issue to Dyar was that the "30 days vacation in Summer" for fieldwork fell far short of what he was accustomed to.[25]

Earlier, Dyar also expressed reservations about the National Museum Building (see Figure 9.1a), noting he expected a new, larger, and fire-proof building to be built to house the collection, given his potential stake in it. He was concerned about other issues, including "the present crowded condition," which he blamed for 'the very poor impression that [scientific] visitors get," and, of course, the need for "a Lepidopterist to keep the collections up to date." Dyar also believed they required space for an expected donation from Mr. William Schaus (see Plate 9c,d) "who owns one of the finest collections of South American Lep[idoptera] whose object always was to leave it at his death to the National Museum."[26]

Howard's reply shows a willingness to accommodate Dyar: "If you can tell me how I can take any steps to overcome criticism, I shall receive them with the greatest pleasure. In meantime, your letter will help me to give the new Director an idea of conditions and needs." Dyar, the realist, mentioning the problem he encountered because the collection was split between the USDA and the USNM, responded: "I do not see that you can very well make any change to correct the impression to which I referred as matters stand now."[27]

On July 1, 1897, the Division of Insects of the National Museum, Smithsonian Institution hired William Harris Ashmead (1855–1908) as Assistant Curator (see Figure 9.1b) and Rolla P. Currie as Aid (1875–1960),[28] and when Dyar arrived in D.C. he accepted an unsalaried position.[29] Hiring Ashmead and Currie was an important step toward expanding and consolidating the Department of Agriculture and the National Museum collections, as well as facilitating a move to more spacious quarters across the National Mall. The two hires through the USNM represented a net gain of one position, and Howard was hoping to get half-time support for Dyar through either the Division of Insects (USNM) or the Division of Entomology (USDA), which would enable Dyar to continue his intensive fieldwork, including rearing. The USDA also had lepidopterist August Busck (1870–1944) (see Plate 9a,b), who arrived at the museum in 1896. Dyar, however, was the anchor to Howard's ambitious plan for the Lepidoptera collections, and Andrew Nelson Caudell (1872–1936) (see Plate 9e,f and Figures 9.1b and 11.3a), a grasshopper specialist employed by USDA, was assigned to Dyar with tasks ranging from cataloging, preparing larval specimens, and fieldwork.

# PART II

# Beginning a New Life at the USNM and in Washington

CHAPTER 8

⊂\/⊃

# Life in the District of Columbia and Wellesca Pollock

*And the story about our grandfather playing the piano . . . One night while he was playing, he observed a mouse sitting under the piano, and he reached down and scooped it up. He was quite certain he caught it. There was nothing in his hand . . . no sign of this mouse. He was wearing a dressing gown with, of course, wide sleeves, and there in the sleeve, looking up at him, was this little mouse. Apparently, it came out every evening and listened to him, and, of course, he didn't have the heart to do it any harm. He just put it back down on the floor.*[1]

Roberta Hill about grandfather Harrison G. Dyar, Jr., August 3, 1999

In 1898, Dyar purchased a brownstone at 1512 Twenty-first Street NW, Washington, D.C., several miles north of the museum but near Dupont Circle, one of the capital's most fashionable residential districts (Figure 8.1a).[2] To the rear of the home was a vacant lot, which Dyar soon purchased for a garden, between his home and the Larz Anderson Mansion on P Street. On the corner across the street Thomas Walsh, an Irish immigrant who made a fortune from his Camp Bird Gold Mine in Ouray, Colorado, would build his mansion in a few years hence (Figure 8.1b).[3] Walsh's daughter Evalyn soon married Edward B. McLean, the owner of both the *Washington Post*—a paper Dyar would frequently appear in—and the Hope Diamond, whose future home was Dyar's workplace, the Smithsonian.

When Dyar settled into the house along with Zella and his daughter Dorothy, his mother-in-law, Harriet M. Peabody (Plate 11b), became a part-time resident.[4] According to Roberta Hill, Dyar's granddaughter, a certain amount of entomophobia (fear of insects) followed the family's taking up residence in the house: "Mama [Dorothy] said that the black nanny used to be convinced that there were insects. . . . [L]iving in the same house with an entomologist . . . gave her this notion. She was convinced that there were creatures crawling on her back."[5]

Many of the family stories relate to Dyar's love of insects and other small creatures. Included among them is one about Chesapeake Bay crabs he was supposed to

**Figure 8.1** Dyar's home in an exclusive neighborhood in Dupont Circle (*a*) and the Walsh Mansion, across the street (now the Indonesian Embassy) (*b*).

cook: "[When he put] the lid on the pot [it] did not fit securely. The crabs all managed to come out . . . and there they were all running all around the kitchen floor, and he couldn't sentence these crabs to double jeopardy. He couldn't put them back in the pot. They'd survived. He gathered them all up and took them back down to Chesapeake Bay and turned them loose."[6]

The Dyars purchased the house adjoining their own in November 1904 and connected the main floors. Thereafter, the children were said to roam freely between the dwellings. The extra space allowed Dyar more room to rear moth and mosquito larvae and also to get live-in seamstresses—adding to the resident staff of a maid and cook—to make and fit the family with clothes. "The most exciting thing was the decision to buy a second house." Dorothy's daughter Roberta recalled being told. "At first, they had people who came once a year and did clothes for them, and they'd stay over in the house."[7] Two years later, Dyar acquired property behind his home from Franklin T. Sanner for $17,910. The property had "a frontage of 120 feet on P Street and a depth of 110 feet," which Dyar intended to use "to enlarge [his] rear gardens" (see Figure 22.1c). Dyar also wanted to make a "fine improvement" of the alleyway but needed the neighbors' approval and later built a small, single unit apartment behind the residence and across the alley.[8]

Dyar's home was also a meeting place. His active participation in the Entomological Society of Washington led to at least four meetings at the residence. The first was hosted in February 1899. At the January 1900 meeting Dyar showed the communal larval cases of the madrone butterfly caterpillar (*Eucheira socialis* Westwood) from Mexico and gave a talk about how to identify adult males of saddleback caterpillars in the Americas. "Saddleback" refers to a white bull's eye oval or green blanket saddle pattern in the middle of the back surrounded by spines on horns (Plate 1c). This group was obviously one of Dyar's favorites because he was to name many *Sibine* species for women in his family.[9]

Soon after the January 1900 Entomology Society meeting, the Dyars took a winter trip to Palm Beach. Zella was pregnant, and they returned home after two months. The Dyar's second child, Otis Peabody Dyar, was born on May 13, 1900. Upon hearing of the new arrival, Dyar's uncle Joe, now living in Syracuse, wrote, "Dear Harry Yours of yesterday just to hand . . . We congratulate you & the mother on the advent of the 'baby boy.' May he 'live long & prosper.' Aunt Bessie says you (Zella) can thank her for it being a boy—as she dreamed the night before the event it was to be a boy. Dr Knopf will attest the prediction. Who says there is nothing in dreams now?"[10] Surely Bessie's (Elizabeth) prediction resonated in a family of mediums. Also, the mention of Dr. Knopf, Perle's husband, attests to the closeness between relatives at the time. The new baby provided challenges for Dyar, as illustrated by a cautionary note to Strecker when inviting him to stay at the Dyar home: "I shall be glad to have you come down any time after Oct 1. . . . There is plenty of room in the house but the baby's noise is penetrating."[11]

## STONY MAN AND WELLESCA POLLOCK

In July 1900 the Dyar family went to stay in a cabin at Stony Man Camp in the Blue Ridge Mountains near Luray, Virginia. The proprietor of the Camp, George Freeman Pollock, transformed it into a Chautauqua, a program that provided the upper-class patrons with lectures, elaborate balls, and entertainment, including jousts, tournaments, and productions with cakewalks and pageantry. During camp festivities, Dyar was out in the field on Stony Man Mountain and found some tiny caterpillars of the crenulate moth (Epipleminae: *Callizzia amorata*). Dyar had become acquainted with these caterpillars in the Adirondacks in 1894 on wild honeysuckle bush. A more significant acquaintance for the Dyar family that summer was George F. Pollock's sister Wellesca, a D.C. kindergarten teacher with wavy auburn hair, brown eyes, fair complexion, and who stood 5 ft. 2 ½ in.[12] The siblings grew up at the camp, which was purchased by their father George H. Pollock (Figures 8.2a, b). Dyar began developing a friendship with Wellesca, who, along with her sisters Loue and Uila, was a regular attendee at the "many new social novelties," picnics, and parties that took place during the season and at which Dyar was sure to find himself. He, Wellesca, and Uila took the roles of wood sprites in a play at the Stony Man Cake Walk and Carnival on August 25, 1900.[13]

The Pollocks' mother, Louise Plessner Pollock (1832–1901), was a famous educator and among the first to bring kindergarten to the United States. Born in Prussia, she and her husband from Boston started a family in the early 1850s—Susan, was the oldest of eight, and the youngest, Wellesca, was born in Weston, Massachusetts, on February 12, 1871. In Washington, D.C., Louise would head the National Kindergarten and founded the Pollock Normal School, along with her daughter Susan: a school for teaching education.[14]

Dyar would draw Wellesca further into his world, teaching her how to find crenulate moth caterpillars while at Stony Man in 1900. Her success would be credited upon Dyar's publication of a paper on the caterpillar: "The next season

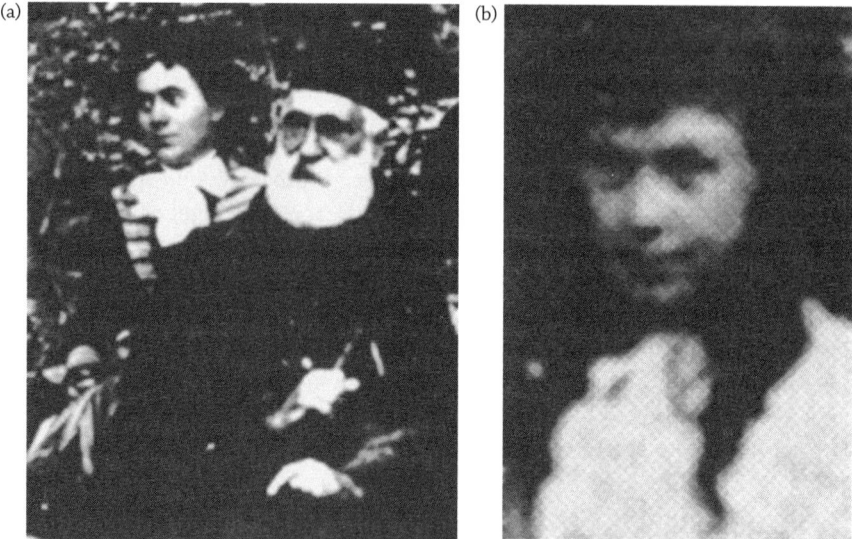

**Figure 8.2** The Pollock family at Stony Man, Virginia, in 1888: George H. Pollock and his son, George Freeman Pollock (*a*), and daughter, Wellesca Pollock (*b*).

[1901] Miss Wellesca Pollock," Dyar wrote, "kindly undertook to get some more from the mountain and she successfully carried them to the pupa stage."[15] He also employed her as a typist in the fall, and she helped Andrew Caudell complete Dyar's "List of Lepidoptera," a catalogue of all the known North American species. Her job at the museum began each afternoon after she finished teaching at the Pollock kindergarten.[16]

Dyar then decided to rename *Parasa prasina*, a species of limacodid from Mexico and Central America, *Parasa wellesca*. Dyar had named the moth *Parasa prasina* in 1898, but the name, or a homonym of it, had been used for a species from Asia. Dyar wrote on top of a reprinted copy of his 1898 article in *Psyche* "Parasa//prasina Dyar*" and on the bottom, linked by the asterisk, "*call it wellesca." Dyar quietly published the new name in the November 1900 issue of *The Canadian Entomologist*, not calling attention to the recipient of the honor, his usual practice even for species he named for Zella or Dorothy. In a 1905 article, however, Dyar noted that the *wellesca* species was "[n]amed in honor of Miss Wellesca Pollock of Washington, District of Columbia," suggesting that passions had grown between them.[17]

## ZELLA DYAR, HER MOTHER, AND PLACE IN WASHINGTON SOCIETY

Zella, part of Washington high society, given her wealth, education, and New England connections, hosted social gatherings as she had in Boston, including one in March 1907 for the Washington Bahá'í community. In April 1909 Zella and her mother co-hosted a tea "in honor of Mrs. Isabelle B. Chase and Mrs. William McFarland [at The Brunswick]. . . . The tea table was presided over by Mrs. Swasey, wife of the new representative from Maine," and others. The attendance by Mrs. Swasey suggests earlier connections with Zella and Harriet Peabody, given that Swasey's husband was born in their hometown of Canton, Maine, in 1839. He certainly knew Mrs. Peabody's father Caleb Holland, who served in the Maine state legislature.[18]

Zella and her mother hosted another evening at the Twenty-first Street home in May 1910, this time a musical gathering featuring several singers, including a contralto from New York, Miss Thompson, a pianist/violist from Chevy Chase, Maryland, and Mrs. Leland O. Howard. Mrs. Howard gave a "talk on the folk songs of the South, and sang three of the old melodies . . . , showing how closely allied to the Egyptian music the crooning style of the African plantation songs are."[19]

Zella probably could not hear the music or conversation at this and other gatherings. She had suffered severe hearing loss around the time of Otis's birth,[20] hindering her ability to enjoy music, though she did attend concerts with her daughter, using an ear trumpet to aid her enjoyment.[21] According to Dyar, at a time when Zella was nearly deaf, she would "entertain [him] evenings by reading stories . . . so he didn't have to go out. However, as the children got bigger and were put to bed later, she would begin her reading later and later and at last she stopped it altogether . . . [when] . . . getting out of the habit of it and the fact that the children now . . . [ate] up the whole evening."[22]

Although Dyar's wealth could comfortably provide for the family, Zella needed her mother's help to raise the children, though Harriet Peabody's work as a social activists did keep her out of Washington, D.C. for long stretches. Her work among the poor dated back to the early 1880s in Boston, when she became the matron of the Children's Mission on Tremont Street. In the 1890s she traveled the country for speaking engagements, fundraisers, and the founding of the Ethical Club of the Twentieth Century in Los Angeles, California, and the New England Peabody Home for Crippled Children in Newtown, Massachusetts, in 1894, a home that remained in operation until 1961.[23] From 1899 to 1905 Peabody taught at the Navajo school in Aneth, Utah, in the southeastern part of the state from 1899 to 1905. Her work with the Navajo made her "famous all over the country" because she not only taught but also established Navajo schools. She aided in the preservation of traditional Navajo craft, that is, "the making of blankets by hand with hand-dyed wool," and facilitated Navajo commerce by organizing a fair that "gave modest cash prizes for best blankets, corn, melons, etc., and roused much hopefulness among the Indians" (Plate 11b).[24]

Peabody was as concerned for her daughter as she was for the downtrodden, as evidenced by a November 1902 letter to Dyar: "I hope you will see that Zella attends strictly eating and the inhaling business. Got here on time last evening. Tell Zella I am all right." Peabody was also investing in western mines, telling Dyar about "Little Louise Mine" in Leadville, Colorado: "I am here and have closed my deal with Mr. Kunge . . . and send you the certificates to care for. Dr. Kingsby of White Plains is to have 10,000 Der and Ex. and 10,000 Little Louise. Wish you had as much." Six months later, she started a reading room for miners of Poland, Arizona, where she found moths for Dyar, as well as in Bluff, Utah, in 1903. Perhaps the gift of specimens was quid pro quo for investment advice, though Dyar's reward for the Utah specimens was to name an owlet moth *Noctua phyrophiloides* variety *peabodyae* after her. It has since been raised to the level of species *Pronoctua peabodyae*.[25]

Harriet Peabody also had an interest in religion, or as Dyar wrote, she "had a wide experience in things spiritual, and she was fond of investigating each new idea ... as it appeared."[26] One of these "ideas" was the Bahá'í Faith; in fact, she had met Dr. Ibrahim G. Kheiralla when he first brought the nascent religion to the United States in 1892. Her familiarity with Bahá'í philosophy led to Harrison Dyar's first acquaintance with it: He later adopted it in his own way.

## CHAPTER 9

# Beginning a Career Building the National Collection of Lepidoptera

The third Secretary of the Smithsonian Institution, Samuel Pierpont Langley (1836–1906), appointed Harrison Dyar Honorary Custodian of the Section of Lepidoptera of the United States National Museum (USNM) (Figure 9.1a) on November 12, 1897,[1] allowing Dyar to travel at his leisure and seek new insect finds. Meanwhile, Howard began his efforts to compensate Dyar for his work.

Once Dyar arrived in Washington in October he wasted little time taking charge of the USNM Lepidoptera collections, "rearranging the Lepidoptera . . . , which are in considerable confusion," he wrote Lintner.[2] Howard also needed Dyar to make authoritative identifications of these insects, particularly those that threatened the nation's agriculture and forestry.[3]

Dyar returned to New York City for Christmas, but continued thinking of his work, writing Strecker to ask for a detailed description of a rare limacodid moth (*Kronea minuta*) on Christmas Eve. Although Dyar surely would have preferred the moth itself, he wrote: "Will you be so good as to do me a Christmas favor? . . . I must ask you to examine your specimen with a lens and note whether or not the palpi reach above [the] vertex of head . . . , [the] hind tibiae [have] . . . two pair of leg spurs, [and if the male] antennae [are] simple. I want to know these points quite exactly."[4] Dyar also continued to identify samples sent by post from Howard while on his Christmas holiday: for instance, forty-three moths from a lamp in Texas on December 30, 1897.[5]

Shortly after returning to the National Museum, Dyar began to show himself as the cantankerous Custodian of Lepidoptera, expressing dissatisfaction about inadequacies in the collection. Dyar complained to Howard that he was "humble pie" about the collection not having specimens of "birdwings." That term referred to Ornithoptera, the world's largest butterflies with a natural range from Southeast Asian archipelagos, New Guinea, and Australia, which Dyar felt should be represented in a collection of the caliber and reputation he sought to build.[6]

(a)

(b)

**Figure 9.1** Old National Museum building (now Arts and Industries), where the Smithsonian insect collection was housed when Dyar arrived in 1897 (a), and its entomology workers inside including (*right* to *left*): R. Shannon, E.A. Schwarz, Dyar, W.H. Ashmead, O. Heidemann, D.W. Coquillett (partially hidden), J.C. Crawford, A.N. Caudell, C.V. Locke, and unidentified (c. 1905) (b).

Dyar's tenure at the museum began during a time of unprecedented growth of the collection. Insects donated by Dr. William L. Abbott of Philadelphia were notable during the 1898 fiscal year: "940 specimens, . . . including many Lepidoptera from 'Trong, Lower Siam'" in Asia.[7] One of Dyar's early tasks was to put names on Abbott's exotic Lepidoptera specimens. For this he enlisted the help of his friend Herman

Strecker. When Strecker also offered to relax and mount the dry specimens kept in envelopes, Dyar, although appreciative, wanted it done in house, as he wrote: "Mr. Caudell is doing it under my supervision. A few of the last lot from Abbott were mounted last summer [1899] while I was away in the old style and several good things ruined."[8] Among the Abbott specimens was a new species of map butterfly (*Cyrestis aza*) described by Strecker from Trong. Dyar also donated roughly 300 of his own specimens; half of them reared sawflies, and the remainder Lepidoptera.

Ashmead (Figure 9.1b), the Assistant Curator, wrote Dyar the following letter of appreciation in June 1898 about the growth of the USNM Lepidoptera collection since his arrival eight months earlier: "The past year has been phenomenal in the history of the Department of Insects and I believe the authorities are working up to its importance & growth." The letter also reveals that an office move was taking place in the National Museum building, for Ashmead wanted Dyar, at his cottage in Bellport, Long Island, to send back his office key so Dyar's "desk [could be] moved to the room below."[9]

Although Howard and Ashmead appreciated Dyar's efforts to expand and improve the collection, the museum administration was disinclined to pay for his services. Still, Dyar wanted a salary, and Howard was keenly aware he had assured Dyar one. After Dyar's first two years at the museum, Howard pushed to get him back pay, writing, in 1899, and similarly in later requests, "The value of the voluntary services . . . rendered . . . cannot be overestimated. He is one of the foremost living workers of the world in the order Lepidoptera. His energy and working capacity seem unlimited. In [a] . . . short time he has entirely re-arranged the collection of Lepidoptera, has added to it very largely by donations from his own collections, has published a number of papers based upon Museum material, . . . [made] exchanges with foreign Lepidopterists, and . . . made his portion of the collection, which was prior to his advent lamentably weak, in many respects, a credit to the Museum."[10]

Howard was attempting, apparently unsuccessfully, to get Dyar paid $900.00 for work he completed during his first two years as a contract and continued, in 1900, to seek payment for Dyar, observing that Dyar was "discouraged at the apparent lack of appreciation which the Museum has shown" and that he [Howard] was "very anxious that this feeling of discouragement should not increase."[11] Howard had dual interests in retaining Dyar: He was needed to build the nation's Lepidoptera collection at the Smithsonian (USNM) along with protecting the nation's agriculture through accurate identifications of agriculture and forestry pests (the mandate of the USDA).

Howard, notably, did not seek compensation for Dyar's personal collecting trips, which provided specimens for the USNM. In fact he proposed that Dyar have a six-month appointment to allow him the travel he desired on his own dime.

## DYAR'S "LIST" AND EXPANDING LEPIDOPTERA INTERESTS

*If any of us are inclined to regret that a man like Dr. Dyar, one of the most original and gifted investigators in America, should spend his time in preparing a catalogue, we may console ourselves by recollecting the character of some other catalogues, prepared by men of less ability. In truth, the thing was well worthwhile, and its value to students of American Lepidoptera can hardly be overestimated.*[12]

T.D.A. Cockerell on Dyar's "List of Lepidoptera," *Science*, 1903

One of Dyar's first major accomplishments at the museum was writing "A List of North American Lepidoptera" (Figure 9.2a) to replace J.B. Smith's 1891 "Checklist of Lepidoptera of Boreal America." Ordered to reflect his recent classification of Lepidoptera, Dyar's "List" was a referenced catalogue, not, as Smith's was, a simple checklist of current species. The project apparently had its genesis at an October 1896 meeting of the New York Entomological Society when a motion, which was voted down due to "lack of funds,"[13] was raised to have Dyar edit such a list. Dyar, who was apparently already working on one, continued to pursue the goal.

Achieving the goal was problematic. The principal experts on two groups of Lepidoptera with abundant species, Prof. Fernald ("micros") and Prof. Smith (owlet moths), were not enthusiastic supporters of the project. That fact seems to have led Dyar to postpone, or consider postponing, his work, writing, also in October 1896, to Fernald:

> Your withdrawal will render the preparation of a good list impossible, so we should have to give it up. I have had a somewhat similar letter from Prof Smith in regard to the fact that there are copies of the 1891 list still unsold, which he regards as an objection to a new list at this time. I should like to prepare a list to illustrate my classification, but if the present time is not suitable, shall be obliged to postpone the matter.[14]

Two years after he withdrew from participation Fernald rejoined, providing information on most subfamilies of snout moths (Pyralidae) and the leaf rollers (Tortricidae).[15] August Busck (see Plate 9a,b) focused on the other "micros" including Gelechiidae, while Hulst worked on Geometridae and Phycitinae of the Pyralidae. Dyar had put together his team of taxonomists for writing his annotated

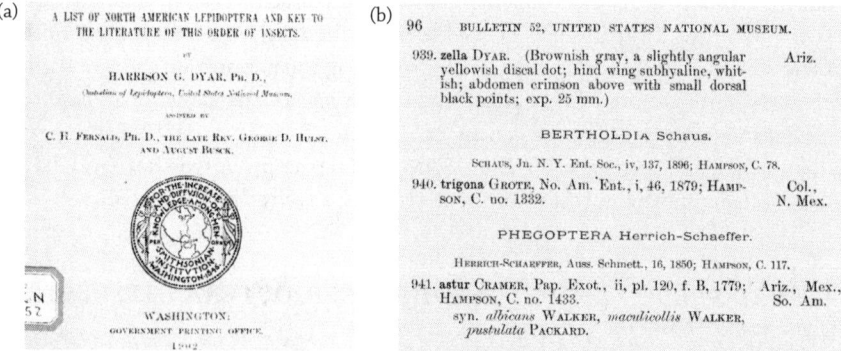

**Figure 9.2** Title page of "A List of North American Lepidoptera" (printed 1902) (*a*) that included Dyar's description of *Calidota* (now *Euchaetes*) *zella* from Arizona (top), named for his wife (*b*).

"List" of the North American species, which included synonyms and multiple references to literature on each species.[16]

To help Busck, Dyar wanted all of the USNM Gelechiidae borrowed by Fernald to be returned; at the same time, he wished to borrow Fernald's own type specimens (those originally used to name and describe species). Dyar feared that Fernald would not comply to such an asymmetrical request, so he asked the more diplomatic Howard to contact Fernald, who, according to Dyar, was "more likely to reply favorably . . . than to any one else."[17]

Fernald did not react approvingly to Dyar's support of Busck to become a "serious student" of "Microlepidoptera." In January 1900, Dyar wrote Howard, "I think the learned professor . . . can't refuse access to. . . . [types] to students even though they may not be his equals in age . . . The slur on Mr. Busck is quite unnecessary but characteristically Fernaldesque. It was this sort of thing that disgusted me with Prof. Fernald personally." Dyar was opposed to sending specimens to Fernald because Busck would not gain the knowledge needed to name them for his "List." Given Fernald's attitude, Dyar went on to suggest that Busck work at other museums such as the Museum of Comparative Zoology (Harvard) and the Academy of Natural Sciences in Philadelphia.[18]

∽

Dyar's "A list of North American Lepidoptera and key to the literature of this order of insects" was printed on December 1, 1902, in the *Bulletin U.S. National Museum*: its official publication date was March 1903 (when printed copies reached the scientific community). In the preface Dyar stated he organized the moths and butterflies known to occur in North America by "somewhere near a natural classification . . . based largely on larval characters in . . . the arrangement of the family and superfamily groups." He went on to compare his system with Edward Meyrick's for the British Museum, though in somewhat different order. In general Dyar's "List" goes from groups he believed to be most derived (advanced), the butterflies and "macro" moths, to those most primitive (the "micro" moths).[19] In addition to his primary collaborators Fernald, the late Rev. Hulst, and Busck, Dyar acknowledged advice from J.B. Smith and A.R. Grote for the largest group of moths, the Noctuidae. Dyar's names for owlet moths closely followed those in Smith's 1893 noctuid catalogue: roughly a third of all Lepidoptera reported in the publication. Although Fernald was not acknowledged for the names used for superfamilies (groups of related families), he and Dyar had extensive correspondence on this topic.[20]

Dyar's "List" included the same species names as in Skinner's catalogue for the butterflies but used Samuel H. Scudder's earlier names of genera, noting: "I have not been able to give the matter full study, so that some of the genera are doubtless not pure, while there are probably too many of them; but as they stand they seem more in harmony with generic ideas, as understood in the rest of the order, than they have been in any other general list, and are at least an improvement on these." He arranged the different butterfly families in deference to Grote, having placed "Papilionidae (swallowtails) first, not because I regard them as higher than the Nymphalidae (brushfoots)[,] the reverse is the case[,] but because the studies of Prof. A. R. Grote seem[ed] to show that they can not be interpolated anywhere, having a very distinct phylogeny."[21]

The catalogue had one instance wherein Dyar wrote a very brief, even for him who was often spare, description of a new species of the tiger moth *Calidota zella* from Arizona, which was named for Zella (now *Euchaetes zella*) (Figure 9.2b).[22] The new species (collected by E.A. Schwarz and H.S. Barber in Arizona) apparently came to Dyar's attention during the latter stages of writing the manuscript. Perhaps Dyar chose to name it for Zella because of her history with tiger moths, collecting a colorful but common species, *Arctia docta* (now *Notarctia proxima*), before they got married. Dyar soon wrote a more complete description of the *zella* species.[23]

Dyar's "List of Lepidoptera" was written about in such newspapers as the *Washington Times*, whose headline read "Over Six Thousand Kinds of Butterflies"[24] while T.D.A. Cockerell's professional review appeared in March 1903. Comparing Dyar's "List" to Staudinger's European catalogue and contrasted it with J. B. Smith's "List of 1891," which was a list rather than "a detailed catalogue, including full references to literature and brief indications of localities ... searched with extraordinary care." Cockerell praised Dyar for making use of "the most recent advances in our knowledge of the classification of the Lepidoptera, many of them due to Dr. Dyar himself." Cockerell, however, found fault with Dyar's practice for the butterflies of lumping together "names of aberrations and some names of geographical races" with "pure synonyms" (same species, different name), and in other cases arbitrarily listing subspecies or forms as valid.[25]

A review of Dyar's "List" in *The Canadian Entomologist* was far less positive, expressing disappointment in the marked increase of generic names, over twice the number, for butterflies to go with only a small increase in the number of species. Dyar presented 158 butterfly genera for 652 species versus the 65 and 74 genera for 645 and 640 species found in earlier lists by Skinner (1898) and Smith (1891). Much of the increase was the result of Dyar resurrecting old Hübner names, also used by Scudder, on the basis of priority. The review suggested that "the ordinary student will feel much hesitation in adopting this List as his guide" because so many names, already learned, would have to be dropped, and the student "will naturally be inclined to think that the List cannot be final, and that it will be safer ... to wait for further developments" before changing collection cabinet labels and filling "notebooks with new names."[26]

Smith published an updated version of his 1891 "Checklist" in June 1903, borrowing liberally from Dyar's catalogue.[27]

⁂

In Washington, Dyar's interests expanded to other groups of moths. Among the first of these were inchworms (Geometridae) (Plate 7g–j). Dyar, setting out to correct mistakes in preparing his "List," took on this large moth family, poorly known for its larvae, the same year he completed his main body of work on the New York slug caterpillars. Published in *Psyche*, the inchworm series through 1907, comprised of sixty-eight publications, one per species, that described the caterpillars' various instars and, when available, the egg and pupal stages.[28] The decades-old work of

Benjamin Walsh on the caterpillar of *Aplodes* (now *Nemoria*) *mimosaria* was among the earlier descriptions Dyar found fault with, observing Walsh's "description is not only brief, but erroneous, as the larva is entirely without 'short velvety hairs' and has none of the structure of *Phobetron*."[29]

Dyar also described many new species and genera—35 species in 1904 alone— more than twice as many as in previous years combined. The death of Hulst, a specialist on inchworms and snout moths (Pyralidae), each having among the largest number of species in Lepidoptera, may have influenced Dyar to focus on them. Although Dyar maintained his interest in families formerly associated in the Bombyces (e.g., Saturniidae, Limacodidae, Megalopygidae), these had far fewer species than the influx of arctiines, noctuids, pyraloids, and geometrids from Mexico and Panama. The Mexican material came in large part from purchases from Roberto Müller and specimens from Panama were collected by August Busck, who had gone on six-month trips there in 1911 and 1912 (see Figure 13.2c).[30]

Dyar described many new species collected by New York native William Schaus, Jr. (1858–1942) (Plate 9c,d), one of the most significant contributors to the Lepidoptera collections at both the National and American Museums of Natural History. These came in large part from Schaus's expeditions throughout Latin America, particularly from Mexico, Costa Rica, and the Guianas. Along with donating hundreds of thousands of specimens to the USNM, Schaus described over 5,000 species of Lepidoptera. Schaus—who spent most of his time in England or collecting in Latin America or the Caribbean with his valet John "Jack" Barnes (1871–1949) from 1895 to 1914 (see Plate 9d)—would eventually become, in 1918, formally associated with the National Museum, where he would place his collection thanks in part to Dyar's efforts.[31]

The new moths that Dyar described from the American tropics often belonged to the largest moth families (those with the most species). Between the years 1904 and 1919 many of the new genera and species Dyar would describe were owlet moths: Noctuidae (600). However, he was also prolific with snout moth families Pyralidae (471) and Crambidae (311).[32] Dyar would describe and name roughly 88% of the new species and genera of snout moths by 1919 after Fernald had openly encouraged him to work on Crambidae in 1906,"[33] and continued into the 1920s on a collaborative effort with Carl Heinrich.[34] His affinity for these moths led him to refer to them as a pet group.[35]

## DYAR JOINS THE ENTOMOLOGICAL SOCIETY OF WASHINGTON

Dyar was an active participant of the Entomological Society of Washington during his first fifteen years in D.C., becoming a member at his first meeting in mid-October 1897. He regularly presented papers until 1912, including on sawflies and limacodids in the first years, his primary interests at the time, and later mosquitoes. These entomology meetings brought up intriguing topics, including show and tell (dead or alive), which led to discussion by the attendees.

At the January 1898 meeting Dyar brought a sawfly specimen he named for F.C. Pratt (*Lophyrus pratti*), a "young Englishman" adept at rearing, who had found the larva on arbovitae in Woodstock, Virginia, and at the March meeting he exhibited live slug caterpillars feeding on mangrove leaves from a recent Florida trip, while presenting a talk about identifying limacodid caterpillars from Townend Glover's 1872 "Illustrations of North American Entomology."[36]

In October 1898, after the summer break, Dyar spoke on *Acronycta* moths (now *Acronicta*), discussing work he had done with John B. Smith that was soon to appear as a monograph in *Proceedings of the United States National Museum*. At the November meeting Dyar engaged the patriarch of the society, Eugene Amadeus Schwarz (1844–1928) (Figure 9.1b), in a discussion about the uncanny resemblance between a burnet moth and a net-winged beetle in southern Arizona. Schwarz thought the beetles were "deceived by the appearance of the moths," whereas Dyar thought it could be a more complex "mimetic relationship" involving several moth species resembling the beetle.[37] In January 1899 Dyar gave a talk on the phylogeny of tent caterpillar moths, based on their larval features and adult wing venation, similar to talks over the previous five years on giant silk moths, limacodids, and other moth families. Perhaps most intriguing was his hypothesis that the profiles of these woolly caterpillars protect them from predators by obscuring shadows and suggested that long warts of the most primitive tent caterpillars, lost through evolution, were replaced by specialized wool.[38]

Howard gave an account of the colorful primordial meetings back in 1884 when "entomology and beer went together." He explained the reason: "Marx, Schwarz, Heidemann, Pergande, Lugger, Schoenborn, Ulke, were all Teutons ... [and] John B. Smith's real name is Johann Schmidt. At that period the German university idea dominated scientific America." Howard regaled that after the meetings "the refreshments were unlimited in quantity but limited in kind; you could have light beer or dark beer, and that was about the extent of the variation. It was my custom to order two cases of beer, each of 24 bottles, for an average attendance of 7 or 8."

By 1909, the meetings became more moderate, as Howard, who "would not vote the prohibition ticket," noted a change in refreshments, when Nathan Banks "gave us hot lemonade and cold lemonade and some very excellent raisin cake." Bureau Chief Howard acknowledged that whereas "a few glasses of beer will make a stupid remark sound witty ... there was no necessity for any such stimulus to the imagination in the old days, because all of the remarks were witty."

Meetings of the Society were cordial, something foreign to visitors, as Howard noted: "On one occasion years ago a member of a prominent northern entomological society ... attended a meeting of our society ... After adjournment I asked him 'Well, what do you [think] of our society?' 'Huh,' he said, 'I don't think it is an entomological society at all.' 'Why?' I asked.' 'There wasn't a bit of quarreling,' he replied."[39]

Such cordiality may have been breeched when Dyar openly disagreed with J.B. Smith at the first meeting held at the Dyar residence in February 1899. Not unusual among lepidopterists, the two held opposing views on whether a tiger moth was to be considered an actual species or a form. According to the minutes of the meeting: "'Professor Smith showed typical specimens of a new species of *Haploa* found in New Jersey, for which he proposes the name *H. triangularis* on account of a peculiarity in the markings' "Dyar stated that he had been familiar "with the form for some years" but considered it "a southern form of *H. confusa* Lyman," not a species. Dyar went on to attribute the forms of the *Haploa* to local isolation, "which complicated the definition of the species."[40] Dyar's opinion on the matter has stood the test of time.

Although Dyar was expressing his professional opinion, a rivalry with Smith was brewing while his personal relationship with him was on a downward spiral.

# CHAPTER 10

# Battle of the Titans Smith and Dyar, and Their New Love—the "Skeets"

*... The story that we had to tell*
*Of bee and butterfly,*
*Our story—have we told it well,*
*With love and earnestly?*
*O, with the lapse of years, how small*
*Do all our quarrels seem!*
*Like children's play, or like the fall*
*Of shadows on a stream!*
*This story of the spider's nest,*
*Of beetles, black or gray.*
*Is but a story, at the best,*
*Told by ephemera!*
*Still is it the pursuit of truth*
*Where all the pleasure lies,*
*A perfect knowledge—that, in sooth*
*Is hidden from our eyes.*
*Upon this quest our little barque*
*Has bravely held its way,*
*On board a crew of men of mark*
*As e'er sailed for Cathay. . . .*

A.R. Grote, January, 1894[1]

Similar to the "bone wars," a feud between paleontologists Edward Drinker Cope and Orthniel Charles Marsh over who named Dinosaurs first, entomologists fought over names of Lepidoptera. Combatants included A.R. Grote, C.V. Riley, H. Strecker (see Plate 9 for all three), J.B. Smith (Figure 10.1b), and Rev. G. Hulst (see Figure 10.2a). However, their battles were often about whether or not

**Figure 10.1** Dyar panning for mosquito larvae at Herzog Island, Maryland (March 1905) (*a*) and Smith in his typical field attire (perhaps near Plummers Island, Maryland (*b*).

antiquated names from the early 1800s should be given priority over new ones, including claiming earlier dates of publication of their own species.

One dispute that spanned decades, involving Grote in the early stages and later Dyar, was over the validity of names in Jacob Hübner's *Tentamen* (1806). A small number of taxonomists adopted *Tentamen* names because they predated others, but were opposed by C.V. Riley and most North American lepidopterists in the 1870s and 1880s because they preferred to keep Lepidoptera names they knew rather than those unfamiliar, including 228 genera that contained only one species accepted by S.H. Scudder.[2] Other feuds were over allegations of stealing specimens, an accusation made against Strecker by lepidopterists including Grote,[3] who "had a part in [many] personal disputes" with North American lepidopterists,[4] some of whom Dyar also battled, including, most notably, John B. Smith.

༒

Dyar and Grote would develop a friendship, exchanging a barrage of letters in the mid- to late-1890s though apparently never meeting, and Dyar would show his admiration for Grote by dedicating his journal *Insecutor Inscitiae Menstruus* to him and modeling it on Grote's short-lived *North American Entomologist* (1879–1880).[5] Grote seems to have contacted Dyar first, writing him from Bremen, Germany, on June 15, 1894, to praise Dyar's work classifying Lepidoptera but cautioning him: "It seems to me that the time is coming nearer when our classification of the Lepidoptera will accord with the results of our phylogenetic studies. All such studies as yours will greatly assist, but no one or single character must be considered infallible." The "single character" referred to the caterpillar setae Dyar used in his classification.[6]

Grote also suggested Dyar was too much of a "lumper," that is, one who grouped together "species" that displayed minor differences: "You are going too far in the reduction of species. The northern forms I think are different from the southern in Parorgyia etc." Although he favored the use of *Tentamen* names, Grote went on to suggest a different approach to nomenclatural priority: "You go back to what appears the earliest name & neglect the priority ... to every subsequent writer in restricting the use of the name. Where two names are synonymous I think the succeeding authority should be respected in his choice."[7]

Grote also drew attention to his antipathy for Smith, complimenting Dyar for not being "rude like Smith but hav[ing] some ... consideration and conscience."[8] Among Grote's problems with Smith at the time, as well as with Hulst, was Smith's rejection of Hübner's *Tentamen* names. This was in agreement with AAAS nomenclature committee of 1894, who stipulated that names for genera and species should reflect biology and evolution rather than strict adherence to priority.[9] Dyar's allegiance to the *Tentamen* seems implausible, given his attempts to build a classification of Lepidoptera supported by evolutionary evidence, but Dyar's view may have been influenced by his early exposure to Samuel Scudder, Boston's leading proponent of the *Tentamen*.

## BATTLES BETWEEN DYAR AND SMITH

Although the fabled "battle of the names" between Dyar and Smith never took place, they did each name a moth for the other in unflattering ways. Smith named an owlet moth *Euclidia dyari* in 1903 and Dyar followed with *Protorthodes smithii* in 1904. The names themselves were neutral, but the purpose for their selection was not: Each author claimed the other made some taxonomic blunder and used a replacement named after the alleged perpetrator.[10] Although this is indeed what a taxonomist does when having to come up with a new name for a homonym, it is not hard to imagine each enjoyed pointing out the other's foible.

The relationship between Dyar and Smith appeared cordial during the 1890s. The first hint of a conflict between the two dates to February 1892, when Dyar published a critique of Smith's "List of Lepidoptera of Boreal America" in *The Canadian Entomologist*. Dyar acknowledged the importance of Smith's work but pointed to "errors of omission" and just plain errors.[11] Smith responded: "There be criticisms and criticisms; those intended as friendly and those intended as destructive in character, and sometimes one is as unwittingly unjust as the other may be intentionally so.... Mr. Dyar is evidently a friendly critic, and I feel obliged for his kind words; but some of the "inaccuracies and omissions" are misleading. The List went to the printer in June, the Bombycids were printed in August, and Mr. [George] Hudson's descriptions of *Dasychira* and *Cerura* did not appear until September or October."[12]

In a harsher critique of Smith's work in *Psyche* later the same year, Dyar wrote: "The species of *Cerura* have been so badly mixed up in Prof. Smith's new list, that I will give a catalogue of them."[13] In many ways what Dyar did was reasonable, as he tidied up the taxonomy of *Cerura* in his review, but the problem was his tone. Rebukes in print between Dyar and Smith would eventually escalate to a similar degree as those between Grote and Smith in the 1880s, and it is not always easy to judge how the combatants felt toward each other. For example, Grote disparaged Dyar's use of the larval stage in classification, yet they had mutual respect and friendly correspondence. Smith, in fact, was predisposed to view Dyar positively, as he, along with other notables, assisted in particular sections of Smith's 1891 List, the preface of which notes that Smith had consulted, along with others, Dyar, who "furnished a list [for the Limacodidae] which is very closely followed."[14]

Smith welcomed Dyar's arrival in D.C., writing, in his capacity as the New Jersey State Entomologist, to Howard, "Glad to hear Dyar is with you. I have heard nothing for months & wondered what had become of him."[15] On New Year's 1898, Smith sent Dyar literature on the window-winged moths (Thyrididae) and enthusiastically discussed Comstock's work at Cornell, writing, "I saw Comstock at Ithaca a few days ago & he has completed a demonstration of the origin and homologies of the wing venation in all orders. It is one of the finest & most complete bits of work I ever saw, & its possibilities in tracing the phylogeny of insects seems immense."[16]

Around the time Dyar became established at the USNM he partnered with Smith on a monograph of *Acronycta* (now *Acronicta*) (Noctuidae)—a revision of the genus and other allied ones. Completed in late 1898 and published in *Proceedings of the United States National Museum*, it was originally to have been written by C.V. Riley along with Smith and to include "all the breeding notes and records accumulated for many years in the U.S. Department of Agriculture."[17] Riley's widow gave Smith "the entire mass of papers and notes, together with all the original drawings and sketches [of] *Acronycta* [caterpillars],"[18] but Smith needed someone with Dyar's expertise to complete the project.

From his own perspective Dyar was unimpressed by Riley's records that he received from Smith, writing in their coauthored work, "Professor Smith had practically no useful notes [i.e., Riley's] on early stages to turn over to me, but there was considerable material of the common species on which I already had notes, and only a few of the specimens were of service. Fortunately, my notes were rather full in the Acronyctid forms."[19] Furthermore, Smith and Dyar held different views of the *Acronicta* classification. Smith found problems in defining *Acronicta* because of inconsistencies between adults and larvae. Dyar thought the classification "coincided in a remarkable manner."[20]

By stating these contrary views it is unclear whether Dyar offended Smith; however, the relations between the two had soon deteriorated, becoming apparent in 1901 when Smith changed concepts of *Acronicta* found in their revision without consultation. In a flippant letter to *The Canadian Entomologist*, Dyar suggested that special treatment be given to certain Guenée and Walker type specimens (i.e., original specimens used to describe species): "Prof. Smith has been to London, and now radically changes the synonymy in the genus *Acronycta*, which I had hoped was to have been finally settled in the revision which was published by him and myself. The changes involve the identification of three of Guenée's species and one of Walker's. As to *impleta* Walk., we must accept Smith's identification as *luteicoma*, G. & R. *I would suggest that the type ought now to be destroyed, lest future changes in the synonymy result* [italics mine]."[21]

Dyar's remark about destroying the type specimen "was looked upon as a good joke" by his Washington cohorts.[22] Joking aside, part of Dyar's sensitivity to Smith's taxonomic changes was they created confusion over which caterpillars went with which adults. In these owlet moths, the adults are difficult to separate from each other, whereas the caterpillars are quite distinctive (see Plate 5a,c)—a situation Dyar and Knab would later encounter with mosquitoes. However, Smith took Dyar literally, replying: "He accepts my identification. . . . in so grudging a spirit that he suggests destroying the type—of *impleta*, I presume. . . . It is to be fact that there are several hundred types in his charge, the suggestion is unpleasant reading. It is a somewhat startling method of securing stability of nomenclature!" Smith then wrote: while "a species is entirely represented only by all its stages and both sexes of the adults" the name of the species is still applied to an adult in order "to fix a type."[23]

Offended by the rebuttal, Dyar drafted a response to Smith and sent it to Howard for his opinion in May 1901:

> Dear Sir, Your justified remarks cannot be justified. You did not see my joke; very well, we are not all alike and what is funny to me may not appeal to another. Therefore I will admit that you could assume that I suggested it would be well if Walker's type was destroyed . . . You have no shadow of right to imply that I would destroy types or injure the collection in any manner, whatever my personal views. Such statement cannot fail to injure the museum. . . . This is entirely aside from the matter of the *Acronycta* synonymy, which I will discuss further in a friendly spirit.

Dyar noted to Howard: "Of course I don't have to explain to you that my suggestion to destroy Walker's type was a merry jest."[24]

Dyar's published reply in *The Canadian Entomologist* July issue was as pointed as the draft. Smith, it reads:

> implies that I might be led by personal views to an improper treatment of the collection in my charge. This implication I indignantly repudiate, and leave Prof. Smith to explain his breach of etiquette as best he may. While Prof. Smith's lack of humour has led him to misunderstand my views, he has no right to imply that . . . I would not properly conserve the National Collection . . . rapidly becoming the finest in the country [and] will continue to be conserved with the greatest care.

Dyar then became more constructive, discussing whether the specimens referred to as types were in fact the specimens originally labeled as such by Guenée.[25]

Although these debates were vitriolic, Smith, a trained lawyer, seemed to enjoy them and gave up some ground to Dyar in the next *Canadian Entomologist*: "In his note . . . Dr. Dyar raises an interesting question, concerning which I would like a general expression of opinion for my own guidance. . . . Concerning *Acronycta*, it is certain that Guenée has mixed things [between adults and larvae], and he may have done so in two or three different ways." Now realizing it was a joke, Smith wrote, "it is, of course, a serious deprivation to be without a sense of humour, but at the risk of losing all reputation in that direction, I must yet confess an utter inability to see anything funny in Dr. Dyar's original note concerning types. . . . I have no apology to make." Again giving ground, he wrote:

> [I]n justice to Dr. Dyar and to myself, that I did not really believe that he would actually or in any way neglect or allow harm to come to any of the types . . . I had too much regard for and confidence in him as a man to believe that; but I did believe that he gave expression to a conviction that the importance of types had been overrated, and that nomenclature would be more stable were there none to be referred to. In which, after all, he may be right.[26]

Dyar was also angered by Smith's naming "some dozen" *Acronicta* species in *Entomological News* rather than in their monograph. Smith, in turn, was upset with Dyar for inserting a larval description into one of his articles in the *Proceedings of the USNM*. Recalling the incident to Howard, Smith wrote, "he shoved in several descriptions of larvae after the proof had been in my hands and without saying a word to me about it. Mind, I do not object to the descriptions of the larvae in connection with the insect. It would have been common courtesy, however, to tell me that he intended to do so."[27] From Dyar's recollection, he "then regarded Smith as a friend and thought the feeling reciprocated. Therefore I did not suppose he suspected me of inserting in his article 'matter that might conceivably be objectionable to him.' This is a mean insinuation of his; uncandid, suspicious, ungenerous."[28]

⚜

Dyar's criticism of Rev. George Hulst (Figure 10.2a) and his work, even if substantive, furthered his conflicts with Smith, who had a long-standing personal and professional association with Hulst, the New Jersey State Entomologist before him. Perhaps most detrimental to Dyar's relationship with Smith was that Dyar had openly critiqued Hulst's identifications of geometrid (inchworm) moths in Dyar's publication on the Lepidoptera collected during the Harriman Alaska Expedition of 1899 (Figure 10.2b); identifications that he requested from Hulst himself.[29] This was bad form to begin with, but appeared incredibly mean spirited because the publication came out after Hulst's death.

Naturally, Smith condemned Dyar's remarks about his deceased friend. Dyar explained the Hulst situation in this note he sent to Howard:

> I thought a little prodding might help him [Hulst]. His untimely death occurred after the paper was out of my control. '*De mortuis nil nisi bonum*' [Speak no ill of the dead] is a good maxim, though not necessarily be applied literally . . . and I should have made my corrections to Dr. Hulst's determination different if they had not been for his own perusal.

In other words, he deemed the remarks appropriate had Hulst survived to see them.[30]

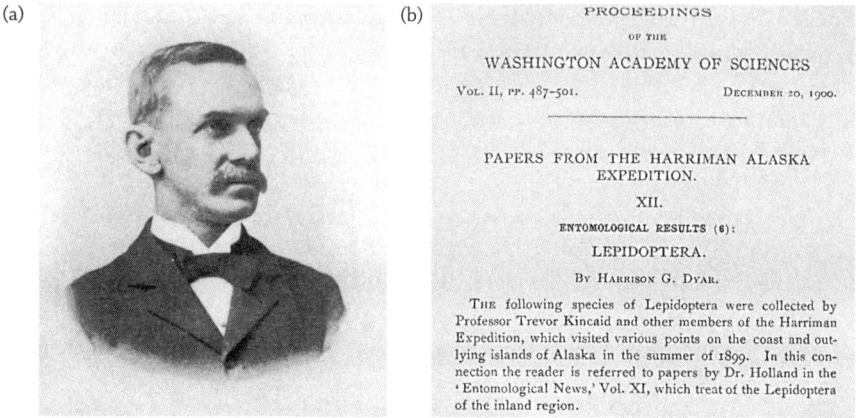

**Figure 10.2** Rev. George D. Hulst (*a*) and Dyar's paper in the *Proceedings of the Washington Academy of Sciences* (1900) that criticized Hulst's identifications prior to his death, but was printed afterward (*b*).

Dyar felt remorse not because of his published remarks about Hulst but because Smith refused to donate Hulst's types to the USNM collection. Smith wrote Dyar and Howard about his intention to "make no effort" to donate Hulst's types. Smith thought the specimens would not receive "considerate treatment" and Dyar would use them to berate Hulst's work in revisions.[31] Dyar hoped Howard could help resolve the issue, writing: "Now I have let off steam to you, I await instructions. I am very sorry to have unwittingly messed the Museum's prospects but perhaps you can rescue us yet with your valuable diplomacy. I am no specialist on *Homo sapiens* (or the var. *insapiens*) and retire in favor of more competent authority."[32]

Seven months before his death Hulst wrote Dyar that the types were not his to give: "As to giving my types to the National Museum.—First, I have no collection—in a proper sense—of my own. All my collection was given years ago to Rutgers College, which may or may not have been wise, but its done, and all I get now, is simply an addition to that collection. Then, as a member of the Brooklyn Institute, I am in a way bound in honor to do what I can with duplicate material to help along there."[33]

To be fair to Dyar, Smith had other issues in regard to giving his own types to the USNM. Whereas Riley forbade types from leaving the museum, the museum abandoned that policy, allowing types to be lent to "certain specialists whose competence and care is without question." Smith was concerned his donated types could be unavailable for him to compare with unidentified specimens he would bring on future visits to Washington.[34]

Although Dyar remained interested in obtaining Hulst and Smith's specimens from Rutgers, he continued criticizing their work. At his second and last presidential address to the Entomological Society of Washington, Dyar, speaking on recent work on North American Lepidoptera at his Dupont Circle home in January 1903, characterized Hulst's work as "undoubtedly brilliant in certain respects, . . . [but]

seriously marred by his habitual carelessness. Nothing ... can be absolutely relied upon, for fear that a thing, apparently most evident, may be found to be vitiated by some blunder that he knew much better than to commit." Dyar ranted that Hulst's use of "secondary sexual characters" in his identification tables (keys) was misguided because one sex or the other was unknown. New genera, Dyar observed, should not be based on solely the female sex of a species, especially because, when doing so, Hulst "simply supplied the missing male characters from his fertile imagination." Nonetheless, Dyar ended admitted, "in spite of defects, Dr. Hulst is badly missed, for he leaves no successor in the study of the Geometridae."[35]

In the same address, Dyar spoke negatively about Smith, an acknowledged expert on owlet moths (Noctuidae). He made unflattering comparisons between Grote—whose descriptions were "clear and concise statements" and had "almost intuitive perception of specific characters"—and Smith—who, despite being "a patient careful man," wrote numerous "lengthy" descriptions "often vague from the very effort at completeness."[36] Furthermore, Dyar found Smith had not "cultivated a knowledge of larval forms, and his work is not checked by breeding," leading "him to describe as species forms not entitled to that rank." In other words, without knowledge from bred adults, it would be easy to mistake variants for actual species.

It seems inconceivable Smith would support donating Hulst's specimens to the USNM given Dyar's callousness. Indeed, Smith even told Dyar, "If you object that I am allowing personal consideration to retard scientific knowledge I will plead guilty."[37] The Rutgers Fire of April 25, 1903, however, gave Dyar new hope.[38] In the aftermath, Dyar wrote Smith: "You ought to have put those types in the National collection long ago as I have repeatedly urged before this warning was directed against you. In the next fire you may not fare so well."[39] To this Smith replied, sending a copy to Howard:

> Under conditions as they exist at the present time, not one specimen of either the Hulst or my own collection will go to the National Museum except in direct exchange. If I find it necessary to get a safer place than we have, the insects will be very much more accessible to me at New York City or in Philadelphia. I do not feel that the National Museum has any claims upon me, though up to the present year I have done everything that was within my power to increase its collections in Lepidoptera.[40]

Ironically, before the fire, Smith had considered donating at least some of Hulst's snout moths (Pyralidae) to the National Museum, and he wrote that he was seeing Dr. Austin Scott, the president of Rutgers, "to have a definite proposition of some kind for the benefit of the National Museum,"[41] but the Rutgers board of trustees voted against moving the specimens.[42] Dyar's critical attitude was a liability that would continue to hinder his goal of gaining a permanent position and cost the USNM valuable scientific specimens. "No one," Smith wrote, "can have a greater regard for Doctor Dyar's ability than I; but there certainly are some things about his way of doing that I do not like."[43]

## DYAR AND SMITH'S RULE OF THE MOSQUITOES

Dyar developed an interest in mosquitoes after the idea that they were disease vectors gained acceptance in the late nineteenth and early twentieth century. Returning from a collecting trip in Colorado to his cottage in Bellport, Long Island, in August 1901, Dyar noted that the mosquitoes "were abundant."[44] At the time he was under the influence of L.O. Howard's mosquito book, particularly its "illustrations of the larvae of *Anopheles*." According to Howard, the onset of Dyar's newfound passion was finding mosquito larvae, known as wrigglers, "in a horse trough [and] studied their characteristic markings and structure."[45] (see Figures 10.1a, 10.3c, Plate 12a–b). Only a few short months later Dyar presented his mosquito observations from Long Island, including novel identification keys to the larvae, at the meeting of the Entomological Society of Washington. John B. Smith, who happened to be at the meeting discussed his own experience with the biology of the same species, though not all in agreement with Dyar's.[46]

Knowledge of mosquitoes became paramount to U.S. interests when it took over construction of the Panama Canal in 1904, and Howard, who had long been interested in them, encouraged Dyar and Frederick Knab (see Figure 19.3a) to redouble their efforts on mosquitoes.[47] Dyar thus began writing a series of papers with Knab, a hired assistant at the Bureau of Entomology, USDA; their first joint effort in 1904 showed the larval stages could be helpful in diagnosing species with similar adults, and through breeding, they were able to better associate females, the blood feeding sex, with the males.[48]

The volume of Dyar's mosquito work makes it seem as if he abandoned his Lepidoptera research early in the twentieth century to focus on mosquitoes. However, Dyar would name most of the 3,200 Lepidoptera from his career after he began researching mosquitoes and name 700 Diptera (mosquitoes and about 100 other types of flies).[49] Still, he did make a significant turn toward the mosquito, something that led to further tension between him and Smith, who had begun mosquito research in the late 1890s. They approached mosquito work from different directions: Smith was a leader of the mosquito control movement in New Jersey (Figure 10.3a, b), whereas Dyar was intrigued by the study of mosquito larvae and adults as independent organisms. Indeed his approach was reminiscent of his approach to Lepidoptera classification.

In the fall of 1901, Smith, confused about the larvae of the New Jersey mosquitoes that he was trying to control, turned to Dyar, asking, "What are [*Culex*] *territans, sylvestris* & *confinis*! With the exception of [*Aedes*] *sollicitans* [the saltmarsh mosquito] you have an altogether different line of names from my own." To remedy the confusion, Smith gave Dyar access to his specimens "for your personal information & for the Washington Society for publication in its due course" and asked to have one of Dyar's identification tables (keys) for the *Culex* larvae as well as specimens in alcohol, live larvae, and adults.[50]

The apparent truce such exchanges suggest began to unravel in March 1903, when Dyar wrote to Smith asking him to "save all stages" of any new mosquitoes for the Carnegie project. This was a monograph on New World mosquitoes that Dyar

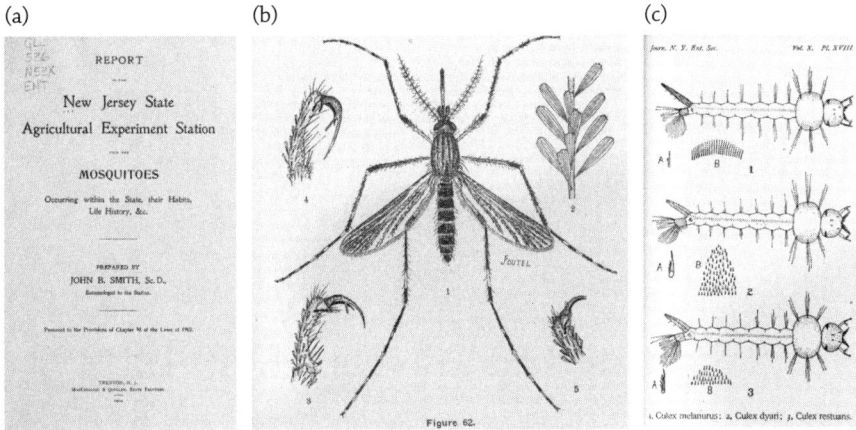

**Figure 10.3** Smith's report on New Jersey mosquitoes (1904) (*a*), illustration of mosquito from Smith's New Jersey report by Joutel (*b*), and illustrations of mosquito larvae by Dyar's sister Perle N. Knopf from his first mosquito publication in the *Journal of the New York Entomological Society* (1901) (*c*).

was working on with Howard, whom Dyar assumed had told Smith,[51] and that the new Carnegie Institution of Washington was funding.

Smith wrote back, saying he knew nothing about the monograph and expressing dismay that Dyar had published a larval description of *Culex signifer* from Smith's specimens, implying his motives for requesting them were deceptive, claiming he would have done the description himself had he known,[52] and demanding his specimens be used "for Dr. Howard's work [the Carnegie monograph]," not Dyar's independent research. Dyar retorted, "I did not suppose that the mosquito larvae you gave me had 'a string to them.' I shall probably never get used to your extremely personal point of view."[53] In other words, it was Dyar's belief that science, including his own scientific work and his criticism of others was "beyond personalities."

In 1905 Smith objected to Dyar making a request for larval specimens through his assistant John A. Grossbeck or other collectors under "my employ," mentioning once again the restrictions he placed on his specimens. Dyar wrote Smith that his "views are a matter of absolute indifference" because he was already following Smith's policy about his specimens.[54] Patterson, in his book *The Mosquito Crusades*, wrote that Dyar's failure to follow this directive by contacting Grossbeck led Smith to vow "to have no further relations with the National Museum so long as Dyar remained" and "not one specimen of either the Hulst or my own collection . . . [will] go to the National Museum."[55] There are several flaws in this account and interpretation, particularly the importance given to Dyar's request to Grossbeck. It appears the root cause of "Smith's vows" likely had more to do with Dyar's scolding letter after the Rutgers fire and other issues relating to Lepidoptera, including the critiques of Hulst.

It does seem curious after all of the rancor between these two men that they still continued to work together. Perhaps Smith gave the best explanation to Howard, but it applies to Dyar as well: "I am more interested in mosquitoes than in quarreling."[56]

## CHAPTER 11

֍

# Collecting and Travels: 1898–1909

*Zella and I went to Sierra Madre and we lunched and walked up the cañon and Zella fell on a rock and sprained her ankle. Mrs. Peabody and I went to Pasadena in the evening to hear Wellesca speak in Mrs. Hammond's house.*[1]

H.G. Dyar recalling a trip to California in 1906 in October 1916

Extended family holidays for the Dyars, replete with collecting and rearing of Lepidoptera and other insects, were commonplace events in the early 1900s. The trips were facilitated by the dramatic increases in rail service, particularly on the transcontinental system. Not always crisscrossing the country, photographs from around 1902 at an East Coast beach show that perhaps his favorite caterpillars remained on his mind when he was away from the museum. Dyar is building a sand castle for his son Otis (see Figure 11.1 and Plate 10a); at first glance the castle looks like a mountain with trees on it, but closer inspection reveals something very different, a representation of a limacodid caterpillar such as the spiny oak (*Euclea delphinii*) (Plate 1c) or rose slugs (*Parasa indetermina*) (Plate 1f) found on bayberry bushes near Atlantic shores. Family photos, taken between 1902 and 1906 (see Plate 10), were from beach vacations to Bellport (perhaps at Fire Island), Weekapaug, Rhode Island (1904),[2] or at Chesapeake Beach, Maryland (1904).[3] Specimens from these trips were donated to the museum as part of Dyar's effort to build the collection.[4]

Other summer trips during Dyar's first decade in Washington included Long Island, Boston, and Plattsburgh (1898)[5]; Long Island (1899)[6]; Summit, New Jersey,[7] and Stony Man Mountain, Virginia (1900)[8]; Colorado (1901)[9]; Center Harbor, New Hampshire (1902)[10]; Kaslo, British Columbia (1903)[11]; Tryon, North Carolina (1904)[12]; England, northern New York, and Massachusetts (1905)[13]; Southern California to British Columbia (1906)[14]; North Carolina and Boston (1907)[15]; Lincolnville, Maine (1908)[16]; and several brief trips, one to Plattsburgh, New York. and two separate visits to Dublin, New Hampshire (May and August, 1909).[17] There was also one spring trip to Florida and Southern Georgia (1905).[18]

**Figure 11.1** H.G. Dyar, Jr., with his son Otis and a sand sculpture reminiscent of a spiny limacodid caterpillar, on an Atlantic coast beach (c. 1902).

The general travel pattern for his trips, at least in the early 1900s, was to make short excursions in the Washington vicinity, before or after longer rail trips with or without the family to the Northeast, West, and Northwest. Once he arrived at his primary destination, Dyar would rely on trolleys, coaches, and automobiles to reach his collecting sites, though he also collected at the lights at train depots, as he had done on his first trip out west.

⚬⚬⚬

After moving to Washington, the Dyars made winter escapes to Florida from the damp cold. The first was in January 1898 to Miami and the Keys, with a side trip to the Bahamas.[19] At the Miami River and the cemetery at Key West he found slug caterpillars on mangrove leaves, as he had the previous winter, and a sawfly on live oak.[20] The Dyars departed for the Bahamas in late January and stayed around Nassau, lodging at the Royal Victoria Hotel in February (Figure 11.2).[21]

The Dyars' trip to Florida in early 1900 was restricted to the Palm Beach area, perhaps due to Zella's pregnancy.[22] Most notable about the trip from an entomological standpoint were thirty-two new moth species—all named by Busck—that resulted from caterpillars Dyar found.[23] Between the two Florida trips the family spent a summer in Bellport, Long Island, during which Dyar took morning treks to the New York swamps where he found a new nolid moth (*Nola clethrae*) on sweet pepper bush (*Clethra*).[24]

⚬⚬⚬

Dyar and his assistant Andrew Caudell (Figure 11.3a, Plate 9e,f) made their first of several productive expeditions in May 1901 to Colorado. Caudell would show a

**Figure 11.2** Royal Victoria Hotel in Nassau, Bahamas where the Dyars stayed during their trip in 1898.

**Figure 11.3** Dyar and Andrew Caudell collected and reared Lepidoptera in Colorado in 1901. Caudell and his wife Penelope ("Poodle") (a) collected on the western slope, and Wellesca Pollock (b) visited in July or August.

talent for indexing entomological information, including the literature on caterpillar food plants and grasshoppers.[25] His personality certainly added an element of levity to the trips, as it was said Caudell had a "quiet native humor and genial kindness" and when he spoke at entomology meetings "everybody was smilingly attentive, because we knew he had something worthwhile to say and that he would present it in his own droll original manner."[26]

The purpose of the Colorado trip was to collect and rear western caterpillars, poorly represented in the National Museum collection. The USNM paid Dyar's expenses, whereas the USDA funded Caudell. Soon after arriving in Denver, their base of operation, in early May, they began collecting "mainly on the prairies and foothills" in the Front Range of the Rocky Mountains "within 20 miles" of the city.

They visited Platte Canyon and sites around Golden, including Chimney Gulch and the summit of Lookout Mountain, known as the Golden Summit.

Among the new species they found around Golden was a snout moth, *Sarata caudellella*, which Dyar named for Caudell. Though Caudell's main priority on the trip was to assist Dyar with finding Lepidoptera, he was able to collect many grasshoppers, including new species. Dyar and Caudell "christened" one common grasshopper they found in Golden the "green fool" (*Acrolophitus hirtipes*) because the species was so easy to collect.[27] Other sites frequented in the eastern foothills of the Rockies included Sedalia, Pine Grove (known today as Pine), and Boulder Canyon.[28]

Dyar had one problem, finding it impossible in the dry climate "to move about ... without losing the larvae already collected ... [and] keeping a fresh supply of food plants."[29] The local collector Ernest Oslar, however, proved helpful, assisting Dyar and Caudell by locating fruitful sites, scouting the conditions on the western slope as he had for others. He also brought larvae or eggs to Dyar for rearing.[30] In July Dyar decided to expand his reach by having Caudell go off to the western slope to search for caterpillars, visiting localities such as Rifle and Grand Junction,[31] but Caudell was unable to find larvae. This led Caudell to write "My Dear Dr. Dyar Lepidopteralogically speaking this place is Bum."[32]

While Caudell was away Dyar collected on the eastern slope at Platte Canyon and Pine Grove.[33] On July 19 they joined forces again at Platte Canyon. Caudell, undeterred by his experience with poison ivy, continued collecting caterpillars on the plant and found two new moth species, one a snout moth (*Sarata rhoiella*) and the other a gelechiid (*Gelechia ocellella*). Two days later they went to the summit of Pikes Peak and brought back 170 Lepidoptera, and at the halfway house below the timberline they found caterpillars on aspen and fireweed.[34]

For Dyar, the trip to Colorado was disappointing, writing his sister: "I could go on collecting but things are going on now ... I want to get back, even if I haven't gotten what I wanted to." In late July he was anxious to return to the East Coast because Busck's scheduled departure to Cuba would leave Dyar without someone at the museum to forward letters, receive specimens, and perform identifications.[35]

As Dyar headed back on August 1, Caudell stayed behind to continue collecting. He also shipped western food plants back for Dyar's live caterpillars in need of special food sources. Sounding a bit timid, Caudell replied to a letter from Dyar on August 9:

> this morning at 7:30 and at 8 your letter arrived saying the food plant was rotten. How sorry I am and especially scared as I fear it will all be that way for I do most earnestly assure you that I took the best of care to send the stuff away in good order. ... *Less* showed me the suit of close [sic] you presented her. I don't know if she has worn them or not.[36]

The "Less" mentioned by Caudell was likely to be Wellesca Pollock (Figure 11.3b), who was commonly called "Lesca,"[37] and the letter suggests that Pollock met Dyar before his departure. While the timing of Wellesca's arrival in Colorado remains unclear, Dyar reported that she was at Stony Man Mountain that summer, raising the same epiplemid caterpillar he had found the previous August, and may have remained until after her mother, Louise, died on July 24.[38]

On his return to Bellport and his family, Dyar collected and reared his first mosquitoes, and other aquatic flies including two species of no-see-ums (ceratapogonids), including *Tanypus dyari*, named for him by Daniel Coquillett the following year.[39] Several months later, Dyar wrote Herman Strecker, who died a month later,[40] about the Colorado trip and life at the museum:

> I am back and hard at work again. In Colorado I got only about 2000 specimens as most of the effort was put in life histories. I got some 200 larvae of which half have been bred and I have hopes of more. I am in considerable confusion at the museum. 200 new drawers ordered and they dont come. Meanwhile piles of new material blocking up the space.

◊

During the field season of 1902, Dyar remained with his family, collecting moths from late April through mid-June at his Dupont Circle home and in Rock Creek Park nearby.[41] This was a year for Brood X (=10) of the periodical cicada, which produced numerous holes in the ground when they emerged during the second week of May on the National Mall (see Plate 9b).[42] The family escaped the noise of the cicadas, going to Center Harbor, New Hampshire, in June. Dyar collected moths and mosquitoes, and Zella assisted her husband, finding a caterpillar he was able to rear.[43] In early August Dyar also visited Red Hill, Moultonborough, and Durham, New Hampshire, and the following month collected on the summit of Mt. Washington and on the piazza of the Profile House Hotel before returning to Center Harbor in mid-September.[44]

In early October Dyar made one of his few known trips to the C & O Canal near Plummers Island, Maryland, hallowed collecting grounds for naturalists in the region. Along with Herbert Spencer Barber (see Plate 13), he collected *Culex* mosquito larvae from "puddles besides the canal on tow path opposite Plummers Is." [45]

While in the Northeast, Dyar told Caudell to be on the lookout for limacodid caterpillars around Washington, D.C., but he instead found a related flannel moth, *Megalopyge opercularis* (Megalopygidae), leading Caudell to seek guidance in his own inimitable way:

> Dear Dr. Dyar:—I am back in Washington. I miss you. I want you. But as I can not have you will you kindly write me a letter of instructions. I am informed [the] very rare and interesting Limacode I sent you was *opercularis*. Well I am sorry. It was on Schwarz's tree and . . . I got another species in considerable numbers on another tree, I blew [inflated] some and have a dozen or more cocoons. I hope some . . . prove valuable.[46]

◊

Attracting and feeding moths with a concoction that included rum resulted in Dyar's most prolific Lepidoptera collecting trip, over 20,000 specimens at Kaslo, located in south-central British Columbia on Kootenay Lake near the U.S.-Canada

**Figure 11.4** S.S. Moyie and S.S. Kuskanook racing on Kootenay Lake (1908).

border during the summer of 1903 (see Figure 11.4). Dyar received help from Mr. J. William Cockle, part-time proprietor of the Kaslo Hotel, Lepidoptera collector, and regional guide, who provided his own local brew of moth bait, which along with a spot of rum included sugar, beer, and molasses. The bait gathered 17,000 specimens, or roughly 85% of the moths captured. Arriving in Kaslo Lake in late May, along with Dyar, was his family, Caudell, and Rolla Currie.[47]

At dusk the bait was brushed onto fence boards, stumps, and telegraph poles on the edge of town. The collectors left at 9:30 p.m. to visit each brushed spot with a lantern, large and small cyanide jars, vials of alcohol, and "two large muslin sacks" with several hundred "empty paper pill boxes." "Macro" moths were captured on bait "by clapping the cyanide jar over them and, partially overcome by the fumes," they were transferred to pillboxes. Caudell devised a method to capture moths off the bait, primarily to prevent them from falling into the grass after being knocked from their perches, fitting a cloth funnel "around the mouth" of the jar, brushing or blowing the moths into it, and corking it. He also used a pocketed canvas apron to carry a number of these bottles. Because the "macros" were only stunned from the cyanide, the group sorted them next morning, keeping the females alive to lay eggs in glass jars for rearing and describing their life histories.[48]

There were times when the collecting crew had little success using bait, so they walked a circuit of electric lights in Kaslo but didn't find large numbers because of cool conditions. Late in the trip they put up a "large white sheet . . . and placed a good lantern and reflector behind it," but met with little success. In late July Currie, Caudell, who gave the following account, and Dyar went to the London Hill Mine on horseback, nabbing a "*Parnassius* [butterfly] and a few moths" below the summit. Overnight they enjoyed "the customary hearty western hospitality," presumably with food and drink, which helped make up for poor collecting.[49]

Although still a novice on mosquitoes on the Kaslo trip, Dyar collected all their life stages and worked at associating larva and adult. This work became part of a larger project on North American mosquitoes supported by the Carnegie

Institution. Dyar enthusiastically praised Cockle for finding larvae near Bear Lake at 7,000 ft: "There he saw many small wrigglers in an old dirty tin pan ... filled with water from rain coming through a hole in the roof ... a pure culture of [*Aedes*] *varipalpus*." Given the success with mosquitoes, Dyar made a request to Howard for Caudell to stay longer than "the three months allowed."[50]

In the publication following the trip Dyar acknowledged Daniel William Coquillett, then Honorary Custodian of Diptera at the USNM, for identifying "1,238 specimens" of British Columbian mosquitoes.[51] He also discussed how mosquito life cycles were affected by climate in mountainous British Columbia, which requires early breading, rapid larval development, and hibernating eggs to survive because it becomes "generally dry" by midsummer and "most natural breeding places disappear." Dyar noted only one *Anopheles* (the infamous malarial mosquitoes) was found and instead of the common house mosquito, *Culex pipiens*, he found *Culex* [now *Culiseta*] *incidens* in "rain barrels and other stagnant water."[52]

Near Kaslo Creek Zella, as she had in New England, showed a knack for nabbing interesting insects as she collected a "single example" of mosquito *Culex curriei* Coquillett.[53] A letter from Schaus several years later shows Dyar was known to use his wife as "a mosquito trap" to collect mosquito specimens by net.[54] Likewise, Dorothy, according to family lore, helped her father collect mosquitoes by similar methods in British Columbia. She was instructed to allow mosquitoes to begin biting her before Dyar would either force the insect into a small kill vial or drop a net over her head. These techniques were reported to have been used by the father-daughter team in other places, including the District of Columbia.[55]

Caudell and the Dyars departed Kaslo-Kootenay in late August, as Dyar collected moths through the first week of September on Vancouver Island at Shawnigan Lake and Victoria before going to the Canadian interior to Revekstoke and Glacier, British Columbia; Field and Banff, Alberta; and finally, Gravenhurst, Ontario.[56]

During the stops at Glacier and Field, Alberta, Dyar collected female geometer moths, he considered to be a new variety of a species found in Europe. He named it after Otis, *Mesoleuca simulata* Hübner var. *otisi* (now *Thera otisi*), only three at the time, "... who assisted me in collecting the specimens."[57]

⚘

In 1904 Dyar began two years of fieldwork restricted to the East. In May, he used a caged female moth to attract a male in his yard, and the next month, he collected mosquitoes with Caudell in Baltimore, at Chesapeake Beach, and Grassymead near Mount Vernon, Virginia.[58] Dyar's next fieldwork, mostly collecting mosquitoes, was at Weekapaug, Rhode Island from July through mid September. In August he went to Tupper Lake and Moody, New York, and visited Lake Champlain, where he collected a sawfly larva on soapberry in late August, before returning to Rhode Island.[59]

The following year, at the beginning of March 1905, Dyar and Caudell collected mosquitoes in southern Georgia and Florida.[60] In Florida they found larvae "in temporary pools of fresh water" in "recently dug holes along the railroad, ... in the pines, ... in pools in swampy land, ... in ditch, ... in a hole with old tin cans and rotten wood" and in southern Georgia "in the pine barrens ... in a puddle by the railroad at a siding."

After the trip Dyar "had the pleasure to name" a Florida mosquito *Culex mitchellae* (Figure 11.5) after "Miss Evelyn G. Mitchell" (see Figure 12.2), a graduate student who previously held "a field and lab assistant mosquito position" in Louisiana. In 1905, Mitchell (then twenty-six), began as a contract illustrator for "The Mosquitoes of North and Central America" at the USNM while also making illustrations for her thesis.[61] Dyar's pleasantries with Mitchell would soon cease (see Chapter 12).

In April 1905 Dyar met up with Frederick Knab, now with the Bureau of Entomology, for mosquito collecting in Massachusetts at West Springfield, Mount Holyoke, and Longmeadow.[62] This was around the time their first paper appeared on mosquito larvae in *Proceedings of the Entomological Society of Washington*. Dyar also made solo stops in Plattsburgh and Elizabethtown, New York, reconnecting with Knab in Hartford, Connecticut.[63] In August 1905, after returning from England with Zella, Dyar collected mosquitoes in northern New York. Nine-year-old Dorothy assisted: Dyar wrote, "Tuppers Lake, Moody, N.Y. 3 larvae in a rock pool [on] a Island in Lake near water. Dorothy found them. Took one." In late September at Tryon, North Carolina, Dyar hunted for limacodid caterpillars, finding *Lithacodes fiskeana* (the only ever reported).[64]

⟡

The next Dyar family trip was about Dorothy's health problems, as Zella, Harriet Peabody, and the children visited the Los Angeles Basin from fall 1905 through the summer of 1906. Dyar did not join the family until the following May, long after their arrival and later than expected. After staying with the family in Santa Monica

**Figure 11.5** Type specimen of *Culex mitchellae*, which Dyar named after Evelyn G. Mitchell soon after his return from collecting it in Florida and Georgia in 1905.

for about a week, Dyar wrote: "My little girl had an operation at the hospital" in Los Angeles. Dorothy had mastoiditis, which required a hole to be drilled in her cranium to relieve the pressure. Dyar was her constant companion throughout the operation and recovery.[65]

June 1906 in Los Angeles was a month of mishap and happenstance for the Dyars. Besides Dorothy's operation, "Zella fell on a rock and sprained her ankle"[66] while collecting with her husband at Sierra Madre. Then Mrs. Peabody encountered a surprise. According to Zella, her mother told her: " 'You cannot guess whom I found in Pasadena.' . . . I said, 'I didn't know; I had no idea.' She said, 'It was Wellesca.' " Thereafter, the family invited her to visit, and "[s]he came in a great many times, or at least a number of times."[67]

Prior to Dorothy's operation, Dyar collected with Caudell, nabbing mosquitoes in the Pasadena area and Lepidoptera at Arroyo Seco in May, and both groups of insects at Gardena and Los Angeles on May 30.[68] In a report on the mosquitoes of coastal California, Dyar wrote: "[In 1906, I] visited California to make collections for the United States National Museum with the idea . . .[that] few species of mosquitoes [were] to be found there, and those mostly well known. The larvae of a few were desired, and these it was hoped to find. Most of the time . . . was spent in the vicinity of Los Angeles, after the seasonal rains were over."[69]

Dyar and Caudell collected mosquitoes in San Diego and Sweetwater Junction, California, in early June. After collecting in National City, they crossed over to Tijuana, making Dyar's only excursion to Mexico, even though he described many new moth species from that country.[70] During the following weeks, the two returned to the U.S. border towns and Dyar later collected mosquitoes in Los Angeles and in southern California at Laguna, San Onofre, Pasadena, including Ostrich Farm, and Carpinteria for the remainder of June and early July.[71]

In late July 1906, Dyar again met up with Caudell and travelled to northern California to collect mosquitoes along the Southern Pacific Railway.[72] They went to Sisson (present day Shasta City) at the base of Mt. Shasta, and Thrall. Collecting in Oregon, Washington, and on Vancouver Island followed.[73] After completing their stay in the Pacific Northwest in August, Dyar and Caudell took the train east through the Canadian Rockies, stopping to collect mosquitoes in Alberta at Lake Louise, Calgary, and Medicine Hat, and in Saskatchewan at Moose Jaw and North Portal, Saskatoon.[74]

While in Washington State Caudell and Dyar encountered a katydid, in Longmire's Springs at the base of Mt. Rainier. Caudell named the particular species, *Cyphoderris piperi* (now *C. monstrosa piperi*).

> Both Dr. Dyar and I readily located its apparent position . . . not over eight feet from us. . . . I cautiously approached, but when I reached the spot . . . the sound no longer seemed to proceed from that point, but from an old stump a dozen feet further on. I now . . . follow that spooky note from point to point, sometimes straight ahead and sometimes to one side or the other, till a distance of over two hundred yards was traversed. Dr. Dyar, lacking the enthusiasm of an Orthopterist in a quest of this nature, strolled on, leaving me to pursue my ignis fatuus [illusion] alone . . .

[Unsuccessful] I . . . was soon pouring my tale of woe into the unsympathetic ears of Dr. Dyar. . . . the next evening, armed with a very dirty lantern, kindly loaned us by an accommodating host, we sallied forth to capture one of the songsters. . .[75]

⌘

In mid-July 1907, Dyar requested to suspend telephone service, suggesting that the family was away from Washington, D.C. Zella and the children may have gone to Maine to visit her family.[76] Dyar went on a moth-collecting trip to Tryon, North Carolina, for a few days in August, followed by a stop in the Washington, D.C., area where he collected noctuids on water lilies in Hyattsville, Maryland.[77] He then went to Boston to attend the 7th International Congress of Zoology (August 19–24) as a representative of the National Museum: He spoke on the distribution of North American mosquitoes, emphasizing their dependence on stagnant water for breeding.[78]

In Boston Dyar found, on a pear tree, caterpillars of the oriental limacodid (*Monema flavescens*). The species was accidently imported on fruit trees to Boston and elsewhere around 1900.[79] To Dyar, the introduction to the United States of this colorful caterpillar with blue dumbbell pattern on its back and striking cocoons was a happy accident. Enamored with having an addition to the sparse limacodid fauna in the region overrode his concern about it defoliating shade and fruit trees. Even though strict quarantine laws had yet to be enacted by Congress, Dyar's views about deliberately introducing an exotic species, or "invasive species," may seem odd, as he was opposed to control efforts in Dorchester:

Conditions . . . favorable for the continued existence of this interesting species in America ["open spaces with trees and shrubbery"] but unfortunately Dr. H. [Henry] T. Fernald [C.H. Fernald's son], in an excess of economic zeal, which we consider premature, destroyed large numbers of the cocoons, for fear that the insect might become a pest. . . . It is to be hoped that the moth has not been exterminated.[80]

Dyar, the entomophile, later at odds with his boss over tactics of mosquito control, displayed fondness for the oriental limacodid moth by attempting to establish it in Washington, D.C. After seeing Dyar's article about the moth in the *Proceedings*, a shocked L.O. Howard wrote: "I have just seen for the first time your article . . . which you speak of liberating the oriental *Cnidocampa* [now *Monema*] in the open in Washington." He was concerned about criticism toward the USDA from those who read the article and wanted to know if there was "any chance that it still exists as the result of this introduction." An unrepentant Dyar responded with a handwritten note at the bottom of Howard's letter: "not any, I am sorry to say. The experiment was entirely unsuccessful."[81]

During the summer of 1908 Dyar was again in New England with his family, finding mosquitoes in Lincolnville, Maine, including *Aedes cantator* Coquillett, in August.[82] A call to duty relating to mosquito control brought Dyar to Dublin, New Hampshire, in 1908 and 1909. However, his visits were mistimed each year: He was either too late or too early to locate their breeding sites. In late July 1909 Dyar collected Lepidoptera, apparently for the last time, in Plattsburgh, N.Y., where he had been going at least since 1885, and on Valcor Island in Lake Champlain, where his friend George Hudson hunted fossils on the island each summer.[83]

The extent of Dyar's travels and the pace of his moth and mosquito collecting had dramatically slowed toward the end of the first decade of the twentieth century. Perhaps his focus had shifted from collecting to naming the new species now arriving at the National Museum. However, an escalation of conflicts with other entomologists, new editorial responsibilities, a salaried government position, and increasing amounts of time visiting Mrs. Wellesca Allen may also have restricted his traveling outside of New England and the mid-Atlantic Region.

## CHAPTER 12

## Literature Wars and Last Battles with Smith: 1904–1911

*Meeting ... Held at the American Museum of Natural History, Tuesday evening. President C.H. Roberts in the chair with thirteen members and two visitors present ... Mr. [Charles W.] Leng, of the publication committee, reported that at a meeting of the committee held this evening Dr. Harrison G. Dyar had been elected editor of the Journal for the coming year.*[1]

Minutes from the meeting of the *New York Entomological Society*, February 2, 1904

In the early twentieth century Dyar's caustic remarks—directed at those in his field and even featured in the national press—were awash along the eastern seaboard in entomology journals. A touchstone event for these vituperations took place during a meeting of the New York Society at the American Museum of Natural History on Groundhog's Day, 1904. Dyar, who had maintained ties with New York City's entomologists, was elected editor of the *Journal of the New York Entomological Society* (*JNYES*),[2] a journal on whose editorial board he already sat (Figure 12.1a). As editor in chief until 1908, Dyar's criticisms of entomologists, amateur and professional, brought his print battles to a new level of intensity.

Among Dyar's primary targets was Henry Skinner (Figure 12.1b), editor of the Philadelphia based *Entomological News*. The *JNYES* catered to a more professional entomologist than *Entomological News*, and Dyar (Figure 12.1c) showed little restraint in critiquing its editor or his more amateur venue. In the December 1905 issue of the *JNYES*, for instance, Dyar reviewed Skinner's "supplement to his catalogue of butterflies", writing:

> It is somewhat bristling with typographical errors and blunders, but we are used to that sort of thing from Philadelphia.... The generic names have not been brought up to date ... [as Skinner] is "not interested" in the subject, which he is pleased to

**Figure 12.1** Cover of March 1905 issue of the *Journal of the New York Entomological Society* with Dyar now as the editor (*a*); Henry Skinner, editor of *Entomological News*, collecting butterflies and skippers (1890s) (*b*); H.G. Dyar (c. 1905) (*c*).

designate as "generic fantasies." This is, we think, a fault. It is easy to stigmatize what one will not take the trouble to understand; but a good opportunity of correcting the antiquated nomenclature of the North American butterflies has here been lost.[3]

Skinner rebutted the harsh critique with "A Review of a Review" in *Entomological News*, lambasting Dyar's own taxonomic paper on skippers (Hesperiidae), a group of day-flying Lepidoptera closely related to butterflies.[4] "The Washington editor of the New York journal," Skinner wrote, "could have used the space to better advantage by pointing out those blunders and the entomological world would have been the gainer thereby. Review by innuendo is of no use to anyone except to vent spleen." Skinner went on to criticize Dyar's own "generic fantasies," arguing contemporary genus classifications were too vague to be useful.[5] Dyar posted by mail:

I notice your "review of a review" and am rather surprised you should have allowed yourself such a hysterical outburst. Better take a sedative. I explained to you that your discoveries of error in detail [i.e., nomenclature] were no just criticism of the system of classification [i.e., based on Darwinian principles]. I hope your studies will result in a new and original system, but it must be based on structural characters to receive any attention. I still think you do not understand the subject.[6]

Skinner responded to Dyar three days later, noting that he liked highballs, not sedatives, and implied that Dyar enjoyed *Entomological News* by mentioning that his rival editor had renewed his subscription.[7]

Dyar confronted other contributors to *Entomological News* in a November 1905 letter to the editor, "New facts that are not new." He observes two unoriginal examples from the previous issue of the *News*: Caroline Gray Soule's observation of the "stemmed cocoons" of the polyphemus moth, whose silk was spun around a twig or leaf stem, different from those she had observed for over twenty years in Massachusetts, and a note from Professor Franklin Sherman, Jr. on the wood pupal chamber of the owlet moth *Harrisimemna*.[8] Both Sherman and Soule responded to Dyar. Soule claimed that he missed the point of her inquiry, which was to "learn in what parts of the country" the polyphemus had stemmed or unstemmed cocoons, and where one type or the other predominated or was absent, and Sherman expressed "a feeling of relief upon seeing Dr. Dyar taken to task in the last issue of the NEWS ... for it shows that I am not alone in my dislike for unnecessary and caustic rebukes."[9]

The battle did not end. Skinner wrote to Dyar, urging him to withdraw his response:

> Now do you really wish me to publish your note in the News just rec'd? It is very illogical and shows how you jump at conclusions. On what ground do you assume that the Editor of the News was ignorant of the facts given by Sherman and Soule being old? Why do you say 'The officers of the U.S. National Museum'—why make that plural? There can be but one Entomological Pope. You drink too much ice water and your blood is too frigid to understand all that goes in the News. I will publish your note if you wish, along with my answer.[10]

Dyar's response to Soule, Sherman, and Skinner was published in the January 1906 *Entomological News*:

> It is a maxim that ignorance of the law excuses no one. Ignorance of previous work should result, not in hasty publication, but in consultation of someone better posted, or in discreet silence. *The officers of the U.S. National Museum* [italics mine] will always reply to questioners seeking information of this nature. We would reply to Miss Soule that we have no objection to "popular" articles that are frankly such and give proper credit to antecedent work.

Dyar also blasted Henry Skinner: "We have criticized authors for hasty and uncritical work; but there is another aspect of the case. What is the condition of editorial responsibility in a journal that accepts these articles without question?"[11] Skinner responded in print by stating he "published the fact [of stemmed and unstemmed cocoons] six years before the citation given by Dr. Dyar" and knew of earlier literature still. In response to criticism about his "editorial responsibility," Skinner wrote, "We publish ... what we think of interest to ... readers, and the assumption that everything is new and [if not] ... is due to ignorance of the facts on the part of the editors is preposterous."[12] Dyar posted Skinner by mail in January 1906, snidely apologizing for thinking Skinner's journal should be judged by scientific standards when its standards were "wholly commercial." Ironically, Dyar then went on to praise Skinner's financial success of *Entomological News* and for his getting "so many new people" interested in the field.[13]

The end of the letter suggests Dyar was envious and had an ulterior motive for being so kind: "I would like to call the attention of your numerous readers to the *Journ. N.Y. ent. Soc.*, as some might like to subscribe. Will you, please, have the enclosed advertisement inserted in a conspicuous place, for a year, and send the bill to me."[14] The battles between the two continued in the April 1906 issue of *Entomological News*. Skinner criticized Dyar's taxonomic review of the U.S. skippers. He refused to use the names of genera in his work, "because in many cases they are unscientific, illogical and untenable" and observed, "I have named specimens for many years from all over the country and the species is the unit of classification, and a multiplicity of bad genera—a Tower of Babel."[15] In a footnote, Skinner called attention to Dyar's fallibility, noting that a recent *JNYES* "gives us *cupracens*, instead of *cuprascens*, as a specific name in *Cicindela* [and other examples]."[16]

Dyar wrote Skinner in June 1906: "I am delighted with the little sarcastic paragraphs you have been putting in the 'News.' Good—keep it up! As you gain experience you might add humorous illustrations and gradually come to rival 'Puck' and 'Life', of course in an entomological way."[17]

⁂

Dyar's editorial conflicts with Skinner dated back to the 1890s, becoming heated when Skinner, commenting on naming of new North American butterflies, wrote "I may say right here that I believe the imago [adult stage], the culmination of nature's efforts, and that while studies of transformations are most valuable, they will not solve the problem of specific difference or identity."[18] Dyar took this statement as an indictment of use of larval stages in classification. Ironically, Skinner did not direct his remarks at Dyar's work, but Dyar chose to "protest against Dr. Skinner's remarks" not in a letter to the editor, but in a note on the phylogeny of giant silk moths in *The Canadian Entomologist*. Following the quote from Skinner about "the imago," Dyar stated: "This is not the view of a careful student of the subject, but reads like an excuse for neglecting studies of the early stages. As if the *larva* were not often the 'culmination of nature's effort,' as in Apatela [= *Acronicta*] or the Limacodidae, or as if the forces determining the struggle for existence must always impinge *most* strongly on the same stage in all species."[19] Responding by personal letter in late January 1897, Skinner admonished Dyar for remarks,

> which seem to me entirely uncalled for. Your protest and difference of opinion are all right and to them I can take no exception. Your remaining remarks are unnecessary and look like a gratuitous insult and certainly have no weight as scientific argument. It is to be deplored that such things get into print as it precludes the possibility of the same kind of a reply from a gentleman.[20]

Bickering with Skinner three years later in March 1900 returned again to typographical errors. In a humorous example, Skinner had misspelled Dyar's surname in a January *Entomological News* article in which he described a new species of limacodid (*Lithacodes fiskeana*). Dyar wrote, "I must really protest against having an article on new species credited to ... *Dyer*, unknown to me."[21] Dyar, however, failed to notice the title of his article had an error. His new limacodid was from the "Pale*aratic*" rather than Palearctic region.

These personal exchanges, although cutting, had a friendlier tone when the parties were on speaking terms. In December 1900 Dyar visited Skinner in Philadelphia, and after he returned to Washington he sent Skinner a letter that discussed, among other things, a wager between the two over whether there would be errors in Skinner's latest *Entomological News* article. Dyar wrote: "I owe you a quarter since Sept. 15th which I have added to the check for Ent News. I found my description of the *Callidryas* in it all right."[22]

## BATTLES WITH MOSQUITO WORKERS AND THE MITCHELL LAWSUIT

The United States' beginning construction of the Panama Canal raised concerns over workers contracting mosquito-borne disease; therefore the need for a monograph on tropical mosquitoes of the Western Hemisphere was identified. L.O. Howard obtained his first grant from the new Carnegie Institute of Washington in 1903 to write such a monograph, and by 1907 he had completed "a large amount of the manuscript."Ced Dyar, Frederick Knab, and initially Daniel W. Coquillett (1856–1911), at the USNM, were involved in the project.[23]

Dyar, as he had done with Lepidoptera, used a larval approach to classify mosquitoes, and he and Knab published "The larvae of Culicidae (mosquitoes) classified as independent organisms," an article in which they defied taxonomic convention, naming 56 species and 4 genera based on larval rather than adult characteristics. The article began with the German proverb "Wer A sagt muss auch B sagen" (The one who says A must say B too), or "what you say is what you do," meaning that if the larval stage best defines a mosquito species, then use the larva to name it.[24]

By treating "the larvae as independent organisms and classify[ing] them separately" from the adults, Dyar and Knab could compare larva and adult-based classifications to "throw light on the phylogeny of the group and indicate the more reliable distinctions [between species]." They knew that this approach would cause taxonomic issues but observed:

> a synonym [two names, one species] is easily dealt with, whereas a misidentification or confusion of two species under one name is really more troublesome . . . [or if] we have named the larvae of previously described species, we believe that less difficulty will be experienced than if we had left them nameless, or doubtfully referred them to known species.[25]

Dyar and Knab named one of the species from the larval stage for Andrew Caudell: *Mochlostyrax caudelli*. Twenty of the new species named from larvae ended with "ator" (all in the genus *Culex*). Quite interestingly, the new names of these "ator" species had nothing to do with larval characteristics. Rather, the authors appear to have thought more of *Homo sapiens*, most of who have little fondness for mosquitoes. The "ator" ending may have actually been a pun, given the authors' credo of "what you say is what you do" because "ator" means "one that does."

The "ator" names are quite funny, especially because a number of them refer to those in the mosquito business, either the scientists or the controllers. For example, someone involved in taxonomy, or who likes mosquitoes, can be referred to as a *Culex investigator, educator, conservator, coronator, proclamator, declarator, elevator* (one who raises a variety to a species), *derivator, mutator,* and *simulator* (looks like another species). Those afraid or are not particularly fond or who wish to control mosquitoes can be known as a *Culex mortificator, lamentator, inhibitator, extricator, gravitator,* and *regulator.* There are a few others more ambiguous, perhaps poking fun at early twentieth-century women including *Culex lactator, habilitator,* and *decorator.*

Skinner, in the May 1906 *Entomological News*, ridiculed the Dyar/Knab article, noting the mosquito names published by D.W. Coquillett earlier the same year had precedence. Skinner went on to write, "It is not our intention to review this paper but only to refer to the dates and the fact that species are described from larvae alone. . . . we have come to the conclusion that the future synonymy, etc., will be somewhat like a Chinese puzzle . . . We expect a paper shortly describing species from the egg or pupa alone."[26] Over half of the 56 species names remain in use, while none from Coquillett have survived.[27]

⚭

The "independent organisms" paper openly disparaged Coquillett's identifications of adult mosquitoes—which were made at Dyar and Knab's request—yet they named a species after him (*Uranotaenia coquilletti*), even though Coquillett misidentified it as *Uranotaenia socialis*. "We dedicate the species to Mr. Coquillett, who has certainly performed a vast amount of labor on a difficult subject, whatever we may think of his results." Another example was *Melanoconion humilis* from Mexico, which they saw "no reason to accept this determination," giving it another name (*mutator*) because they didn't believe *humilis*, described from Brazil, could be the same species. Although sounding harsh, Dyar and Knab's goal was to show that adult mosquitoes have unreliable characteristics rather than the taxonomists themselves.

Dyar once again misjudged the line between his scientific goals and personal relations, and Coquillett responded in the June 1906 issue of *Entomological News* titled "Dr. Dyar's Square Dealing," an exposé on the difficulties of working on the Carnegie monograph with him. According to Coquillett, the conflict began after he was given mosquito larvae to associate with bred adults from the West Indies, but Dyar demanded they be "turned over . . . *at once*" (for work on the monograph) and was "so persistent and vehement . . . an order was issued . . . to immediately place this material in his possession." Coquillett provisionally named the samples, but was upset to find Dyar had published the names, at times incorrectly, or omitted annotations (e.g., question marks).[28]

In his published response, Dyar wrote that Coquillett was allowed to examine the larval skins but was expected to work with the adults and "was quite unaware [he] had changed Mr. Coquillett's marks of doubt from species to genus, and, if so, it was purely by inadvertence and without any such object as I am charged with." Dyar coldly wrote: "I have tried to deal with Mr. Coquillett's work as squarely as possible, and if I am obliged to condemn it unreservedly, it is without any personal animosity [and] the Carnegie Monograph ... has been finally clarified by removing Mr. Coquillett from any connection with it, which is now in my hands."[29]

⁂

Dyar critiqued two works on mosquitoes in late 1907 and early 1908, the first coauthored with Knab concerned F.V. Theobald's "A monograph of the Culicidae of the World, volume 4" and the second regarded Evelyn Groesbeeck Mitchell's (1879–1964) popular *Mosquito Life* (see Figures 12.2a, b).[30] Their jab at Theobald's poor classification is a historical footnote, but the other, a claim that Mitchell had siphoned off information and drawings from the Carnegie Monograph for her own book, got Dyar into legal trouble.

On the surface, some of Dyar's antipathy toward Mitchell may have originated in a spat she had with Knab in 1907 over an article he wrote in *Psyche* about *Deinocerites cancer*. Mitchell accused him of being "possessed of certain delusions."[31] However, during Dyar's 1906 exchange with Coquillett, she and Coquillett had already formed a working relationship. Further, Mitchell and Louisiana Surgeon General Dr. James Dupree had an understanding that after his death she could illustrate his specimens and copyright the drawings for her own purposes. She could redraw illustrations for the Carnegie monograph of species shared by the two projects and be compensated for them. Mitchell believed Dyar wanted to prevent her from publishing *Mosquito Life* as Dupree's family told her there had been a request for his notes to be turned over for the monograph, but she refused to comply "on the ground that the material was already promised to me."[32]

Dyar began his review of *Mosquito Life* in a complimentary tone (Figure 12.2c). "Miss Mitchell's original keys for the ... species will, no doubt, prove convenient to field workers and physicians, as she has largely avoided the use of microscopical structures." However, Dyar lambasted Mitchell for giving inadequate credit to the mosquito workers at the National Museum, at best, or stealing their work. This included drawings he believed were intended exclusively for the Carnegie

**Figure 12.2** Evelyn G. Mitchell in articles about her lawsuit against Dyar (*a,b*); Dyar's critique of Mitchell's "Mosquito Life" (*c*) and (*d*) her rebuttal.

Monograph she claimed were copyrighted for her use. Dyar stated: "Miss Mitchell has played the part of a feminine *Psorophora* among the scientific Aedids of Washington." In other words, she was like a *Psorophora* mosquito, whose larvae are predatory on other mosquito larvae, and the "scientific Aedids of Washington" he alluded to were Dyar, Knab, and Coquillet.[33]

Coquillett, already at odds with Dyar over the Carnegie Monograph, had a very different opinion of Mitchell and her work. He stated that Mitchell didn't receive credit in Dyar's work published in the *JNYES*, and she told Coquillett she would "resign rather than continue working under the unpleasant existing conditions" under Dyar and would continue "her work on the drawings" under Coquillett's supervision in his office.[34] In April 1908 Mitchell sued Dyar for libel, seeking $35,000 in damages.

Mitchell said:

> I have had no help whatever from Dr. Dyar in the preparation of my book, or in that of my thesis except in the one instance.... The book was written at my home in New Jersey and Mr Coquillett never saw it until I had everything settled with the publishers.... If my book reads like a second edition of Dr. Howard's ... beyond treating of the same general subject, I fail to see any comparison in plan, style, or text.[35]

"In Dr. Dyar's review," wrote Mitchell in *The Canadian Entomologist* (Figure 12.2d), "he not only seems unable to say anything against [*Mosquito Life*], but, on the other hand, to so admire it, that he has become possessed of the strange idea that he is actually the author of some portion of it." Her response to being compared with a mosquito whose larvae are predatory on those of others showed she had a better sense of humor than her critic: "I feel rather flattered at the comparison to *Psorophora*, since this insect is large, beautiful, not a frequent nuisance, but an exterminator of common and pestiferous "Aedids." The libel suit was "dismissed for failure to prosecute."[36] Dyar got his own revenge: In 1915 he wrote the story "Taming of a Suffragette" over the letter about the Mitchell case from his attorney A.S. Worthington. Not surprisingly, Mitchell favored women's suffrage and Dyar, Shakespeare.[37]

## LAST BATTLES WITH SMITH

In 1909, at a time when John B. Smith was an established leader of mosquito control, having earned "the distinction of having cleared the Jersey marshes of mosquitoes,"[38] Dyar was invited by Dr. E.C. Stowell of the Dublin Chemical and Pathological Laboratory to the town of Dublin, New Hampshire, to find the breeding wetlands of the salt and pepper mosquito, *Mansonia* (*Coquillettidia*) *perturbans*.

According to Dyar, the abundant mosquito began flight in late June and entered houses, even when screened, through chimneys, and bit "viciously and are altogether disreputable in their behavior.... widely dispersed through the shelter of the woods" and if "abated ... the prosperity of the place would be promoted by the influx of newcomers." He stressed that these mosquitoes posed little threat to human health because the "locality is far out of the range of tropical disease carriers." He did not completely rule out the possibility of malaria in Dublin from two species of *Anopheles*, but he stated there was "no danger whatever."[39]

Dyar's report on Dublin's mosquitoes recalled unsuccessful past attempts at discovering the breeding sites in 1908 and with Caudell in 1909, as well as those by Busck and locally based Stowell and his assistant. Curiously absent among entomologists mentioned was John B. Smith and assistant John A. Grossbeck. A lepidopterist of some renown, Grossbeck did many excellent drawings for Smith in his "Mosquitoes of New Jersey." Dyar credited the "timely" discovery of the "breeding habits" of this voraciously biting mosquito to Mr. J.T. Brakeley, and Dyar located its "breeding place" himself in August 1909. Dyar's remedy "for the mosquito plague in Dublin" was "[l]owering the water level 2 feet" to "destroy the breeding conditions."[40]

Dyar's article perturbed John B. Smith, having himself been to the mosquito breeding sites of Dublin, because he was not mentioned among the visitors. In a critique of Dyar's account, Smith described a jolly adventure of how he and Dyar came to be on the same expedition. "Dr. Dyar decided to break into his vacation and to visit Dublin on exactly that same day, and still more extraordinarily, his choice fell on exactly the same train that carried me," Smith wrote. "We therefore arrived at Dublin at the same time and Dr. Stowell [of the Dublin laboratory] did not seem much surprised to see Dr. Dyar; certainly not so much as I was when he came to me in the train soon after we were well away from Boston."

Then the two rivals and others were each extended invitations by Stowell for an automobile ride to the "suspected breeding places." Dyar remained for the week and Grossbeck, Smith's assistant, found him back at the swamp that Smith purportedly had "pointed out as an ideal breeding place" and found the mosquito's "egg boat and three pupal shells." Smith was offended that Dyar credited Brakeley with the discovery and didn't acknowledge himself and Grossbeck.[41]

Predictably insulted, Dyar fired back in January 1910:

> I am astonished that Dr. Smith should attempt to divert credit to himself from a most generous and warm-hearted, if nonpublishing, friend [Brakeley]. For myself, who am neither generous, nor, I fear, at present particularly friendly toward Dr. Smith, the attempt to annex credit is explicable on the ground that Dr. Smith's point of view is too self-centered to allow him to read the situation in its true aspect.[42]

Smith countered in the January 10, 1910, *Entomological News*. He quoted Stowell, who objected to no credit being given Smith and Grossbeck for discovering the breeding site in Dyar's report and noted he "begged" Smith to help "because ... our problem was so complicated that only the eye of special experience would be able to solve it.'" At the letter's close and wanting the last word Smith wrote: "And the fact remains that after two seasons of work Dr. Dyar 'discovered' a breeding place of perturbans the day after I had pointed it out and told how to look for the insects."[43]

Having spent a decade or more dealing with Dyar directly or on the sidelines, Henry Skinner, in *Entomological News*, took editorial advantage over Dyar's latest battle with Smith, placing the two entomologists in the Arctic as Cook and Peary, who were in the news for their reaching the North Pole in 1909 and 1910:

ON AN EPISODE.

THE CHARGE.
Said Doctor Smith to Doctor Dyar: I'm amused and yet I'm sad;
I found the pond where wiggletails disport themselves, and had
Assigned myself the arduous task of telling to all men,
How such things court, and what they eat, and how they feed, and when.
Then you appeared upon the scene ('Twas strange to find you there!)
I welcomed you and smiled on you and showed you where they were;
Yet when I looked your paper through, my Coddington [magnifier] in hand,
I failed to find that you had this sufficiently explained.
Your presence there was passing strange, but this was stranger still;
I've wondered ever since if you intend to treat me ill.

THE RETURN.
Said Doctor Dyar to Doctor Smith: You surely must be leery;
Your conduct quite reminds me of our townsmen, Cook and Peary;
Your name's not in my paper and the search you may forego;
I left it out on purpose and am free to tell you so.
If I had been a blind man and deaf and dumb beside,
A roaming round the country with no kind friend to guide,
And you had chanced upon me and grasped me by the hand.
And led me round the boggy ground right to the small pond's side,
Your name had been emblazoned in my paper without doubt,
In letters large and very black, a 'steenth inch tall about;
But since I see like anything, indeed I saw through you,
My perspicacity alone had brought the pond to view;
Besides, I'll have to tell you, and with purpose most pacific,
(Though this journal more's the pity's not so very scientific).
Neither your name nor my own name should by good right be there,
Since another found the wigglers and showed us both their lair.
AU REVOIR.
Said Doctor Smith to Doctor Dyar: I'm grieved, but yet I'm cheery.
Said Doctor Dyar to Doctor Smith: You're Cook and I am Peary.

At the end, Skinner wrote: "My steed is a little lame and needs currying. In truth for fifteen odd years she has been tied in her stall and has had only dry feed. Please make some allowance for this.—THE AUTHOR."[44]

༺༻

The final battle between Smith and Dyar was set off in 1910 by seemingly inconsequential statements in an obscure journal from Pomona College, Claremont,

California. Dyar described a new species of owlet moth, *Pleonectyptera* (now *Hemeroplanis*) *cumulalis*, while suggesting that Smith had earlier misidentified *cumulalis* as *P. finitima*.⁴⁵

Smith disagreed with Dyar's assertion in the September 1911 *JNYES*, stating that Dyar had created a synonym by describing the species *cumulalis* in the first place. Although history sided with Smith's contention that Dyar's *cumulalis* is a synonym, Smith's practice of writing original descriptions from multiple specimens and not clearly designating one as the type, such as in this instance, caused problems for Dyar and future workers. The problem was compounded because Smith kept two cotype specimens of *finitima* in his own collection in New Jersey; the other two, those available to Dyar, which looked somewhat different, were at the USNM.⁴⁶

☙

Smith died a little over five months after the aforementioned controversy with Dyar. A special meeting of the Entomological Society of Washington was held on March 13, 1912, at the Bureau of Entomology, four days after his death from chronic heart trouble complicated by Bright's disease.⁴⁷ Noticeably absent from the Smith memorial and the remaining meetings in April through June before the summer recess was Dyar.⁴⁸

Conceivably Dyar did not wish to hear more about Smith, including the society's resolution to write Smith's biography. However, the self-styled curmudgeon may have memorialized his rival in his August 1914 story "The Man Who Did Return," whose central character was John Smith. A quote from "The Rubaiyat" of Omar Khayyam preceded the story: *"Strange, is it not? That of the myriads who Before us passed the Door of Darkness through, Which, to discover, we must travel too."* The John Smith in the story was a man who had died and made a bargain to return to the living: *"John Smith died and went—. Where did John Smith go? Wouldn't we all like to know? Do you suppose there's any way to find out? John Smith himself had no doubts. You see, John Smith had had experience; John Smith had died before. This was the way of it."*

Although in the end the real John Smith wasn't a friend, perhaps Dyar missed their conflicts. The fictional Smith's pact to have a large family was something Dyar, the eugenicist, advocated, as illustrated by the story's ending: "And now the name of Smith is a common one in that part of the country and seems likely to remain so, for all these Smiths that I have met, are blessed with unusually large families for these days of personal responsibility and woman's rights."⁴⁹

Perhaps Henry Skinner said it best:

AU REVOIR.
Said Doctor Smith to Doctor Dyar: I'm grieved, but yet I'm cheery.
Said Doctor Dyar to Doctor Smith: You're Cook and I am Peary.⁵⁰

## 13

# Gains and Losses as a Professional Lepidopterist: 1907–1914

Dyar would begin skipping meetings of the Entomological Society of Washington after John B. Smith's death, in March 1912. Dyar had been a force in the Society since his arrival in Washington. He regularly attended meetings, published papers given at the meetings in the Society's *Proceedings*, served as the president, and held positions on executive and editorial boards.

Dyar did return to a Society meeting in October 1912, when he and gave an oral book review of *Moths of Limberlost* by Gene Stratton Porter, a successful nature writer and wildlife photographer. Dyar's review was given in typical "Dyarian" style. It began with a somewhat complimentary tone: "This is a book intended for 'nature lovers,' not for students or even beginners. It gives a general account of the appearance and habits of some of our larger moths and of the doings of the author and members of her family in relation to these moths, all very entertainingly told." Dyar then chided Porter for not only neglecting various sources, including popular ones such as "Harris's 'Insects Injurious to Vegetation'" but also for using those he deemed less appropriate, including Holland's "Moth Book" and Packard's "Guide to the Study of Insects." He then opined, "the author expresses the opinion that *Citheronia regalis* [the regal moth] is 'beyond all question' of tropical origin. This is true, of course, of the genus, but the author is speaking of the species, which is confined to the United States east of the Plains, as is well known."[1]

Changes in how the Society ran its journal were announced at the November meeting at Saengerbrund Hall. The publication committee of the *Proceedings*, which included Dyar, was to be replaced by a single editor. Dyar was not present then but was there the following December, when James Chamberlain Crawford (1880–1950) was named editor. During the December meeting, Dyar gave the editor's report, noting he had put out three numbers that year and that a fourth was with the printer. The meeting was to be his last as a member.[2] Dyar does not seem to

**Figure 13.1** Dyar moved with the nation's insect collection across the National Mall in 1909 to the new building known today as the National Museum of Natural History (1912).

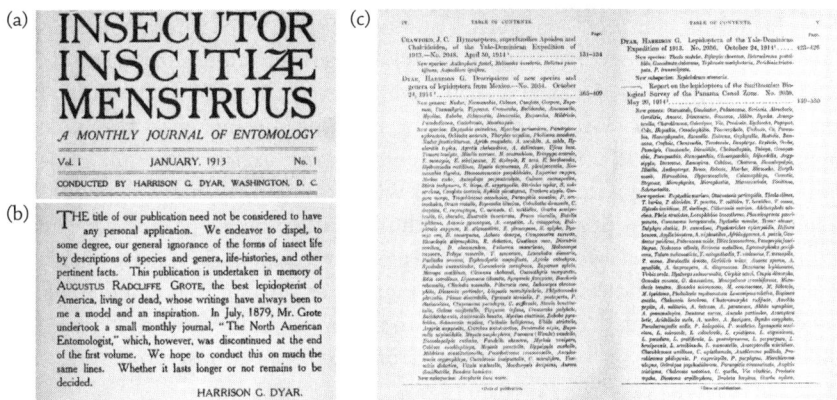

**Figure 13.2** The first issue of Dyar's journal *Insecutor Inscitiae Menstruus* (a) and dedication to A.R. Grote (b); summary of among hundreds of species and many genera described by Dyar from Mexico and Panama (c) in *Proceedings of the U.S. National Museum* (1913–1914).

have been upset with the Society, as he donated the separates collection (reprints) of his own work to it. He just decided to put his energy elsewhere.[3]

In January 1913 Dyar began his own journal, *Insecutor Inscitiae Menstruus* (Figure 13.2a), the title of which means "persecutor of ignorance monthly".[4] Given his previous attacks of entomologists, the entomological community might have had cause for concern about Dyar's journal. To allay their fears, Dyar explained that the journal name didn't have "any personal application," and its purpose was "to dispel, to some degree, our general ignorance of the forms of insect life by descriptions of species and genera, life-histories, and other pertinent facts."

Dyar dedicated *Insecutor* to his friend A.R. Grote, who had died ten years earlier (Figure 13.2b). The journal was to be about insects in general, as Grote's journal *North American Entomologist* had been.[5] Dyar had earlier praised Grote's succinct style in describing new species, which was evident in *North American Entomologist*. The entire first issue was Dyar's "Notes on Cotton Moths"—over thirty species of owlet moths. Dyar's choice of subject was not coincidental, as Grote did extensive work on the cotton worm (*Alabama argillacea* (Hübner)) in the 1870s. However, there was no mention of Grote or his work on the subject.[6]

Entomology lore at the Smithsonian has suggested Dyar began his journal so he could publish on moths and mosquitoes at will. Dyar, after all, had always been frustrated with lengthy delays between completed manuscripts and printing. The problem was a practical one: Delayed publication could mean the names given to newly discovered insects by others would become the valid ones because of their getting published first. For example, the Carnegie Institute took several years to print a completed *Mosquitoes of North and Central America*. Likewise, the government printing office was often slow in printing *Proceedings of the United States National Museum*: the dates of publication on articles could be a year earlier than the date of printing.

Dyar's and Knab's Lepidoptera and Diptera articles filled the seminal issues of *Insecutor*. However, Dyar's major contributions to entomology during this time, including naming a staggering number of new Lepidoptera species, were published in the *Proceedings* (Figure 13.2c). His goals for *Insecutor* may not have been for larger works. Dyar may also have believed the USNM should bear the cost of publications closely linked to the flood of specimens to the museum. Furthermore, Howard wanted him to publish a large article to form a series on Panama Lepidoptera with Busck. At odds with his encouragement to publish the article, Howard wrote of a bottleneck in the printing process for *Proceedings of the United States National Museum*: "I learn from conversation with Doctor Benjamin that Museum publications are in a very stagnant condition . . . not able to get the Government Printing Office to go ahead in the present glut of work."[7] Dyar later observed that most of his publications from the onset of *Insecutor* were either placed in his journal or the *Proceedings*.[8]

The first of Dyar's large Lepidoptera works in *Proceedings of the United States National Museum* during this period involved the Biological Survey of the Panama Canal Zone, published in May 1914, and included many new species and genera from the region (Figure 13.2c). The new species named were divided between his tome and Busck's. Dyar described the small-sized "macros", as stated in his introduction:

> Most of the specimens . . . were collected by Mr. August Busck, who went [to Panama] primarily to collect 'Micros' and took the 'Macros' only as a side issue. Consequently the . . . little 'Macros,' especially the small Noctuidae, Lithosiidae, and Pyralidae . . . [are] . . . unusually well represented, many hitherto

undiscovered species being among them. There are reported on here 8,254 specimens in 1,713 species.⁹

Dyar wrote a second large *Proceedings* paper on tropical Lepidoptera in 1914 describing species collected in Mexico by Schaus and Roberto Müller (135 new species, 20 new genera) (see Figure 13.2c).¹⁰ Why Schaus did not describe these species of "macros" from Mexico on his own is an interesting question. One clue is that he took special care to collect particular favorites of Dyar's such as Limacodidae. Maintaining good relations with Dyar, the man in charge of the National Lepidoptera collection, was important to Schaus, particularly because he no longer had complete control over the several hundred thousand specimens he donated. Now living in D.C., in close proximity of these specimens, Schaus chose other Lepidoptera he could name beside Dyar's pet groups.

Combining the large Panama and Mexico works during these years, Dyar had described over ninety genera, roughly a fourth of his entire career (342) and approximately a fifth of the species (680). Dyar would describe around 3,000 species for his entire career, far less than Edward Meyrick (1854–1938), who was the king, at approximately 15,000 species and subspecies, and Schaus with over 5,000 species.¹¹ Of course, Dyar also worked on the taxonomy of two other insect groups, the mosquitoes and the sawflies, while amassing information on the life histories of all three groups of interest.

Some would question Dyar's taxonomic productivity because he wrote extremely brief descriptions of new species and genera. In a paper on the Mexican fauna, for example, a description of a gypsy moth relative *Leuculodes dianaria* is particularly brief due to the absence of a pattern on the front wing:

> Genus LEUCULODES Dyar.
>
> LEUCULODES DIANARIA, new species.
>
> Translucent white; costa of fore wing black at base; vertex of head ocher; pectinations of antennae yellowish. Expanse, 25 mm. *Cotypes.*—Two males, No. 16500, U.S.N.M.; Zacualpan, Mexico, July, 1913 (R. Müller).
>
> Close to *D. lacteolaria* Hulst, smaller and without lines on the fore Wing.¹²

Toward the end of Dyar's career, Jeane Daniel Gunder asked him about what constituted an ideal description of a new species. He answered that a "model description would do more harm than good" and the

> value . . . resides in the author's intelligence, discrimination and experience, qualities which can never be learned. . . . I have seen descriptions three lines long in which the author appreciated every striking differential character and mentioned it, and others comprising pages which were practically worthless. . . . A poor student will muddle his description no matter what standard you set up for him, and an astute student will only be handicapped by a set form.¹³

Along the same lines and also in the late 1920s, he wrote:

> Aside from generic difficulties and type fixation, it is clear when we get down to fundamentals, that the species is the ultimate unit. Nothing can be done in nomenclature without a specific name, and no specific name can be held valid without a description. A description may become inadequate in the light of future researches; but it is adequate at the time it was proposed, provided the author is competent to describe at all.[14]

Dyar made a valid point that long descriptions without diagnostic information have little value. However, at times he went too far in the other direction. A product of his times, Dyar like many of his contemporaries, rarely described genitalia, today considered important in defining a species. Although Dyar may have preferred succinct descriptions, he was also working as rapidly as possible to have an ordered collection—one without large sections of unsorted specimens. This approach led to creating synonyms (e.g., the limacodid *Metraga costilinea* is now *Euclea zygia*), but to Dyar that was something to be worked out later by him or others.

Dyar favored properly labeling type specimens, often a series of "cotypes." These all had red labels with USNM numbers on them. He clearly wanted to avoid the problems the British Museum had of poorly labeled types. Dyar's cotype series were from a number of localities, including foreign countries. Given the possibility that cotypes were a mix of more than one species, Dyar intended to allow the future taxonomist some latitude in determining the identity of a species, as he wrote Cockerell: "I never gave much thought to the question of 'type' localities. The localities mentioned are all 'cotypes' . . . I let the 'next reviser' restrict if any restriction is required."[15]

Dyar's contemporaries, including Schaus, on the other hand, were often unclear in the published description about which specimens they examined or from what locality they came, making it difficult for future workers to determine which species were types as well as the true identity of the species. Thus, although Dyar wrote minimal descriptions of his new species compared with those of Schaus, his work is often more useful because so long as the types are well preserved, it is perhaps more important for the future taxonomist to know precisely which specimens are the actual types and where they were collected.

☞

Some of Dyar's contributions during this period have escaped notice because they did not result in authorship, including Dyar's partnership with entomologist Theodore Dru Alyson Cockerell (1866–1948) (see Figure 19.2b) to complete Alpheus Packard's treatise on giant silk moths.[16] Cockerell, a prolific author, has been referred to as a "walking Chautauqua" because of his Renaissance interests.[17] They began a correspondence while Cockerell was in New Mexico in the 1890s and continued in Colorado during the 1910s and 1920s. He supplied Dyar with live western caterpillars, moths, and mosquitoes, and Dyar provided identifications and manuscript reviews. Although there is no known record of them spending time in the field together—each attended the Zoology Congress in Boston in 1907[18]—the two entomologists respected each other's work. This was especially true of Cockerell, who earlier wrote about Dyar's prominence in the field of Lepidoptera classification and would soon praise his work on mosquitoes.

After the death of Alpheus Packard on February 14, 1905 (Dyar's thirty-ninth birthday), Cockerell was tasked by Packard's widow Elizabeth Walcott Packard with editing his last monograph on the Bombycine moths for the National Academy of Sciences. To complete the project, Cockerell, neither a moth expert nor in possession of a major collection of giant silk moths, needed Dyar's help and, in the fall of 1911, began corresponding with Dyar about obtaining photographs of giant silk moths (Saturniidae) from the USNM collection for the work. Dyar replied, criticizing Packard's ideas about genealogical relationships between groups of moths: "[H]e had a silly idea that they [Limacodidae] were related to the Saturnians on account of the stinging species."[19]

Although Dyar did not like Packard's concepts of classification, he was eager to assist Cockerell in producing a publication with numerous color plates of giant silk moth caterpillars, from the hatchling first stages to the final instar before cocooning. Dyar expressed interest in providing useful identification keys for the giant silk moths in the publication because he believed that "[Packard] could not write a synoptic table [and] had not the faintest idea of the principle of the subject." Dyar also hoped Cockerell could get Packard's unfinished work on limacodids, complete with Louis H. Joutel's drawings, published, and he "could help to any extent."[20]

Over the next year Cockerell and Dyar worked on the Packard monograph. Their correspondence in March 1912 gives a hint of the difficulties Dyar had in juggling projects: "My desk is filled up with jobs and the pile seems to grow instead of diminish. The proof reading of the mosquito monograph uses up all the spare time. I'm sure I won't be able to make you any tables." Indeed, W. Barnum of the Carnegie Institution of Washington wanted all originals of the large mosquito manuscript Dyar coauthored with Howard and Knab to be returned, and the proofs arrived several weeks later. In spite of the pressure to complete these projects, in characteristic humor, Dyar wrote Cockerell about a letter he received from his colleague misdated May 7: "[W]hat a fast life you must be living! It is only April in slow old Washington."[21]

By late June Dyar had made progress with selecting the type specimens to be photographed and the identification tables on Latin American *Hylesia*; remaining work was put on hold due to Washington's sultry summer:

> Really, I can't do justice to your queries in the heat," he wrote Cockerell. "If I leave the fan ... the wind destroys the types [specimens]. I am going to file your letter ... and take 2 weeks in the [Blue Ridge] Mts. On return I will make up for the delay by some good information. So please ... excuse the loss of time and write all the questions you want .... Now I'll get out of this inferno and give you some good copy when I am cooled off.[22]

From Skyland on the Blue Ridge, Dyar reiterated to Cockerell by post that he would be prepared to work on his return to the USNM.[23]

## PAID POSITIONS AND PROBLEMS WITH DOCTOR BARNES

While Dyar continued to complete many significant works, he finally began to make some strides towards permanent, salaried employment in 1906: He received a temporary, paid appointment in 1906 and 1907 as a substitute for Ashmead, who had been in poor health, in the Division of Insects.[24] In his new capacity Dyar was designated by Smithsonian Secretary Charles D. Walcott as a representative of the museum at the Seventh International Zoological Congress, which was held in Boston in August, 1907.[25]

Ironically, prior to his getting a temporary post and while he was seeking a salaried position, Dyar was featured in a 1906 article in *Technical World* magazine called "A New Type of Patriot" (see Figure 13.3a for portrait in article); it was about wealthy American scientists who worked for the government essentially without compensation and produced "results both substantial and valuable to the people and to the country at large." The Division of Insects was referred to as the "Bug Department of the National Museum," and the place where Dyar "toils every day for long hours[,] is well known as one of the greatest 'lepidopterists' living, [and] . . . know[s] more about mosquitoes than anybody else in America or abroad." Dyar earned, the article reveals, a "modest stipend" of "[t]wenty-five dollars a month."[26]

Dyar was a difficult man to make the full-time Assistant Curator at the USNM in 1908, as shown by his criticism of management bureaucracy and his fights with members of the scientific community. Sensing he would not be chosen for the position, Dyar wrote Howard, "If Mr. [Richard] Rathbun [and director of the USNM] will not accept me . . . do not appoint anybody, but abolish the office. The Museum can keep me at $160 as custodian of Lepidoptera."[27] Indeed, the museum administration went another direction and hired James Crawford, Jr., a wasp specialist like Ashmead, as Assistant Curator.[28] Dyar would later reminisce: "Years ago I had trouble with Rathbun. I was too executive for him and he prevented me from getting the place which Crawford now fills. But I am not criticizing."[29]

After his failed attempt at get hired at the Smithsonian, the Bureau of Entomology (USDA) hired Dyar, but not until 1913. Dyar's duties were to identify Lepidoptera of economic concern and mosquitoes, although there were changes in appropriations and job titles. At one point Howard asked Dyar to fill out the blank application and advised there would be no questions to answer in the subsequent entomology examination, but "merely of statements and affidavits, concerning education, scientific training, practical experience, and publications or thesis."[30] Dyar remained the "Honorary Custodian of Lepidoptera" at the U.S. National Museum. L.O. Howard oversaw him both as Bureau Chief and Honorary Curator. Dyar's role as Honorary Custodian would lead to conflicts with Dr. William Barnes.

⚭

By 1910, William Barnes, M.D. (1860–1930) (see Figure 20.1c), built the largest private collection of Lepidoptera in North America in Decatur, Illinois (see Figures 13.3b, 24.1a). He employed a series of talented professionals to manage his collection and write scientific papers based on its specimens. Although Barnes coauthored

**Figure 13.3** Dyar's portrait as it appeared in a *Technical World* article about wealthy American scientists who worked for the government gratis (*a*); and Barnes Museum in Decatur, Illinois (*b*).

these papers, other than collecting and borrowing specimens, he did little of the taxonomic research that produced them.[31] Barnes was politically connected and his wife, Charlotte, was from the prominent Gillett family of Illinois. Her sister, Jessie D. Gillett of Elkhart, Illinois, would fund the Barnes and McDunnough "Checklist of Lepidoptera of Boreal America."[32]

An adversarial relationship with Barnes surfaced in the 1910s, when Dyar refused to lend the physician type specimens from the USNM collection to help identify his own. In December 1912 Barnes wrote his congressman from Illinois, Rep. William Brown McKinley, about having the Barnes Collection purchased by the National Museum. The one stipulation: "it will never go there as long as Mr. Dyar has charge [and if he is] at the time I wish to dispose of it, it will probably go to the New York Museum." McKinley served in the house (1905–1913; 1915–1921) and later became a senator (1921–1926). Barnes wanted McKinley to bring his letter about Dyar's fitness to serve at the Museum to the congressional committee in charge of the National Museum. He mentioned the alienation of J.B. Smith as a factor to be rid of Dyar, as well as the USNM losing out on the Frank Merrick Collection. Barnes criticized Dyar for expecting the elderly Merrick of New Brighton, Pennsylvania, to send his collection without packing assistance.[33]

When Dyar allegedly refused to return types to Barnes's own collection, Barnes threatened to bring the issue to John M.T. Finney, a prominent surgeon at Johns Hopkins, who would then "take it up with [President Woodrow] Wilson personally." Howard was keenly aware of the Decatur doctor's intentions to sell his collection and of his bad opinion of Dyar. In August 1913, after he read a draft of Dyar's critique of Barnes and McDunnough's new series "Contributions to the natural history of the Lepidoptera of North America," Howard warned Dyar:

> I ... beg that you will modify it so as not to make Doctor Barnes angry. A mere statement of the synonymies, it seems to me, would answer the purpose perfectly well and with a modification of the introductory paragraph, leaving out the suggestion that they are in haste to anticipate new species, and the omission of the final paragraph, will not hurt in the least and will not be so apt to give serious offense.[34]

Dyar ignored Howard's request to tone down the critique, writing that the work would have been better done by Smith's former assistant John Grossbeck and W.D. Kearfott: Grossbeck had been on Barnes and McDunnough's expedition to southern Florida and had sent Dyar specimens of snout moths to compare with the USNM collection. He went on, "we have been inflicted with an unusually large proportion of needless synonymy." Dyar then gave a lengthy report on the synonyms and other problems, and in closing wrote that about half of the twenty species in the Barnes paper were synonyms and hoped that "the other unknown half are in better fortune." He also chided Barnes for not making use of the "resources" available to him by the National Museum. "Nothing but haste to get ahead of someone else will explain this work, and that aggravates rather than palliates the offence."[35]

After reading Dyar's critique in September 1913, a defensive Barnes responded to Howard:

> His identifications . . . which he throws into the synonymy [are] probably right in some cases, but he has certainly not, in any case, proven he was right and, personally, I have little faith in such identifications. . . . Some of the wildest identifications I have ever had have been made by him, and I have kept quite a number of the actual specimens, with his hand-writing, for curiosity. I thought I would write you this to show you how well Dr. Dyar is holding onto his grudge.[36]

The "grudge" began when Barnes and McDunnough published a paper stating they had been refused help by the National Museum and according to Barnes, Howard concurred. Dyar was then to have written Barnes an angry note "wanting to know whether it was to be war or peace between us." Dyar's pugilistic tone suggests that he was held accountable for the embarrassment he caused the Museum. He may have also known about Barnes attempt to "dethrone" him as Honorary Custodian.

After the fallout from Dyar's critique Howard had attempted to mediate the conflict between Dyar and Barnes by arranging a meeting between the two. Barnes wrote Howard, "The meeting Dr. Dyar and I had . . . was a failure, as I feared it would be, as regards [to] bringing about any more cordial relations between Dr. Dyar and myself." Barnes went on to make the following request for the prohibition of borrowing USNM types to be lifted: "I thought it possible that, due to the fact the other Museums are granting us the privilege of examining their types, that you would extend us the same courtesy." Barnes held the opposite opinion from Smith, who the previous decade was concerned about types being out on loan from the USNM during his visits.[37]

Part and parcel of having Harrison Dyar in charge of the USNM Lepidoptera collection meant there would be schisms between him and well-connected personages outside the museum. Indeed, William Barnes had replaced John Bernhardt Smith as Dyar's chief adversary.

# PART III

## Scandal, Divorces, and Their Aftermath

## CHAPTER 14

⚜

# Wellesca's Bahá'í Faith, New Wealth, and Growing Concerns by Zella as Dyar's Life Begins to Unravel: 1906–1908

*Wilfred knew that my greatest desire in this world was to make the trip to Acca to visit 'Abdu'l Bahá. So he gave me this treat. The boat sailed from New York City, April 1, 1907.*[1]
                                    Wellesca Pollock Allen Dyar to A.B. McDaniel, 1936

Dyar learned about the Bahá'í religion—which seeks to reconcile science with religion, promote universal peace and equality, and see the divine inspiration and unity of many faiths—from his mother-in-law, Harriet M. Peabody, but apparently did not immediately get involved with the movement. Wellesca Pollock (Figure 14.1a), whom Lua Moore Getsinger introduced to Bahá'í in Washington, D.C., in 1901, quickly became active in the religion, beginning a correspondence with 'Abdu'l Bahá Abbas (1844–1921) (Figure 14.1b), the leader of the faith and the son of the founder Abba Bahá'u'lláh,[2] in 1902. Over the next five years, Wellesca received seventeen tablets, all hand-written letters in Persian on parchment from her leader. In one she was given the Persian name Aseyeh (pronounced OzeeAy) after his mother.[3] On at least three occasions between 1905 and 1907 'Abdu'l Bahá, under house arrest, sent answers on tablets discouraging Wellesca's requests to visit him in Acca, Palestine, then part of the Turkish Empire.

Wellesca married Wilfred Preston Allen in Richmond, Virginia, on September 5, 1906,[4] perhaps creating the "happy occasion," which 'Abdu'l Bahá had said she must wait before visiting him. Wellesca, in any case, remained in Washington while her husband was in Philadelphia and did not introduce him to any of her family and friends except Dyar.[5] The following year, Wellesca, as Mrs. Allen, resigned from her teaching position. With the help of her husband, who "knew that my greatest desire in this world was to make the trip to Acca to visit 'Abdu'l Bahá, so he gave

**Figure 14.1** Wellesca, now married to Wilfred Allen (note ring on left hand), on her pilgrimage to Acca (*a*), and Bahá'í leader 'Abdu'l Bahá (*b*).

me this treat,"⁶ sailing from New York City, April 1, 1907. Wellesca spent six days at the home of the Bahá'í leader in Acca,⁷ along with six other Washington pilgrims, including Getsinger who asked 'Abdu'l Bahá, "'if it were possible for the spirit of a departed to materialize through a medium.'" He answered "No, just as the spirit never returns from the Kingdom."⁸ After leaving the Holy Land Mrs. Allen visited other Bahá'í followers in Paris and London.⁹

Wellesca had asked 'Abdu'l Bahá "to pray for Wilfred and I might have a child," and in August 1908, Wellesca wrote Monever, 'Abdu'l Bahá's daughter, that she was expecting a child, who was born on December 29, 1908. The child was named for his father Wilfred, though he received a Persian name "[w]hen a few weeks old at my request, Abdul Baha named him 'Roshan' (which means light—illumined)."¹⁰ However, not everything had gone smoothly for Wellesca. Prior to her trip a celebration at a Bahá'í meeting hosted by Zella at the Dyar home turned sour:

> We had a beautiful meeting at Mrs. Dyar's house, and she played the Pilgrim's March because I was to be the pilgrim going to see the beautiful prophet . . . and she served us with her own hands, lemonade she had made for the occasion; we were all beautiful friends until this lady gossip told her I was getting rich at Dr. Dyar's expense.¹¹

Such gossip was undoubtedly fueled by those reading the real estate sections of the local newspapers, which revealed Wellesca's sales resulted in profits in the tens of thousands of dollars.¹² There were also rumors that Dyar was the real Mr. Allen.

## INVOLVEMENT WITH DYAR IN WASHINGTON REAL ESTATE AND LOANS TO GEORGE F. POLLOCK

The real estate sections in Washington D.C. newspapers of the early twentieth century appear today like the board game *Monopoly* (© Milton Bradley Co.). Apartments, business properties, or lots changed hands within days of purchase, sometimes having been traded in various combinations (e.g., several smaller properties for a large one). Dyar and Zella, through realtors and with the help of advisors, made such deals with well-known builders and architects. Besides the purchases of land adjoining their home in Dupont Circle, the Dyars' real estate deals in 1906 included the sale of a Kalorama Heights property located northwest of Dupont Circle to architect George S. Cooper for $12,500 and part of a trade for the Coopers' luxury apartment house known as the Ashburn, which was located on the corner of Thirteenth and Harvard Streets. The Dyars paid $45,000 to $50,000 for it; the property later was transferred to Wellesca Allen.[13]

At the time, Zella reported, Dyar was keeping a tight budget at home so that he could afford his growing real-estate investments, and Dyar often used friends as proxies in real estate transactions because he believed he could not have outside income and be a government employee at the museum. Among these proxies were Wellesca, her family, and less frequently Frederick Knab. Wellesca, in particular, was gaining wealth in her property exchanges (Figure 14.2a, b).

**Figure 14.2** Real estate exchanges in Wellesca Allen's name in 1911 included the four-story apartment at 1826 M Street NW (*a*), for the Russell Building at 927 G Street NW (*b*)

In July 1907 Charles Early, Dyar's first Washington realtor, was also having difficulties due to Dyar's miserliness with regard to the management of the Ashburn over paying a husband and wife to be janitors. Early wrote:

> On my return from the country this morning I find your letter of July 3rd. As you know, we only pay the janitor at the Ashburn $12.00 a month, with the understanding that his wife is to attend to his duties after he leaves in the morning and up to the time of his return in the evening. You will see that this is the most economical way for us to employ a janitor.[14]

In September 1908 Dyar threatened to remove Charles Early, whom he had difficulties working with, from managing the Ashburn. Early responded that he, "spent a great deal of . . . time, and energy and money in advertising the apartments in the Ashburn and have rented all but one. I therefore, should not be asked to turn over the leases and building to another."[15] Late the following month Early wrote about carpeting stairs at both of Dyar's large apartment houses, including protective rubber treads, which would cost $50. The carpet at the Ashburn was "quite worn" and needed replacement, but Dyar was against making the improvements.[16]

Wellesca also had realtors handle her tenants and properties. She acquired a property on Fourteenth Street, sold by Early, across from Franklin Park in late February 1908. She held it "as an investment." The block was being "rapidly transformed from residences into business houses." "Boss & Phelps, real estate dealers" had "a five-year lease" on the property.[17] However, it appears making money from the building wasn't easy.

Eugene Gough, a friend of the Pollock family, apparently the first realtor to help Wellesca manage her properties, seems to have suggested that her finding a larger firm to manage her properties could provide her with tenants for eight units on Fourteenth Street. In June, after she decided to make Early her agent, Gough sent her what she was owed on the building, writing: "My dear Lesca: Enclosed please find statement of account together with check to your order for $4.30. I regret exceedingly the net amount to you is so small but as you, of course, understand, the coal bill is for coal which has been used during the past several weeks [$64]."[18]

Early also had problems finding tenants for Wellesca's Fourteenth Street apartments, writing to Dyar, which suggests Wellesca was the "ghost owner" of the properties, about the issue. He also discussed Dyar's failure to obtain the title to the Winchester, which according to Early was "no fault of mine" but the fault of "cheap, blackmailing, Congress entering Dist. [of Columbia] without any . . . grounds for so doing." He hired whom he considered to be the "best lawyer in Washington. Mr. Jos[eph] J. Darlington," and agreed to pay for "the matter at my cost, and expense."[19]

In February 1911 Dyar was still working with Early but the relationship remained tense, and by June 1911 Dyar was working with Alfred Higbie, who sold two of Dyar's "lots on Sheridan . . . for $5,250."[20] Higbie had solicited Dyar's business first, but Dyar later said he discovered Higbie, who "wanted to get in business for himself" and claimed he used Mrs. Allen as a proxy for Higbie, as well, because "I was an employee of the Agriculture Dept and thought they might object to my name in business."[21] However, Dyar's work with Higbie and Wellesca predated Dyar's government positions with USDA or the USNM, so perhaps he was anticipating future employment or had other motives.

⚜

Dyar made significant loans to Wellesca's brother George Freeman Pollock, including $7,000 in October 1907 and $15,000 in 1912.[22] None of this money would ever be repaid directly to Dyar.[23] The money helped Pollock manage and pay the mortgage on the Skyland resort atop the Blue Ridge during the decades prior to the establishment of the national park. Indeed, President Franklin D. Roosevelt, who would dedicate a new Shenandoah National Park in 1935, probably owed Dyar a debt of gratitude for his contributions to keep parcels of the land conserved.

Dyar had become a high-profile player in the booming real estate business in the nation's capital and a patron of Skyland in the Blue Ridge during his first decade in Washington, D.C. In terms of his personal life, the question was whether Zella's marriage to Dyar could sustain his relationship and real estate deals with Wellesca.

## ZELLA'S ULTIMATUM AND RETURN OF THE STONES FROM 'ABDU'L BAHÁ

In November 1908 Harrison and Zella Dyar made a curious set of real estate transactions. They sold their Dupont Circle homes on Twenty-first Street NW (Figure 14.3a) and all of the surrounding land to Mrs. Harriet M. Peabody for $43,900, and on the same day, sold Wellesca an 18th Street property for $70,000. One week later Mrs. Peabody sold the Dyar residences back to Zella, and not Harrison, for $43,900.[24] Zella had apparently made an ultimatum that Wellesca pay back what she received from Dyar and put a chunk of assets in Zella's name to insure the future security for herself and their children. If they did not comply it appears Zella would go forward with a very public divorce, naming Mrs. Allen as a corespondent ("the other woman"). The following spring, however, real estate transactions with Wellesca continued.

On May 23, 1909, Wellesca hosted a three-part celebration for her baby Roshan, the Bahá'í "Feast of Declaration of the Bab" (the foretelling of the coming of Bahá'u'lláh), and 'Abdu'l Bahá's 65th birthday (Figure 14.3b). On the invitation, along with a photograph of Roshan, was written, "Roshan is rejoicing in the springtime of life, in the springtime of the year, and especially because we are now living in the springtime of the Day of God, the Reign of Peace, the golden Age, when there is to be no more winter but an everlasting season of Spring."[25]

Zella Dyar took the celebration as an opportunity to write a reproachful note to Wellesca about her husband's generous gifts to her. Perhaps as an attempt to mollify Zella and her mother, Wellesca had given them monogrammed stones she

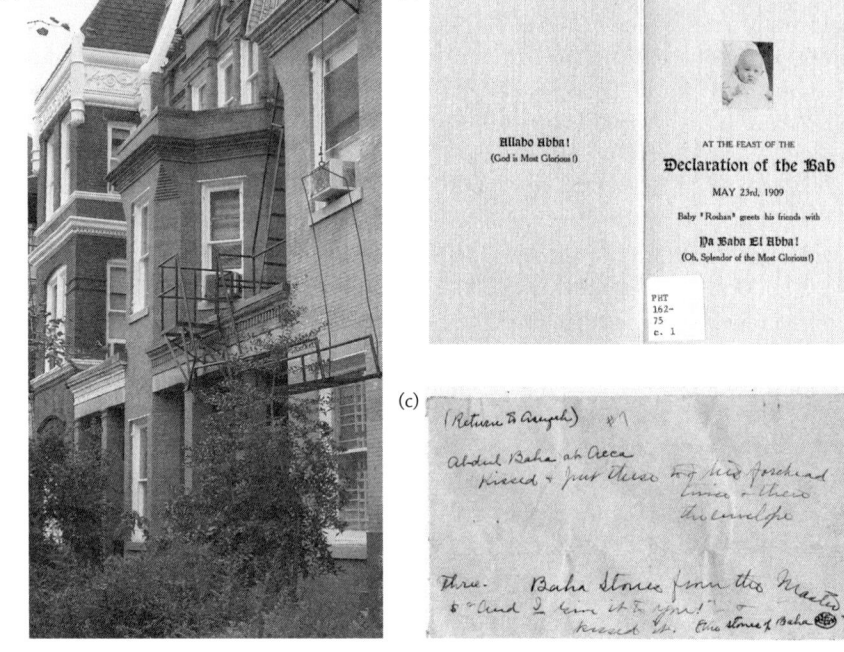

**Figure 14.3** Dyar's Dupont circle home (*a*); announcement of "Declaration of the Bab" and celebration of the birth of Roshan Allen (*b*); the envelope that contained three stones given to Aseyeh (Wellesca) from 'Abdu'l Bahá (*c*).

acquired from 'Abdu'l Bahá in Acca.²⁶ Zella returned them ( Figure 14.3c) and a distraught Wellesca wrote that she,

> would have never have taken any money except on baby's account.... Your husband knew this. He is the best friend I ever had and I trust and honor him next to my own husband. Except for him we would not have had our son, & he is to be 'Uncle Harry' to little Wilfred [Roshan] always. The money is certainly safe. Mr. Dyar has my [financial] notes and if he should die ... I would pay them to his estate. If I should die my will is deposited in the American Security and Trust co., and it provides for the payment of everything and makes it impossible for the property to pass into wrong hands.²⁷

A month earlier, Dyar sent Zella a written ultimatum, demanding that she accept his relationship with Wellesca as "sisterly in nature." Dyar asked her to forgive both him and Mrs. Allen "for any indiscretions that we have committed or that you think we may have committed." He noted that if he were in Mr. Allen's position, he "should be thankful to have someone take as friendly and brotherly an interest in my wife as I do in his and help her out of difficulties."²⁸ He also offered an explanation for Allen's mysterious nature, explaining that Wellesca's large family made him nervous and "Lesca," given her "usual impulsiveness and generosity said 'well you [Mr. Allen] needn't see any of them then.'"[29]

Zella replied:

The time for jealousy [about Wellesca] on my part has long since passed. It is all now a question of right and wrong. You know better than I what you have done and are doing. If you can conscientiously think that you have done no wrong, then your standards of right have changed a great deal in the last ten years.... I shall never call on Mrs. Allen because I do not have as a friend any one who can be so careless of a reputation, as she has of yours. And after all you (and I too,) have done for her. For, as hard as I try, I cannot feel any more that she is innocent in this matter. I shall be only too glad to help you get back into the straight road, and feel that you should be willing to make any sacrifice in order to realize your best capabilities once more.[30]

Writing idealistically about more important things in life than money, Zella suggested her husband work more for the betterment of society than spend idle time with Mrs. Allen:

There are so many vital subjects today to be threshed out for the benefit of society ... it is a pity to lose one possible worker, and one might easily forget his own little pleasures in being actively interested in one of these great world movements for good ... Will you not work with me for something more inspiring and of more lasting value? As it is, I think that you [are] doing wrong, and my 'jealousy' is only for your reputation, among your fellows.

She then reiterated her hope that her husband would reset his moral compass and suggested she would assist him in typing his manuscripts, writing, "I am delighted to see the old type-writer here again: I would be glad to help you any time."[31]

## DYAR'S DECLINE IN COLLECTING

Dyar went with his family to Weld, Maine in 1910, but Dyar increasingly spent his summers with Wellesca and her growing family from 1911 to 1914, when he reported only scant collecting.

Among Dyar's records in 1911 were limacodid moths in July, the first from a visit to Skyland[32] (see specimen label in Figure 14.4). The other was from a trip to Tryon, North Carolina, a pinstriped slug moth (*Monoleuca semifascia*) (Plate 1d)— ending a fifteen-year quest to successfully capture the female moth. Lured into one of Barber's new tent light traps, the moth "deposited eggs and [the hatched] larvae fed normally upon smooth-leaved trees." Dyar described the elusive moth's life history in the last supplement of his "New York Slug Caterpillars."[33] In 1912, while at Stony Man for ten days in July, Dyar collected few Lepidoptera, though he and

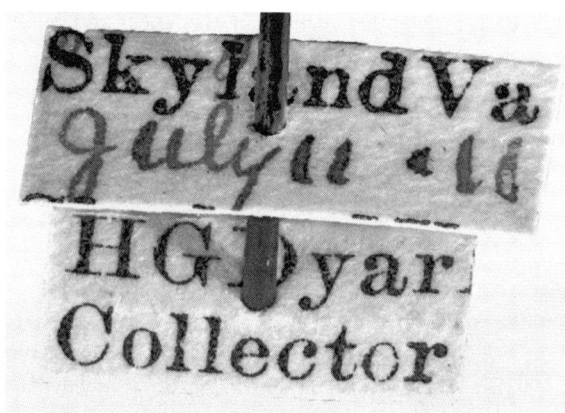

**Figure 14.4** Label on a limacodid moth collected by Dyar in Skyland, Virginia, on July 11, 1911.

Wellesca each found brown katydids (*Atlanticus davisii*). Dyar's only known record for 1913 was an unidentified moth collected from one of his properties in D.C. on May 29.[34]

Perhaps most symbolic of this period of sparse collecting was that Dyar on a visit with Wellesca to the romantic Paramore's Island in July 1914, certain to have abundant mosquitoes, reported collecting only the eastern salt marsh mosquito (*Aedes sollicitans*).[35] The only other report that year was a rock pool mosquito (*Aedes atropalpus*) at Chain Bridge near the District of Columbia in August.[36]

## CHAPTER 15

# Marriage Troubles: 1909–1913

*... Then you built her a house, somewhat strangely located [across from the museum], she has become well off, rich even.*[1]

Zella P. Dyar to H.G. Dyar, 1912

On April 16, 1910, Wellesca purchased a property on B Street SW, a short distance from the flat on 1228 B Street SW that she shared with Mr. Allen, at least according to that year's census (Figure 15.1c).[2] Dyar would soon build Wellesca's home on the lot across from the Old National Museum building (Figures 15.1b, d). Later Wellesca wrote to Zella (Figure 15.1a), telling her the house was paid for with profits from a successful real estate deal made in Wellesca's name, Dyar's *modus operandi* for other real estate purchases and exchanges.[3] The house was intended for a growing family that included young Roshan, born in December 1909 (Figure 15.1e), and Wellesca was again pregnant, though that pregnancy ended in a miscarriage.[4]

Amid gossip fueled by her newfound wealth and once again expecting a child, Wellesca wrote Zella in May 1911, perhaps under Dyar's instruction, after Zella refused an invitation to her home. Wellesca, attempting to ingratiate herself to Zella, inquired after her health—knowing she was unwell, perhaps recovering from typhoid fever—and empathized with her need to nurse her children: "Certainly you had a time of it, first nursing Otis and Dorothy through chicken pox and then Otis through measles. If you had not given out under the strain then I fear you would have if you had been up to care for Dorothy through the last siege of measles."[5]

In an attempt to put Zella at ease, Wellesca explained that her financial dependence on Dyar was only temporary. Mr. Allen was to inherit a "considerable sum" from his aging and wealthy employer, McGrath, who had no heirs. The plan was for him to work in Philadelphia until his employer's death, vaguely estimated to be between two and ten years. She also explained how she came to own the house and how idyllic it was: "Mr. Dyar ... loaned us a sum which we invested in hopes of

( 125 )

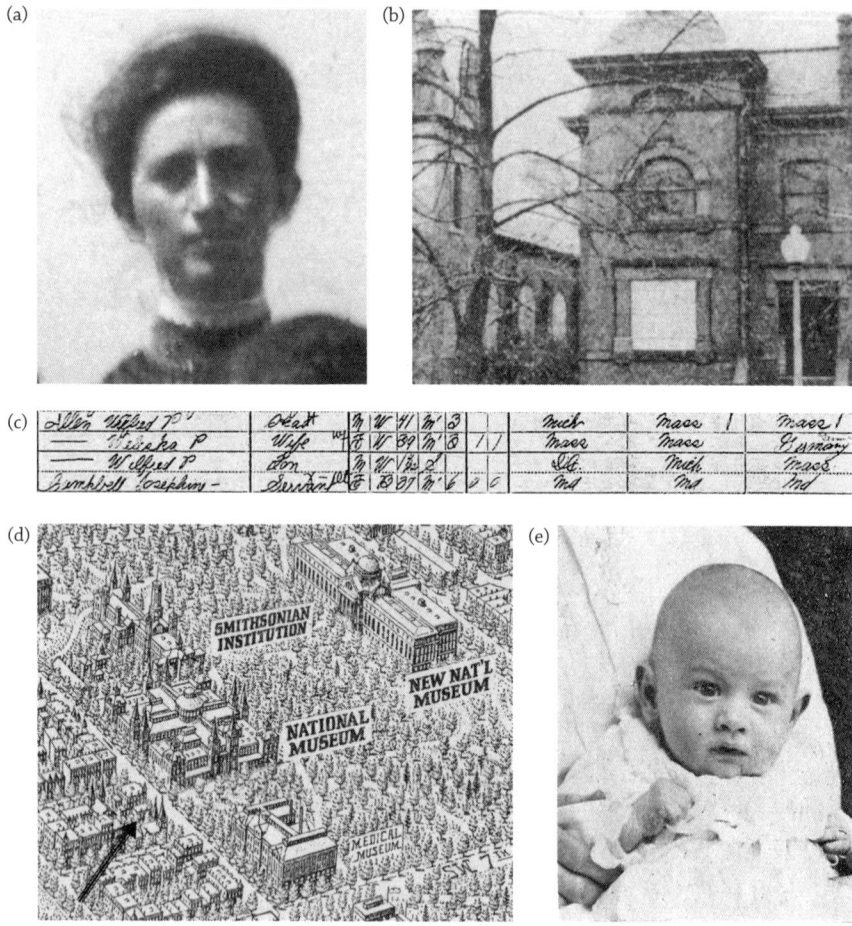

**Figure 15.1** Zella Dyar (*a*); home H.G. Dyar built a home for Wellesca and her growing family at 804 B Street SW (*b*); 1910 U.S. Census of the Allen family and house servant Josephine Campbell at 1228 B Street SW (*c*); location of home on 804 B Street SW across from the National Mall (*arrow*) (*d*); baby Roshan Allen (*e*).

making something. Our hopes were more than met. We made not only enough on real estate deals to pay back 5% interest, but also to build the lovely house at "804." And now, all that was used as investment in houses has been signed back to him, so there is no indebtedness."[6]

Wellesca went on to observe that Dyar's generosity to the Allen family was not without precedent, as he had made similar loans to colleagues at the museum, "Mr. Caudell and Mr. Curry [Rolla Currie] to buy their homes" and "neither of the two ... have yet paid their debt. And Mr. Busck does not even pay interest on the loan he got." She also explained that the home was to become a safe haven for Baháʼí travelers, who needed a hospice now that the Florida one had been vacated. Perhaps with a nod to Zella's mother Harriet Peabody's work for the less fortunate, Wellesca also left the house to the Baháʼí Assembly in her will. It was, she told Zella, "to be used as a hospice or for a home for maiden ladies and orphans."⁷ Her hope was that 'Abdu'l Bahá, when visiting America, would for "at least one night occupy the hospice room."⁸ Although 'Abdu'l Bahá did visit the B Street home in 1912, he did not stay over.⁹

Wellesca also restated Dyar's common refrain that there was "no reason . . . why we should not be friends all along" and "the only way I erred was in not discussing the matter of the first loan from [her husband] with you." Near the close of the letter, Wellesca, playing to Zella's pride, characterized herself as "plebian and sort of common" and Zella as a woman "of refinement and high birth."¹⁰

⚭

During the fall of 1911, after Dyar spent part of a summer with Wellesca and her son Roshan at Virginia Beach and Skyland, Zella wrote him, complaining that their son Otis had become aware of the situation. In search of Dyar, Otis had gone to the museum and could not find him, so he went to the B Street home.¹¹ Zella, therefore, complained:

> It will be necessary to make an appointment with you, so that he shall see no more of this: and you will be at your work to receive him. . . . I had hoped, I suppose because it is natural to hope, that when we returned from Maine, I might find things changed somewhat for the better. . . . You evidently made an effort at first to feel some interest in us all, but it was too hard work, and we soon found ourselves alone as usual.¹²

Dyar continued to spend significant time at Wellesca's home, including on Christmas Eve 1911, when Wellesca's second child was born, and remaining past New Year's Day. Dyar was tending the furnace during the birth, and the attending nurse saw him in the bedroom with mother and child. The newborn was named Harrison after his mother's friend and benefactor. This was an audacious decision given the gossip that Dyar was Mr. Allen. The baby, who was later nicknamed "bunny," received the Persian name Golshan from 'Abdu'l Bahá. With Golshan as a middle name, he was now Harrison G. Allen.¹³

⚭

After 1910 Dyar began sending Zella ultimatums, demanding she publicly accept his friendship with Wellesca. He wanted Zella to keep up appearances by allowing the Dyars and the Allen family to spend time in each other's homes:

> I am not going to let you spoil our friendship. . . . What is it you fear? I am not going to run away with another man's wife. I am 46 and Lesca is 41, and that is rather late in life for any such foolishness . . . We are friends of the Allen family and have rendered them assistance where we could, that's all.¹⁴

From Zella's perspective, the proposal was absurd. Moreover, unless Wilfred Allen materialized, rumors that Dyar was the Allen children's father would continue. Zella's response shows that Wellesca had been a regular visitor at the Dyar home since their first meeting. She knew the attraction between her husband and Mrs. Allen was not "sisterly." Zella countered her husband's assertion that she was jealous and suggested that Dyar had once been jealous of "another man" seeking her affections.[15]

When it became obvious that Wellesca and Dyar were involved in a variety of real estate deals, Zella was bothered that the transactions were occurring at all and that the house on B Street was being built. Zella complained, "I do think that she ought to be satisfied with the princely gifts you have made her and be glad and thankful to go and live with her husband, as other women do . . . [S]he can never recompense you for your lost manhood, for your steady decline morally, for your lack of interest in your work." Dyar replied, "I am not going to give her any more money. What I did give was in exchange for the business relations you wished me to give up, and I have property which may return me the amount. Are you afraid my interest in her family will detract from that in my own? I assure you that the reverse is the case."[16]

If Zella was distraught by Wellesca and Dyar's relationship, Dyar was disappointed by his wife's wish to have only two children, obliging him, as he put it, to "[give] up half my family on your account." Zella defended herself, asserting:

> When I found that motherhood duties were fast rendering me less capable as a mother, . . . I considered it just as much my duty to the two living children as well as those unborn, to refuse further motherhood, as it had been my most sacred duty in the beginning to have children. . . . I realize full well that they have many faults, but you should realize, I think, that had I my sense of hearing, they might be very different. (And that is the principal reason for limiting the size of the family, as I have said before.) . . . It is a hard struggle alone.[17]

Dyar also felt that he should be in charge of the household. Even if he had not been away for lengthy periods, either in the field or with Mrs. Allen, Zella clearly needed help from her mother or someone else given her hearing disability. However, Dyar played the victim card with his wife and mother-in-law, writing, "[I] am helpless while you will not allow me any influence in the direction of the family. It is time there was a change and that my authority was recognized and decisions respected."[18] Zella responded,

> As to you not being the head of your house, you surely do not believe that the man is either qualified to or should in reason, decide all of the matters pertaining to the home. . . . I let your desires override my better judgment in the matter of schooling for Otis, which, as you know, resulted so disastrously at the Force school: so that I had to withdraw him before the term was over.[19]

Living with Zella's mother was often a source of contention. Although clichéd to "blame the mother-in-law" for meddling, both Dyar and Zella found living up to her mother's expectations difficult. "I have endured (more or less patiently)," Dyar complained, "the unspeakable insolence of Grandma for your sake for 20 years;

I have watched Dorothy grow up into a little prig under this influence and now am seeing Otis' naturally sweet disposition being spoiled by too much hectoring from three women [Harriet, Zella, and Dorothy] without any real direction." "You seem to forget that she is the grandmother of your children, beloved and looked up to and revered by them," Zella responded. [She is a] woman of the highest moral character, of sound judgment and ability, and is looked up to with genuine admiration and almost reverence by both men and women.... I know that she thinks I have fallen far short of what I should be now, if I had used all of my opportunities to the utmost."[20]

In spite of all of Dyar's domestic problems, it is surprising he requested to have the children accompany him to England for the 2nd International Congress of Entomology in 1912. Alternatively, he offered to pay for their travel in the western United States, but not his mother-in-law. Zella responded, "I would say, that your suggestion in regard to a European trip, would necessitate my going along with the children, an undertaking which I would not attempt, as Dorothy would probably be very ill, and Otis is not yet old enough to be of much assistance at such a time." "The Rocky Mt. region," she continued, "they are totally unacquainted with. There are many wonderful regions to be visited[:] The Yellowstone, Pike's Peak, and the new Glacier Park in Wyoming [sic]."[21] Dyar apparently took the children on a western trip during the summer of 1912 (Plate 11a), though probably without Zella and Mrs. Peabody.[22]

Dyar continued to try, unsuccessfully, to get Zella to allow Wellesca and her family to visit. Zella's acquiescence, given the gossip surrounding Dyar and Wellesca, was impossible. Still, Dyar persisted: "Now I ask you this in the interest of peace and harmony and I promise you that you shall not regret it. At least you can give it—trial. If you find in a year from now that things are going worse than they would (can they go any worse than at present?), you can quarrel with Lesca again and we can begin the fight where we left it off."[23]

Zella did not mince words about Wellesca:

I would say there would be just as much chance for her to work in Philadelphia among the Bahas [sic] as here. She has broken with all of her relatives here, and over this very question, too. . . . I certainly shall have pity for her on that day when she looks into the calm pure eyes of the Master, where I almost feel that he will be able to read the story that must be written there. . . . [In] the interests of peace and harmony and justice, I ask you to explain, really, the separation of herself [and her] husband. No one wishes or needs peace more than I. The mother of your children. Zella.[24]

In 1913 Dyar's official home address remained in Dupont Circle; Dorothy was college age and her brother Otis, now 13, graduated from Friend's School Intermediate Class that June. We don't know if Dyar attended the ceremony, but perhaps not, given the growing strain caused by visits to Mrs. Allen's.[25]

The prior fall Wellesca made a request to 'Abdu'l Bahá, during one of his two visits to the District of Columbia in 1912, for a third child with Mr. Allen. During their personal interview he was said to have replied: "Already God has granted your request." When M. Alma Stover, a graduate nurse, arrived at the B Street home on July 21, 1913, Wellesca was about to give birth to her third child. Stover found the expectant mother, a "little German nurse girl," a cook, and Dr. Dyar. One person she expected was the father, Wilfred Allen, but he was missing.

Dr. Mary Parsons delivered Wallace Parsons Allen at midnight on July 22, but Dyar, the godfather, was no longer there. The child was named for Wallace McLean, Wellesca's family friend, a lawyer and head of the Morris Plant Company in New York. The middle name was for Dr. Parsons.[26]

Soon after the birth of each child, they received Persian names from 'Abdu'l Bahá. The first and last of the Allen children were given English names with the initials W.P.A., the same as their mother and father. Wilfred Preston Allen, Jr., was later given the Persian name Roushan (illumination and light). Harrison, the middle child, was given the name of Golshan (rose garden or beauty). Wallace Parsons Allen would receive the name Joushan (breastplate and protection), which later became his middle name: Joshan rather than Parsons.[27]

According to Wellesca the spellings of the Persian names Roushan and Joushan "evolved" to match how they were pronounced in Washington. The change in spelling, however, may have resulted from a genealogical twist from Dyar. Given his wordsmanship it would be surprising if he was not involved in tweaking the names to Roshan and Joshan, for the first three letters of the first and last name had significance on Dyar's maternal side, that is, *Jos-han* (*Jos*iah Cushman *Han*num) and *Ros-han* (Eleonora *Ros*ella *Han*num Dyar). Wallace believed his name was entirely from 'Abdu'l Bahá,[28] but given Dyar's genealogy, one is led to wonder if dropping the 'u' from the children's Persian names was a "Dyarian" riddle.

Perhaps Aseyeh requested a name from 'Abdu'l Bahá for her second son to begin with a "G" (Golshan) to make the first and middle initials the same as his godfather, Harrison G. The name Golshan seemingly doesn't fit the genealogical pattern of the other sons, as the root "Gols" is not found among the Hannums. Curiously, the translation of Golshan as "rose garden" alludes to Dyar's mother Eleonora's middle name Rosella.

Dyar attached a baby photo of Wallace Joshan and his nurse along with a note to Perle during Christmas 1913. In a rather curious exchange, the brother wrote: "This was meant for a picture of Mrs. Allen's baby, but its [sic] as good of the nurse and she looks so handsome that I begged the picture to show you as a sample of German art. The girl is a German." Although it seems odd for Dyar to mention the attractiveness of the nurse to his sister, perhaps it was a ruse to divert her attention from Wellesca and the baby. Given Perle's close relationship with Zella, she would have known about her sister-in-law's torment over her brother's mistress. Perle would eventually accuse Harrison of fathering the Allen children.[29]

## CHAPTER 16

# The Separation: 1914–1915

In early 1914 Dyar's irregular appearance at the National Museum, now that he was on the USDA payroll, became more noticeable. He had been on sick leave at Wellesca's. Dr. Parsons said Dyar was chronically ill with the grippe (influenza) and linked his nervous condition to his recovery and a "weak heart."[1] Dyar, taking opiates to get to sleep, described his own condition in the following way:

> The nervous disease like that I could do fairly well during the day time but ... at the time of going to sleep, there is always a reaction when the nerves let go. On going to sleep I would feel restriction in my throat that I could not breathe; set up in bed, beating the air, gasping for breath the heart in violent action; that will last for a few minutes and gradually disappear in possibly, some times [sic] twenty or thirty minutes and then relapse again and would repeat; and if I was very bad there was nothing to do but to take opiates and go to sleep.[2]

To avoid the appearance of impropriety, Dyar instructed Busck and others to hold his mail at work for his courier, but the B Street home was his contact address. In February Dyar wrote to Howard: "Please invent some way so I shall know automatically when my letters have been attended to and sent so I shant have to keep them on my mind and get grayer than I otherwise would." Howard replied, "I will try in future to send you word each time one of your letters is mailed from here."[3]

Concerned for her husband's health, Zella wanted him to come home but would not visit him.[4] Over the summer of 1914, Zella, along with her mother and children, went to Maine, as was her routine, but without her husband. Zella still collected a mosquito for Dyar while in Norcross. The family stopped in New York briefly on their way home to visit Perle and Dr. Knopf, who were certainly aware Dyar was now with Wellesca,[5] with whom Dyar spent part of the summer in Accomack County, Virginia, at Island House in Wachapreague.[6]

When the family returned from Maine in September, Zella wrote Dyar, again asking him and telling him to "[g]et well as fast as you can."[7] In September 1914 it became clear Dyar would not return to his family. August Busck wrote to Howard: "I beg to inform you that Dr. Dyar sent a messenger for all of his official mail last Saturday, and he presumably intends to answer it from his home," that is, 804 B Street. Busck continued: "I shall of course be very glad to help out as far as I can in any group of Lepidoptera, while Dr. Dyar is away, but there are no Department sendings on his desk now."[8]

❧

In December 1914 Dyar went to New York to convalesce at his sister's, although he continued to get work from Howard, who sent him the crop pest, the velvet bean caterpillar (*Anticarsia gemmatalis*), for identification. Knab also wrote to Dyar with updates on the journal, *Insecutor*, which Knab was keeping on a timely publication schedule: "Enclosed is proof of Insecutor. I have the other set, but will hold it until I hear from you. [C.H. Tyler] Townsend devised a scheme for clipping down his article. . . . Searle is also asking for the mss for the Dec. number. Have you any? How are you getting on?"[9]

When Dyar requested government leave to go to a warmer climate for his convalescence, Parsons suggested Dyar go to Atlantic City, whereas Howard recommended Florida.[10] In the end, Howard arranged for Dyar, probably at Dyar's cost, to "stay in the Bahamas for sixty days from January 2nd, collecting. . . . [and] not be charged against your annual leave."[11] Dyar, he noted, required a nurse for "continual services during the nighttime" for his condition, so Wellesca hired Maud Leech to stay with the children at B Street[12] and accompanied Dyar. Although Dyar's need for a nurse seems suspicious, Wellesca reportedly saved his life on numerous occasions during his nighttime attacks.[13]

In late January 1915 Dyar wrote to Howard about mosquito work, noting that mosquitoes

> are very much in evidence, at least the two common kinds. . . . The [*Aedes*] *calopus* rest by night and the [*Culex*] *quinguefasciatus* rest by day; but we, that is to say the specimens of *Homo sapiens*, we have rest neither by night or day. Dr. T. H. Coffin found some curious *Culex* larvae with only two tails somewhere in these islands, which he failed to breed. I shall take a look for these as soon as I feel able. Hoping to be soon restored to duty and on the firing line.

Howard thus sent Dyar additional supplies, including dissection scissors and cyanide bottles,[14] and Dyar made "a few observations . . . worthy of record" of three species of *Culex* found breeding "in holes in coral rock containing water."[15] He and Knab thus published an account of the *Culex* mosquitoes found on the Bahamian island of New Providence in *Insecutor. Culex aseyehae*, one of the new species described in the article, was named after Wellesca, Aseyeh being her Bahá'í name. Part of the description read: "Female. Proboscis moderately long, nearly uniform, brownish black scaled throughout." Dyar was able to find the rare larvae, probably with Wellesca's help, in rock pools.[16]

Wellesca also helped Dyar in another matter. She sent one of Dyar's short stories to the Frank A. Munsey Co., a large publishing house of a number of magazines, including those with pulp detective fiction, and many newspapers. The story was rejected. "We have read the story, 'The Cloud of Fate,'" H.R. Durant replied, "and regret to report that it does not seem to quite fit the needs of any of our publications."[17] The rejection was among other setbacks in Dyar's literary career: The American Literary Bureau had rejected Dyar's novella *Diamonds Going and Coming* the previous month. The feedback, however, was positive, and he later self-published it.[18]

⚜

Zella's decision to remain married to Harrison—beyond the stigma of getting a divorce—was related to her fear of losing support for herself and the children: "If I have tried to influence your financial policy," she wrote, "it was solely on account of protection to your children's inheritance."[19] She did, however, wish to separate from him. During the final days of his trip, Dyar wrote Zella about her proposed separation agreement in which she accused him of desertion.[20] Negotiations for a settlement had begun, though Zella, who agreed with Dyar that they needed to curb bad publicity, still wanted Dyar to return home because of the impropriety of his staying with Wellesca and her children. Wellesca had kept her travel with Dr. Dyar a secret from her friends and associates for the same reason. Zella's plan was for her husband to stay on the first level of the 1512 Twenty-first Street house, while she and the family would live at the 1510 address. She went on to discuss the financial arrangement, agreeing to settle for $250 a month and suggesting that Dyar pay any tuition that needed to be paid for educating his kids.

Zella, responding to Dyar's insistence that she not hold the Twenty-first Street property, revealed her husband's Washington, D.C., real-estate empire was in the red, observing that the property she wanted was "the only piece of unencumbered real estate that you hold now in Washington . . . [if] it is deeded to Mrs. Allen—how long will it be in that condition! As it is, if left under my management, it will be conserved." Her final words show Zella not only blamed Wellesca, but also Alfred Higbie for the financial mess, writing: "Can you wonder that I have very little confidence in Higbie—? On that I feel that even through a reliable agent, that I might do better."[21]

Dyar answered Zella's letter on March 8, 1915, from Florida, writing,

> In my long letter I did not leave any matters for compromise or "meeting half way." Therefore I expect you to agree to all the points except the question of your piano and the time of sending the furniture which are in your discretion. I do not think it wise to take the furniture with you, first because I want your remaining array to seem spontaneous and second because you will not know what to do with the furniture til settled nor what you will need. But these are not cardinal points.

Dyar made two other demands. First, he wanted his mother-in-law, Mrs. Peabody, to leave the house by March 15. As he had in his 1912 ultimatum, he stated his dislike for her, adding that his "nerves" could not "stand her" and that he "occasionally [had] dreams of having fierce fights with her." Dyar went on to complain, "Her attitude, her ideals and her objects and methods are repugnant to me" and noted that if he had authority "over the children she should never associate with them." His second demand was to deed the property in Dupont Circle (square 67) to Wellesca, stating he had expected her to "yield me back my property when asked," as, he claimed, Wellesca did with other property.[22] This was something Zella, it seems wisely, refused.

Dyar's financial condition at this time is unclear, but a letter to Zella on March 8, 1915 reacted to her suggestion that the Twenty-first Street homes be used as collateral to borrow money because he accrued debt on his other properties and was having financial difficulties:

> I am heartily sick of borrowed money and I have done extremely little of it, whatever you may think you have found out. I have borrowed exactly $22000 in all and I have paid off old trusts . . . on the properties to almost that amount and shall soon exceed it with payments I am arranging for. What is the fuss for? I am sick of all the distrust and your setting my judgment aside and supplanting your own. Am I a spend thrift or fond to throw my money away? I have had losses, but they have been due to bad times and causes beyond my control.

In response to Zella's distrust of Higbie, his business partner, Dyar claimed Mrs. Allen's name was used to make real estate purchases, although he was vague about whether that was the case with some or all of them. In fact, Dyar claimed the "Co." in Higbie & Co. was Wellesca.[23] Zella's concern was that if her husband lost everything, it would harm her and her children. Dyar, nonetheless, continued to argue with Zella over her holding the Twenty-first Street property. He conceded he couldn't force her to give it up as it was in her name but threatened that if she refused to do so, he would withhold "support [from her] till the court compels me," would not pay the property taxes due in May, and would "change my will and not leave you or the children anything except your 1/3 of the income for your life." Dyar did give Zella the choice of keeping the Twenty-first Street property, "worth almost $50,000 but not saleable now," but at the cost of losing "the rest of my estate, worth about $400,000." At the close of the letter, he wrote: "Agree to these two points in the Jacksonville letter [of having Mrs. Peabody move out and deeding property to Wellesca] and all goes well. Otherwise we'll see if little old H.G.D. has any fight left in him yet!"[24]

Although his letter was harsh, it closed on a gentle note. Zella had expressed hope for his safe passage on the ship, and Dyar became sentimental about their earlier trip back:

> I had a fairly good passage back . . . But nothing like the blissful passage we remember [from Nassau in 1898]. I have . . . a number of fine mosquito larvae representing two or three new species of *Culex* I believe. By staying longer I could have done some fine collecting, but this was a purely health trip, as I am looking to bring what I have. It is cold in Miami Today.[25]

"The Meeting," Dyar's unpublished short story written several years later, gives his view of the events that led to his separation from Zella (Matilda in the story). The story includes the infirmed protagonist's (Paul) sojourn to a warm climate with his nurse Mrs. Allen (Mrs. Jones):

> He had the effrontery to be sick at the "other woman's" house. Matilda was furious. So was "Mama." They engaged a lawyer and refused to admit Paul [Bartlett] into his own house. But there was no suit for divorce. They possessed no evidence what ever and they knew it. There were plenty of threats of legal action and much talk of possible scandal and gossip. Paul finally signed a separation and his family went west to live. Paul engaged Mrs Jones as nurse until he should recover and he also left town in search of a favorable climate.[26]

In late March 1915, as negotiations for the separation agreement proceeded, Dyar, who was living at the Metropolitan Hotel, a few blocks from the National Museum, wrote to his children, explaining that the children of parents who get a separation can choose which parent to remain with if they are over fourteen. Each letter began: "Your mother contemplates separating herself from me to live in California." Dyar told them he would not support them after the age of twenty-one if they went with their mother. If Otis chose his father, he was given the option to "finish out the Friends School, then go to a college of your choice, preferably Mass. Inst. Tech., where I graduated. Then I will see you established in a profession." Dorothy, now nearly nineteen, was given a different option: "If you stay with me, I will send you to college in New York where you can stay at Aunt Perle's.... I will keep you till your marriage."[27] Both children stayed with their mother, though Dorothy would follow her father's advice for college.

Dyar and his wife Zella signed a separation agreement on May 13, 1915 (Figure 16.1). Zella was to receive $300 per month ($6,640 in today's dollars) in resettlement costs to Berkeley, California until April 1917, when Dorothy would turn twenty-one. Afterward she would receive $250 per month, and have all of the furniture, while her husband would keep some portraits and have visitation rights with the children. In the agreement Wellesca Pollock Allen renounced her separate income from properties in her name (earnings in her name were to be considered part of HGD's earnings), and half of Dyar's net income was to go to Zella.[28]

∞

Dyar had a rough re-entry to the museum. In January 1915, when Dyar had a dual role as Honorary Custodian at the USNM and as a civil service employee with the USDA, he got into a scrap with the Civil Service Commission. His attempt to train a potential future employee to replace James Crawford's wife, Emily Baker (Wellesca's niece) was frowned on. As Dyar wrote Howard in May 1915: "I am sorry to hear that the Civil Service Commissioner does not like me to train in people ... Can you tell me if they have some people on the list ready for Mrs. Crawford's place and if so who they are? If there will be an examination I can let my man take it."[29]

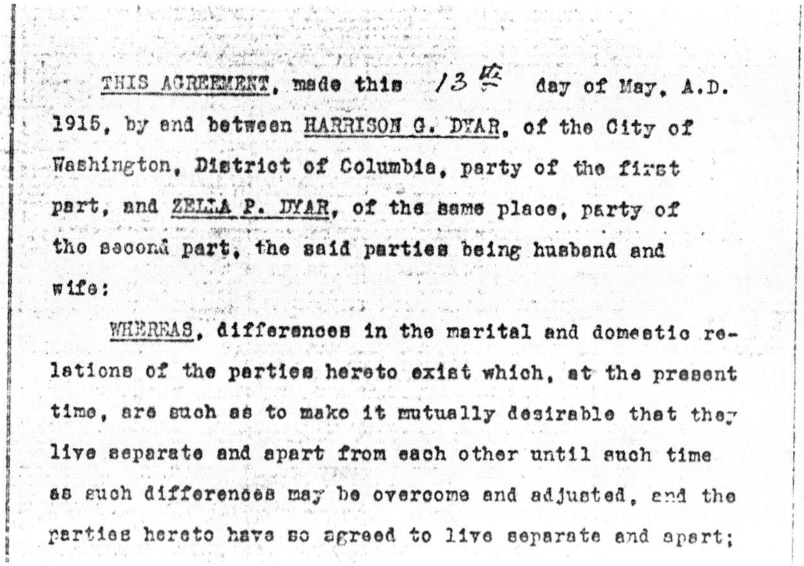

> THIS AGREEMENT, made this 13th day of May, A.D. 1915, by and between HARRISON G. DYAR, of the City of Washington, District of Columbia, party of the first part, and ZELLA P. DYAR, of the same place, party of the second part, the said parties being husband and wife:
>
> WHEREAS, differences in the marital and domestic relations of the parties hereto exist which, at the present time, are such as to make it mutually desirable that they live separate and apart from each other until such time as such differences may be overcome and adjusted, and the parties hereto have so agreed to live separate and apart;

**Figure 16.1** The beginning paragraphs of the separation agreement between Harrison and Zella Dyar, signed on May 13, 1915.

By late April 1915, conflicts surfaced between him and William Ravenel of the USNM administration. Howard stepped in, writing to Dyar:

> In all matters of Museum administration, the custodians and in fact all of the employees of this Bureau who are stationed in the National Museum must obey Museum rules, and it should be the duty of all of us to try to preserve the most friendly relations with the National Museum people. Mr. [James] Crawford, as Associate Curator, is charged with seeing that these Museum rules are followed, and it is perfectly proper for him . . . to hand to you . . . the communication which Mr. Ravenel intended for you as Custodian.[30]

Conflicts such as these were minor, compared to what was on the horizon in both the personal and professional life of Harrison G. Dyar.

## CHAPTER 17

# Divorce Wrangling in Reno: 1915

*Dear Gran: Thanks for your telegram. Zella's presence is necessary. The trip is easy and cannot do her any harm. Take a section on train 6 S.P. [Southern Pacific] and you arrive here at 8:25 A.M., a very convenient time. Take the hotel 'bus to the Riverside. Leave Zella there and take a taxi to 781 Mill St and see me. I think this best because I think I ought not to be seen around the hotel while Zella is there. Her action must seem independent of mine.*[1]

Harrison G. Dyar to Harriet M. Peabody about meeting in Reno
to negotiate a divorce in Nevada, October 21, 1915

In the spring of 1915 Harrison and Zella Dyar made plans for a quick divorce in California to avoid having the potentially high-profile case heard in the Washington D.C. limelight.

Dyar (Figure 17.1a) would move to nearby Reno, Nevada, under the guise of doing government mosquito work (and recovering from various infirmities). His family and Mrs. Peabody (Figure 17.1c) would move to Berkeley, staying with cousins Louise and Zinie Kidder until their household items arrived; Dorothy, as a California resident, would begin college for free at the University of California. Afterward, Dyar would return to work in D.C.[2]

Zella went to Berkeley by rail, whereas Dyar, touting the health benefits of a cruise, traveled from New York to Reno via the Panama Canal to San Francisco on the SS *Finland* (Figure 17.1b). His traveling companions included Wellesca and her three children (now 6, 4, and 2). Dyar and Wellesca would later testify she accompanied him as a paid nurse. Also with the group were Mrs. Maud Leech and her children.[3]

Mrs. Leech, who had bonded with the Allen children while staying with them on B Street the previous January, was hired to assist with the children, including Roshan, who had a bad ear infection. Her husband, Coray Leech, would watch the Allen home while the party was in Reno.[4]

**Figure 17.1** H.G. Dyar (*a*) and his party, including Mrs. Allen, Mrs. Leech, and their children, took the *SS Finland*, through the Panama Canal to San Francisco (the sister ship *SS Kroonland* shown here) (*b*); Mrs. Harriet M. Peabody (*c*), who was in lengthy negotiations with Dyar to reach a divorce settlement with Zella while she and her daughter were in Berkeley, California.

The cruise was not as restful as he had hoped. When Mrs. Leech's three-and-a-half-year-old son, Laurel, was screaming, Dyar, she claimed, threw him down, causing a limp, though Wellesca reported that the child fell after struggling and could walk fine afterward.[5] They arrived in San Francisco on July 5, 1915. Wellesca, seeking adventure, unsuccessfully attempted a late-night visit to the World's Fair with Leech and Roshan.[6]

After reaching Reno, on July 7, Dyar wrote Mr. Leech that his family would be returning sooner than expected. Mrs. Leech was "nervous and far from well"; consequently, "we do not always keep good natured and this is bad for my convalescence." In closing, he noted that there would be no objection "to Mrs. Leech and the children occupying the house [in Washington] with you til fall."[7]

Leech, on her arrival in D.C., gave accounts of Wellesca's amorous relations with Dyar to Zella's D.C. attorney Robinson White, who immediately contacted Zella. She reneged on her agreement with her husband and filed suit in D.C., following the advice of her attorney and the Knopfs. In early August, Wallace McLean wrote of his concerns about Leech to Dyar: "Mr. Ford [Dyar's Washington attorney] has spent the greater part of the day with me [in New York] and decided . . . a course of action that will protect your interests. It smells of blackmail and must be handled carefully. I will be in Wash. in a day or so."[8]

Leech's testimony and the rebuttals from Dyar and Wellesca over the next several years present contrasting accounts of interactions among the three of them. However, it appears Leech and Allen were close confidants. According to Wellesca, Leech suggested a move to Reno for Allen and Dyar divorces, as well as her own from Coray Leech. A letter to Dyar's attorney in Reno from Leech implies that she did not receive adequate compensation to make the return trip to D.C. with her children.[9]

As a result of Maud Leech's contact with Zella's lawyer, Dyar felt a second jolt after a long earthquake in Nevada. The *Washington Post* on November 8, 1915 proclaimed: "wife charges her husband with misconduct and names a corespondent.

The court is asked to restrain the husband from disposing of any of his real estate while the suit is pending"[10] (see Figure 18.1a for a *Washington Herald* article on the same date). When Dyar heard about the D.C. suit he reminded Zella that she had broken their agreement, which would "spare" him "the disgrace and notoriety of a suit in the District where I will live" and that she had accepted his "money [and] good will on this promise." He also noted that Wallace McLean, who apparently helped broker the deal, "assured me you were sincere. And I believed it like a fool."[11]

Dyar demanded Zella telegraph her attorney Mr. White "to withdraw the suit" and immediately divorce him in Reno. He reasoned a quick divorce negotiated by the two parties was legal, but her "foolish attorney did not know." Perhaps alluding to his spiritualist heritage, Dyar wrote:

> You and White are the only people I ever met who would not trust me with their immortal souls or sight. And I assure you that I feel that I could always have returned their souls in as good condition as received. They all knew it. So do you in your heart. But I am not asking you to trust me. Only don't do this nefarious thing since it is not necessary.[12]

Dyar's attorney, Richard A. Ford wrote to Wellesca about Zella's suit, reflecting on Wellesca's concern about her standing in the community:

> Your name has not appeared in any of the newspaper accounts of the filing of the suit, only Dr. Dyar's name . . . with "a corespondent was named." Unless curiosity has prompted persons to examine the bill as filed in court, or they have learned it from some person who has made such an examination, your name is not known in the proceeding. . . . I am still hoping that some way may be found in which the suit here [Washington] may be gotten rid of and the desired result achieved with as little knowledge on the part of the public as possible.[13]

Dyar soon began to attempt getting a quick divorce in Reno, first writing to his mother-in-law in Berkeley in late October to request she and her daughter quietly take the train to Reno. He planned to meet with Harriet Peabody (Figure 17.1c) at his home and reach an agreement to stop the suit in Washington. She would then bring papers to Zella for signing, avoiding the appearance of collusion. A quick divorce would keep more news out of the tabloids, particularly findings from the raid of the B Street home by Zella's attorneys and Leech's potential testimony. According to Leech, Dyar feared he would be arrested for having sexual relations with Wellesca[14]: a violation of the Mann Act.

Following the divorce notices in the Washington newspapers, Dyar wrote "Gran" Peabody to get Zella to wait for the advice of his sister and brother-in-law before pursuing the divorce. After learning the Knopfs' favored a Washington divorce, he wrote:

> Dr. [Knopf] and Perle fail to understand the situation and the vital importance it is to my health. The suit was in the Wash. papers, all my friends are scandalized

and I am nearly prostrated. I want you and Zella to come to my assistance at once. Let Dr. and Perle go and Mr. White and everybody else and I promise you shall not regret it. Bring Zella up here as soon as you can.[15]

Two days later, Dyar wrote Peabody:

Now that I have discovered this short, easy way [to divorce] I feel that every day's delay is one more day of unnecessary suffering for me. Also I cannot tell what my enemies in Washington will do to undermine my plans if I wait. Perle and Doctor are blundering around, innocently betraying me.... I propose that we leave them out of the question here and now, and everybody else as well, except you, Zella and me. We will settle this to our satisfaction and it is nobody else's affair.[16]

The same day, he wrote to Peabody again, saying that Perle and Dr. Knopf asked, "to be relieved of any further part in my affairs." Dyar then asked "Gran" not to "mention to your lawyer [the] ... divorce bill filed in D.C.... Let Zella get her divorce [in Reno] first, then notify White that the case is off. Otherwise White will fight his best to prevent our reaching a settlement to save his personal fee."[17]

By early November Dyar—perturbed that Peabody didn't give immediate attention to the negotiations—responded to a letter she had written expressing distaste for the scandal and stating that the $300 a month settlement Zella had accepted was too low. "Now having vented your spleen, perhaps you may be able to return to reason," he wrote. "[Zella] agreed to sue in California where no property security would accrue to her. I have kept my bargain absolutely. Zella was induced by false representations to break hers, but because she was deceived I am ready to overlook her bad faith if she will act promptly to remedy its effects."[18]

When the Dyars made their divorce plans neither apparently realized the interlocutory decrees in California meant divorce was not final for one year, compared with the six months in Nevada, as Dyar wrote to Mrs. Peabody: "Out of kindness and consideration, I let [Zella] go to California. Now I do not know whether counsel purposely deceived me or were ignorant.... If I had known ... I would never have consented to her going to that state. So she owes me the justice of suing here [Nevada]." She could, he noted, live in California and sue in Nevada.

Professing his innocence of adultery, Dyar wrote,

You have no shred of real evidence against me and I have never in the least admitted your slanderous charges. Ask Zella if I have.... I have not been able to adduce any 'proof of my innocence' since it is very hard to prove that one did not do a thing.... But lately I have come into some remarkable evidence that would convince anybody, judge, jury, prosecution, you, Zella, even perhaps my own counsel! I shall not reveal it except at the proper time but it gives me absolute confidence in my case.

Dyar was likely bluffing, as the only "remarkable evidence" would be to bring Mr. Allen to the witness stand, produce witnesses besides Wellesca who had seen Allen, or had strong circumstantial evidence, such as bank statements.

Dyar again pushed Mrs. Peabody for Zella to agree to a Reno divorce:

> The best way to vindicate myself would be to let the Washington case proceed; but I am not physically able to stand the strain. Moreover it would be a great expense and necessitate the repetition of residence and divorce suits here since Zella would not get hers and I should be obliged to sue. Alas it would submit Zella to scandal and ignominy? and show up her credulity and unjust suspicions. All of which I am willing to spare her. This is written in the hope that you may be able to see reason at the 11th hour.[19]

Zella rejected Dyar's proposal. Desperate to win her trust and that of Perle, Dyar sent an odd letter to Zella, in which he wrote,

> I have been very careful to make [Wellesca] no proposal of marriage. I may decide to do as White hopes and give her the "double cross." There is a nice young girl over in Truckee interested in my line of work who might fill the bill better than Lesca. I am under no obligations to marry Lesca as there have been no illicit relations between us. I only considered the possibility to oblige you and Perle.[20]

The young girl Dyar was referring to was a person of note in the entomological community, Miss Ximena McGlashan. She and her father were active in butterfly farming and in writing the magazine, the *Butterfly Farmer*. A decade later Dyar would name species of limacodids—*Sibine ximenans* and *Sibine zellans*—for her and Zella (see Figure 25.2b).[21]

※

Dr. Dyar was not the only one putting pressure on Mrs. Dyar to end the divorce bill in Washington. In December 1915 Wellesca wrote Zella and suggested that she drop the suit and write him a pacifying note, something like: "*If we must separate, let it be in peace rather than in desperate warfare. What will you do for us, for us all?*" If Zella followed this prescription, Wellesca assured her, Dyar "would run to do the kind and generous thing by you."[22]

In the letter, Wellesca also tried to explain that Zella's suit in Washington was working against Wellesca's attempts to quiet the gossip by keeping her trips with Dyar to Nassau or Reno secret from her friends. She mentioned her disdain that the Washington suit was now in "four regular newspapers . . . , the Law Reporter and that scurrilous 'Bulletin' . . . distributed in stores and saloons where all the juicy morsels were doubtless served up in spicy manner." Wellesca also claimed bad publicity from Zella's counsel led to the news headline: "HARRISON G. DYAR CHARGED WITH DESERTION AND NON SUPPORT." She went on to say because of these headlines Dyar "will be looked upon by the Washington public as a cruel monster."[23]

Wellesca also mentioned the toll the growing scandal and accusations from his family was having on Dyar, stating, "when he would see the postman come or the door bell ring or the telephone ring his heart would go with spells [until] . . . he could not stand it any longer." This was affirmed when Dyar wrote Zella about escaping to an undisclosed location in southern California "to avoid mail and telegrams, which worried my nerves badly."[24]

Wellesca's more threatening prose at the close sounded more like Dyar—who perhaps dictated the letter. She wrote, "A wise lawyer would have been so mild in his charges that we might have let the bill go by default and so you would have gained your object. But now—only over our dead bodies! And what an audience we will have if that suit comes off!"[25]

Later, asked under oath about the letter's goal, Wellesca said it was so Zella "would quickly be friends with Dr. Dyar" and if she trusted him "he would do the most honorable and just things by her but he didn't wish to be compelled by law to do something: he wished to do it for his honor." She went on to affirm that she was hoping to persuade Zella to be agreeable to Dyar "[w]ith as little publicity as possible."[26]

⁂

As anxiety over the divorce and ensuing scandal continued to mount, Dyar became concerned about the contents in his desk at the museum ending up in the wrong hands, telling August Busck,

> No one not connected with the Museum should have access to my desk without my written order. I mention this because the house at 804 B Street, S.W. was thoroughly ransacked last fall by agents of Mrs. Dyar and many things stolen which it is now being attempted to use in that divorce suit. I do not want these people to disturb my papers at the Museum.[27]

Doing little to ease Dyar's worries, Busck replied that if "any really enterprising agent wanted to, he could easily get hold of whatever is in [the desk]; nobody here could accept responsibility for what may be done or what may have been done with that unlocked desk when you are not here, but none shall be or would have been allowed access to it of course, while Mrs. [Jennie] Locke or I am about."[28]

Zella's agents had collected letters from B Street, and Robinson White took depositions from January until early September 1916 to establish there had been adulterous relations between Dyar and Wellesca. The transcripts were also sent to Zella's attorneys Brown and Belford in Reno to be used in her defense. The central witnesses were Maud and Coray Leech, the nurses M. Alma Stover and Catherine Booth, Wellesca's sister Mrs. Amia Louise Baker, and Eugene B. O'Leary of the USDA.[29] Karl Vaupel, who booked the steamship *Findland*, and Joseph Davis, a clerk at the Hotel Arlington, where Dyar's party stayed in New York City before their departure, were also questioned.[30]

## "A FOOL AND HIS FRIEND"

In November 1915 Dyar sent another one of his short stories to Perle. This one was purportedly autobiographical and was intended "to purge her mind of the erroneous idea ... in which she considered that I had been acting wrongfully."[31] In "A Fool and His Friend" Dyar changed the real names to the following pseudonyms: Paul French (Dyar), Mrs. French (Zella), Flossy (Wellesca), Mr. Smith (Wilfred Allen), Mama (Harriet Peabody), and Mrs. Bloodletter (Maud Leech).

Although clearly reflecting Dyar's rather than Zella's point of view, many of the points in the story jibe with accumulated evidence and accounts from family members, including Wellesca's miscarriage of what would have been her second child in 1910. However, the critical point that fails to conform to scrutiny was the existence of Mr. Smith (Wilfred Allen).[32]

Couched with eugenic philosophy, "A Fool and His Friend" suggests middle and upper classes of people should have more children for the betterment of society due to their genetic superiority. Dyar would write other more fictionalized stories that had eugenics-related themes. What he certainly didn't know when he wrote the story was that a complete version or synopsis, including Flossy and Paul French, would appear in court and in the newspapers. As was noted by the Nevada judges, one aspect of "A Fool" did not square with Dyar's own accounts: Mr. French (Dyar) was still among the living. Coincidentally, noted science-fiction writer Isaac Asimov would later use "Paul French" as a pen name for his series of juvenile science fiction books titled *Lucky Star*. Dyar also wrote science fiction, but under his own name. Many of his stories, however, were never published. Eugenics, bigamy, and reincarnation were all common themes.

## CHAPTER 18

## Divorces, Appeals, and Bigamy: 1916–1921

*Is changing wives such a sin these days? They're all doing it. Dan, these marriage vows are an inheritance, a worn-out heritage. They do not fit the times and seasons. Should a man hold fast to a woman whom he has outgrown when he finds better? It does not seem to me that this rigidity suits our times. Divorces are made easy and more and more take advantage of them. Why not? Is this not a necessary adaptation to changed conditions? The orthodox and the old-fashioned rail because they do not understand that things must change, must evolve. You are not living up to your highest, perhaps, after all, Dan.*[1]

Unpublished short story "An Old Manuscript,"
by Harrison G. Dyar, October 23, 1923

Zella filed for divorce in D.C. and the story hit the news there,[2] in Reno, and in San Francisco (Figure 18.1a). Dyar, unable to avoid scandal by getting a quick divorce, countered by filing in Reno.[3] Wellesca also pursued a divorce from Wilfred Allen in Reno, filing for it in January 1916,[4] but in May the judge delayed ruling on it. Zella's attorney had raised objections because of the "alleged connection" between the Allen and Dyar divorce suits. He also suggested "the defendant in the Allen case was . . . imaginary"[5] (see Figure 18.2a).

Dyar's filing as plaintiff for a Reno divorce in March 1916 was also covered in the *Washington Post* and *Reno Gazette* (Figure 18.1b). He was said to be "making a determined fight for a share of his estate worth $150,000" (over 3 million in today's dollars). His properties in New York's business district were "owned jointly with his sister" and "valued at $360,000" and his "property in Washington . . . about $100,000." Dyar was ordered to pay Zella's attorney fees, $1,750, in addition to her monthly support (Figure 18.1c).[6] When he completed the short story "Uxores Defendam" (Support of a Wife) that July in Reno, Dyar was likely thinking about Zella.[7]

(a) (b) (c)

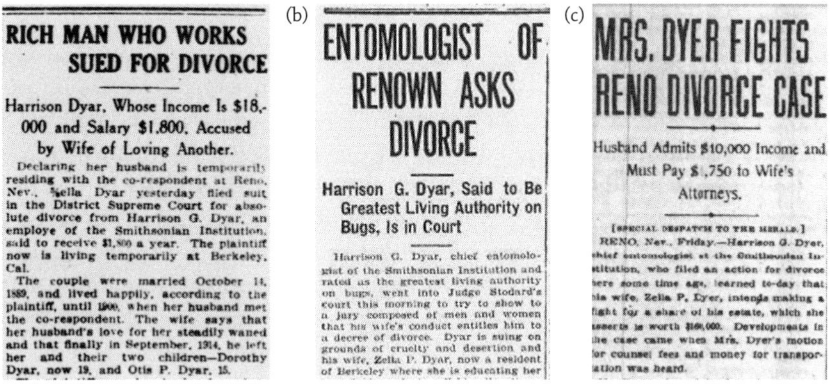

RICH MAN WHO WORKS SUED FOR DIVORCE

Harrison Dyar, Whose Income Is $18,000 and Salary $1,800, Accused by Wife of Loving Another.

Declaring her husband is temporarily residing with the co-respondent at Reno, Nev., Zella Dyar yesterday filed suit in the District Supreme Court for absolute divorce from Harrison G. Dyar, an employe of the Smithsonian Institution, said to receive $1,800 a year. The plaintiff now is living temporarily at Berkeley, Cal.

The couple were married October 14, 1889, and lived happily, according to the plaintiff, until [illegible], when her husband met the co-respondent. The wife says that her husband's love for her steadily waned and that finally in September, 1914, he left her and their two children—Dorothy Dyar, now 19, and Otis P. Dyar, 15.

ENTOMOLOGIST OF RENOWN ASKS DIVORCE

Harrison G. Dyar, Said to Be Greatest Living Authority on Bugs, Is in Court

Harrison G. Dyar, chief entomologist of the Smithsonian Institution and rated as the greatest living authority on bugs, went into Judge Stodard's court this morning to try to show to a jury composed of men and women that his wife's conduct entitles him to a decree of divorce. Dyar is suing on grounds of cruelty and desertion and his wife, Zella P. Dyar, now a resident of Berkeley where she is educating her

MRS. DYER FIGHTS RENO DIVORCE CASE

Husband Admits $10,000 Income and Must Pay $[illegible],750 to Wife's Attorneys.

[SPECIAL DESPATCH TO THE HERALD.]

RENO, Nev., Friday.—Harrison G. Dyer, chief entomologist at the Smithsonian Institution, who filed an action for divorce here some time ago, learned to-day that his wife, Zella P. Dyer, intends making a fight for a share of his estate, which she asserts is worth [illegible]. Developments in the case came when Mrs. Dyer's motion for counsel fees and money for transportation was heard.

**Figure 18.1** *Dyar v. Dyar* newspaper clippings: Zella's divorce bill in Washington, D.C., in November 1915, naming Wellesca as "the other woman" (*a*); Dyar's 1916 divorce bill in Reno (*b*); Zella's attorneys motions for Dyar to pay her legal fees in the Reno case (*c*).

The same *Post* article named Wellesca as "the other woman," noting that Dyar had "transferred property in Washington" to her name "to fulfill a moral obligation." This obligation, stated by Dyar on numerous occasions, was to promote larger families, particularly of New England ancestry. Zella, interviewed by the *San Francisco Chronicle* in Berkeley about contesting her husband's divorce suit in Reno, implicated Wellesca, saying, "My husband has deeded a large share of his property to her. It is a terrible affair, and I do not care to discuss it in detail." But she did say that if she sued as plaintiff, she would name Wellesca as corespondent.[8]

Dyar wrote Busck in March 1916 about being unable to "return to Washington yet for the reason that my divorce suit is on here." Dyar also noted that his "nerves are continually improving, in spite of the vexations of laws." Referring to the Lepidoptera identifications Busck was doing at the museum, particularly those of the larger moths ("macros") when he normally covered the "micros", Dyar requested he "have patience with the Macros a little longer."[9]

Dyar continued to maintain that he was out West for collecting in spite of the publicity back home. Fieldwork in San Diego and Yosemite in the spring and in the Tahoe region most of the summer did serve as a respite from the proceedings in Reno. While they were at Fallen Leaf Lake, California in June, Wellesca (as Aseyeh Allen) gave free lectures of a Bahá'í flavor to folks in the camp.[10] Meanwhile, Zella's attorneys were gathering testimony in D.C. In a strange twist, Zella wrote Howard, "or one in charge" at USDA, for Dyar's administrative leave in July: "Please send Dr. Dyar letter requesting leave of absence for six months to collect west of Rockies. Dr. Howard showed us the letter in Washington. Most important that we have it immediately."[11]

Although Zella may have strived to save her husband from himself, she needed him employed to receive her $300 per month. It is also possible she was gathering evidence to have him declared a nonresident of Nevada.

## DYAR V. DYAR

As the divorce approached, Dyar was ready to put it behind him. He informed Howard: "If no unforeseen accidents occur, my case will be tried here on Sept. 14.... I shall certainly be relieved to have the matter over with at last."[12]

The anticipated divorce suit between Dr. and Mrs. Dyar caused a stir in the Reno press, in part because it was the first of its kind to have women on the jury in Nevada.[13] The case was to be heard "behind closed doors" because the evidence was deemed "unfit for public hearing." The hype was much ado about nothing, as the court dropped the Dyar suit in late September 1916 because he was considered a nonresident of Nevada.[14]

Dyar attempted to establish residency by purchasing a home in Reno, but he was careless, spending considerable time in California. Furthermore, he left a paper trail at the Bureau of Entomology to prove it. Zella's attorneys obtained copies of these documents perhaps to get the suit moved to California. In any case, Dyar decided he would no longer pursue the suit in Nevada. Perhaps he saw a speedier settlement in California, where Zella had established residency.

At some point in 1916 or early 1917, soon after Dyar's divorce attempt in Reno failed, the couple drafted a letter in Zella's voice to end the divorce proceeding in D.C. and move the case to California. The first part of the draft is about Zella's fears about "being left penniless" and her "children's prospects blasted." According to the letter, she pursued a divorce in D.C., following the advice of her mother and attorney White, who Zella "believed to be a man of experience." Leech's affidavit convinced her of her husband's guilt, while White and her in-laws persuaded her that the only way to save her "husband from jail was to file a suit for divorce in Washington, D.C." Zella admitted that her agreement of support from Dyar was based on the "express understanding that there should be no suit in Washington."[15]

There were negotiations between the attorneys Richard Ford (representing Dyar) and Mr. Colloday (Zella) on the "settlement of the Dyar matter" by 1918.[16] However, it took until 1919 for the divorce to move ahead in Alameda County, California. Once again the story hit the papers, this time in San Francisco, because of the tabloid notoriety it had achieved in Reno and in D.C. Unlike the earlier suit, Zella did not name Wellesca as a co-respondent, although the allegations from previous attempts to divorce found their way to the front page of the *Oakland Tribune* in February 1919. There, along with already reported information, it was mentioned that Wellesca was the real-life "Flossie" in the Dyar story "A Fool and His Friend."[17]

On December 1, 1919, Zella received an interlocutory decree of divorce, which would be finalized in a year, as California law required. She appeared as plaintiff; Dyar was represented by an attorney.[18] Wellesca's attorney, A. Grant Miller, said, "Mrs. Dyar obtained her divorce from Dr. Dyar in California by his default after a property settlement," which was brokered by a person trusted by New York attorney Wallace McLean.[19] In the end it took five years to divorce from the time the Dyars moved to the West. Wilfred and Wellesca Allen's divorce, however, remained unsettled.

## ALLEN V. ALLEN

In May 1916 Zella's attorneys moved to hold off Wellesca's divorce before Zella's defense in Reno on grounds that the cases were linked and that Mr. Allen was a paper defendant in the Allen suit (Figure 18.2a). Wellesca's suit, however, was held up again on September 1, 1916 after a sheriff on the East Coast failed to find Wilfred P. Allen to serve him a summons (Figure 18.2b). In October 1916, Judge Moran asked attorney George Brown to stand in as *amici curiae* (friend of the court) and represent Wellesca's missing spouse. Perplexed about his role, Brown told the court:

> I don't represent the defendant; I am attending as a friend of the court ... with the idea of bringing out such matters as the court should know in this case ... which this plaintiff was charged with being co-respondent in which adultery was charged by the wife, ... so I don't know what we could call this; it is not defendant's exhibit, surely.[20]

Brown had a complete grasp of the evidence at his disposal. Especially important was the testimony of Mrs. Leech, who claimed to have been a former confidant of Dyar and Wellesca, about their amorous relations. Brown asked Wellesca to describe his "client," Wilfred Allen, to which she replied, "He ... had dark eyes and a very earnest expression; brown hair.... A deep line in his nose, [which is q]uite straight. [He was a]bout four or five inches taller than I [, and weighed] about 165.... He had a moustache [and] wore a beard at that time."[21]

Wellesca answered two of the attorney's questions nearly verbatim from Dyar's autobiography *A Fool and His Friend*. When asked about her first meeting of Mr. Allen, she said she encountered him as a train conductor at the Chicago depot on the way to Pasadena in 1906. After asking Allen about the time of her rail connection,

**Figure 18.2** Wilfred Allen as an imaginary spouse (a); halt of the Allen case when Mr. Allen could not be served with a summons (b); appeal to the Nevada State Supreme Court (c).

she discussed the Bahá'í Revelation with him, as she would "do anything on earth to get ... [a listener] interested." After this briefest of encounters, they were to have met again in D.C. after her trip to the Grand Canyon and Pasadena. Mr. Allen was to have proposed marriage to her through correspondence, while she was sending him Bahá'í literature.[22]

Although Wellesca said she did not initially accept the proposal, Dyar advised her to do so, she told the court. Answering a follow-up question, Wellesca explained that Wilfred didn't know if he could meet the expenses of having children but that Dyar said: "'All right, you have a fine big family because I always wanted one and never got it and if you ... come to hard times what you are unable to give to your family I will help them out in the line of education and necessities.'" Wellesca went on to say, "He [Dyar] is like [Theodore] Roosevelt, and believes in people of the refined, higher class in contributing to the world of their kind instead of leaving it to the foreign element."[23]

During Brown's cross-examination, Wellesca's answers had many inconsistencies, including whether she would have started a family without Dr. Dyar's financial assistance. Contrary to the letter she wrote to Mrs. Dyar in 1909, which indicated Dyar was already giving her money at the time of Roshan's birth, Wellesca testified she could afford but one child. Her marriage to Mr. Allen seemed even more implausible when Wellesca forgot the month of her purported marriage in Richmond, Virginia in 1906, stating June rather than September.[24]

In response to questions about why Mr. Allen remained in Philadelphia, she answered that he was "marrying her [Wellesca], not her relatives," that he couldn't leave Philadelphia because Mr. McGrath, who employed him as a typist at $65 per month, allowed him only short and infrequent leave, and that McGrath, who had no heirs, would leave an inheritance to Mr. Allen as long as he stayed "right with him for a few years more."

Wellesca testified that the first year of marriage went well, and Wilfred was "perfectly lovely to me" and that her children were conceived on trips to Philadelphia.

Wellesca also offered a biography of Allen. He was from Detroit, Michigan, the son of George, a "high official in a Western R. R. [railroad]," and Susan Allen. Allen desired anonymity, especially in D.C., because during his dark past in the city, he had fallen "into certain bad company and became rather seriously involved financially." Afterward, while in Michigan, he was in a boating accident on Lake Superior and safely swam ashore. However, he was presumed dead and was thus able to escape his past and start over. He claimed that his real middle name was Pollock and that he was related to Wellesca but changed his middle name to Preston to avoid connection with her relatives.

There were questions about there being no handwriting samples of Mr. Allen: He was to have dictated several letters typed by Wellesca. One to Zella was about the home Dr. Dyar built on B Street. Another was sent to Wellesca's childhood friend Mrs. Boyd to prove he existed. The dictated letters had a variety of other purposes. For example, he wrote brother-in-law George Freeman Pollock requesting he approve a visit by Dyar to Skyland with Wellesca, and to 'Abdu'l Bahá and the Washington Bahá'í community expressing his opposition to "mixed-race meetings." Wellesca said Wilfred had good handwriting, but explained she typed the letters because a stroke made it difficult for him to write. His condition didn't seem to affect his typing for Mr. McGrath.[25]

In addition to typing the letters dictated by Mr. Allen, Wellesca admitted writing a letter about "Papa Allen" to her son Roshan while she was in Nassau with Dr. Dyar in early 1915. Mrs. Leech, who was watching the children, was instructed to read it to six-year-old Roshan when his younger brothers were asleep. This was because when his mother was away he was asked about his father, to which he answered: "never seen him." Wellesca felt a letter about "Papa" would quell Roshan's concerns. Wellesca wrote to Roshan from Nassau not to repeat the "big mistake" in his answer about his dad, assuring him that his father was with the family often when he was a child and had lovingly cared for him.[26]

Further evidence that Mr. Allen did not exist was that Dyar, who after initially denying he wrote Wilfred Allen's letters, admitted he had penned a draft about the B Street home. Zella testified that the draft was in her husband's handwriting and had only previously seen the typed version. There were also a number of checks allegedly written by Mr. Allen from the bank in Philadelphia in which "Dyar ... furnished the money for the account on which they were drawn."[27]

Unconvinced of Allen's existence, the judge ruled that Wellesca was unable to divorce him in Nevada. She appealed in July 1917, but given the implausible testimony, the appeal failed, as did a retrial in 1917. Wellesca then took her case to the Nevada Supreme Court (Figure 18.2c), where in 1920 George Brown reprised his role as a "friend of the court."

This time Brown went on a tirade, perhaps to end the proceeding once and for all. Particularly noteworthy was the letter from Wellesca to Roshan and the reference to Mr. Allen calling his son trilobite; perhaps the final blow in convincing the court that Allen didn't exist. In fact, Dyar could not deny under oath he had called Roshan the nickname. In March 1920, the Nevada Supreme Court upheld the District court's decision, agreeing that Mr. Allen was fictitious. In the Appellant's response Wellesca's attorney A. Grant Miller did not mince words: "Then should the State blast the mother of three beautiful children and stamp those children bastards, their mother a harlot, unless the evidence be of the most clear and convincing character?" The final attempt for a rehearing of the case before the Nevada Supreme Court was denied. Judge Coleman concluded, "The evidence in the case, as a whole, shows an utter disregard on the part of Mrs. Allen and Dr. Dyar for the conventionalities of life."[28]

Clearly both Dyar and Wellesca believed it was critical for them to have the victory in court that eluded them. Given their defeat, each needed to convince society that Allen existed and fathered Wellesca's children. The stakes were high: Dyar's standing at the museum was threatened, and Wellesca risked repudiation by the Bahá'í community in Washington and around the world, not to mention by most of her family.

## WELLESCA'S FICTIONAL MARRIAGE AND FICTIONAL DIVORCE

Dyar and Wellesca announced they were married in Reno, Nevada, in April 1921. Wellesca, according to Harriet Peabody, "managed some how to get a divorce in Maryland," making it possible for her to marry Dyar. The betrothal allowed for a more peaceful return to Washington society for the couple and her children. Upon hearing the news, Wellesca's brother George Freeman Pollock wrote her while she was in Oregon: "I received your glorious announcement . . . and you could have heard me singing all over the camp. I was so merry that some of the folks felt sure that I had some moonshine on board. . . . I can tell you [my wife] Addie was delighted too."[29]

Given the fantasy built around the existence of Wilfred Allen, it should not be a surprise that Wellesca's marriage to Dyar (as Dyar) and her divorce from Wilfred Allen were fictitious.[30] That Dyar and Wellesca were married in Reno on April 26, 1921, is indicated in many places, including Dyar's passport application in 1923, biographical accounts, an alumni card at Columbia University (see Epilogue), and even in obituaries. However, no Allen divorce made this possible in Nevada. Indeed, a petition for rehearing Allen v. Allen was denied on April 13, 1921, in Reno, effectively ending the effort. Furthermore, neither Dyar's petition to adopt the Allen children nor Wellesca's consent in April 1922 mentions a divorce; rather, it states that Allen abandoned his wife and children in 1913 and that Dyar and Wellesca married on April 25, a day earlier than mentioned elsewhere, in Reno.[31] Likewise, there is no evidence of a marriage or divorce from records in any other state, including Maryland, the location of the Allen divorce mentioned by Mrs. Peabody. Therefore, nothing but elaborate fiction supports Dyar's marriage to Wellesca or her divorce from Mr. Allen.

## BIGAMY: AN ALLEN MARRIAGE RATHER THAN A DYAR DIVORCE

*Reminds me how Phil Utterly went to law. Ha, Ha, that was a funny suit. I'll wager you never heard of it, Jane, me girl. Phil was a young fellow then and inclined to be frisky as young fellows will, and some old ones too. He'd been courting two girls at once, the rascal, till it came to near the point of popping the question and then he came to a difficulty. He didn't know which he preferred and the thought of possessing one made him sad at the thought of losing the other.*[32]

Short story "A Saucepan for Breakfast" by Harrison G. Dyar, October 27, 1917 (later published as "An Anecdote of the Law—A Story")

Although divorces were much less acceptable in the early twentieth century than they are today,[33] Dyar's choice to remain married and live in a fictitious marriage was an odd one. Perhaps, as the Nevada judges suggested, the deception allowed Wellesca to have children with her true love without the stigma of them being illegitimate.

Wellesca testified she had an operation to bear children in June 1906, as evidence of her stated belief that "marriage without children is a failure."[34] Although the "Allen children" were born several years after her wedding in September 1906, Wellesca could have been pregnant with one of Dyar's children earlier that year, and her admitted operation a cover for an abortion or miscarriage. Indeed, the prospect of children from an unplanned pregnancy may have fueled a desire to start their own family and to fulfill Dyar's eugenic-related beliefs to increase the population of his race and social class.

Evidence that Dyar had found a soul mate in Wellesca is seen in his writing to Zella about her: Dyar contrasts the "relaxation in a change of thought" in philosophical discussions with Wellesca, whereas the same topics had a bad reception with Zella and her mother (eugenics appears to have been one of them).[35] And, according to Dyar family lore, sexual relations ended in Dyar's marriage in the early 1900s, so he sought intimacy outside of the marriage, and mutual attraction between Dyar and Wellesca, as noted by Zella, was a major cause of their affair.[36]

Wellesca surely wanted to avoid being the "other woman" in a publicized Dyar divorce given her place in the Bahá'í community. With the prospect of divorce Dyar, on the other hand, may have feared losing connection with his two children, as well as damaging his relationship with Perle, and his aunt and uncle in New York who were close to Dyar's family. Although Dyar valued Zella's social status, he also appears to have had genuine feelings for her. This is suggested by a photograph taken of the couple around 1926 and Dyar naming a species of one of his favorite moths, a limacodid, after her in 1927 (see Figure 25.2).

Indeed, in Dyar and Wellesca's secret world, the solution to having a family together was an Allen marriage rather than a Dyar divorce.

## INVENTION OF AN ALTER EGO: WILFRED P. ALLEN

The curious process of creating alter ego Wilfred P. Allen by Harrison Dyar and Wellesca Pollock is suggested by similarities the imaginary Allen shared with people in their lives or with their forebears. Dyar's "Preliminary Genealogy," published in 1903, three years before Wellesca "married" Wilfred Allen shows that Dyar's uncle John Wild Dyar and his son-in-law Ephraim Williams Allen (compare "Wilfred Allen"),[37] railroad executives, moved to Michigan from New England like railroad executive George Allen (Wilfred's "father").[38] Furthermore, Dyar, Sr.—several times the president of the New York and New Haven railroad—had similar job experience (see chapter 1), and Wilfred Allen's parents had the same names (Susan and George) as Wellesca's oldest and closest siblings in age.

There were several "Allens" connected with Wellesca's family. Her father George H. Pollock, a "George" like Wilfred's father, had a business partner Stephen M. Allen and her mother Louise Plessner opened a kindergarten in West Newton, Massachusetts on the request of Nathaniel T. Allen.[39] Later in life, Wellesca's attorney Wallace McLean had a close associate Cebas Allen.[40]

Presumably the co-conspirators decision to give Wellesca a married name with the same initials W.P.A. as her new husband and first born was not a coincidence—as Pamela M. Henson has often noted.[41] The name Wilfred, while not in Dyar's genealogy, was mentioned as "Sir Wilfred" in one of Dyar's early stories "The Three Green Sisters" sent to his sister Perle in 1901 (see chapter 2).

CHAPTER 19

# After the Scandal, Dismissal, and Friend Knab: 1917–1918

Dyar and Wellesca went to San Diego in December 1916, just as the USDA—perhaps unbeknown to Dyar—had launched an investigation of his extramarital activities to determine whether to remove him from government service. The following month Dyar headed back to D.C., and by late April 1917 he learned of the USDA's ongoing investigation and noted to Howard that he "worked three months [at the museum] without pay and . . . cleaned up all the jobs assigned to [him] except jobs that require[d] a preparator." Sensing perhaps his dismissal was imminent, Dyar went off to the Cascades to "try and get some more of the early mosquitoes."[1]

A key part of the inquiry involved comparing Dyar's handwriting with Allen's on the marriage documents from Hustings Court in Richmond, Virginia. As the investigation approached its completion, the office of David F. Houston, President Woodrow Wilson's Secretary of Agriculture, wrote the court clerk in Richmond on May 1, 1917: "It is very important that the Department secure for a short time the original signature of Mr. Allen in order to enable certain comparisons with other writings to be made." In addition, Mr. Horrigan, an investigator from the Solicitor's Office, had gone to Reno to make "certain inquiries for the Secretary's Office regarding the record of Dr. Harrison G. Dyar of the Bureau of Entomology." The investigator gave assurances "that he is in possession of abundant material to justify the preferment of charges against Dyar looking toward the latter's removal from the service."[2]

After receiving copies of the marriage documents from Richmond, Virginia, Secretary Houston wrote Dyar. Noting that,

> [t]here is inclosed [sic] a photostat copy of an affidavit filed in the office of the Clerk of the Hustings court, Richmond, Va., in connection with a marriage license issued to Wilfred P. Allen and Wellesca Pollock on September 5, 1906.

The Department will be glad to consider such statement, if any, as you may care to make as to whether or not you wrote the signature 'Wilfred P. Allen' on the original affidavit . . . [or] any evidence tending to support your statement. . . . All statements or evidence furnished should be under oath.[3]

After flatly denying the charges, Dyar wrote Howard from Seattle, "I have replied with the statement" from the Secretary of Agriculture "that I did not sign the affidavit with the signature of Wilfred P. Allen and that I have no knowledge of the whereabouts of this person"[4] (see marriage license without Pollock or Allen's signature in Figure 19.1).

On June 2 Dyar was terminated "because of conduct unbecoming an employee of the government." The following week, Dyar wrote Busck, expressing his hope that he could hold onto his honorary position as Custodian of Lepidoptera and concern over James Halliday McDunnough, William Barnes's curator in Illinois, supplanting him: "I hope you will not agitate for McDunnough or any one [sic] else, as I intend to come back. I still have a string to my bow."[5] The string Dyar was likely referring to was the donation of his private collection to the National Museum through J.C. Crawford on December 18, 1917—including 15,000 Lepidoptera, collected prior to his arrival at the museum and housed there, 1,000 sawflies, and 19,000 mosquitoes and other Diptera (flies) from Western trips over the previous two years.[6]

※

News of Dyar's dismissal spread to detractors William Barnes (see Figures 20.1c, 24.1a) and B. Preston Clark (1860–1939), who quickly lobbied to replace him, as Dyar feared. Clark, a politically connected collaborator at Pittsburgh's Carnegie Museum and soon to be the same at the USNM, favored William Schaus (see Figure 20.1b) over McDunnough. He wrote Howard that Dr. Barnes had heard that Dyar's "relations with the National Museum have been severed" and suggested Schaus as his replacement because he had "the best knowledge of any man in this country of tropical Lepidoptera." He also mentioned that if "Dr. Dyar was in charge, any connection between the National Museum and Mr. Schaus was out of the question, as there would have been too much friction."[7]

Howard wrote Clark:

Doctor Barnes was mistaken; Doctor Dyar's relations with the Department of Agriculture have been severed, but he still holds his custodianship in the National Museum. I do not know how long it will last. I have been thinking about the matter of his possible successor, and the same idea had occurred to me which you express in your letter, but of course the question of such an appointment cannot be seriously considered at present.[8]

Despite being unsure of Dyar's future, Howard still considered him an asset to the museum and could still count on him for critical identifications of mosquitoes and moths related to the USDA mission.

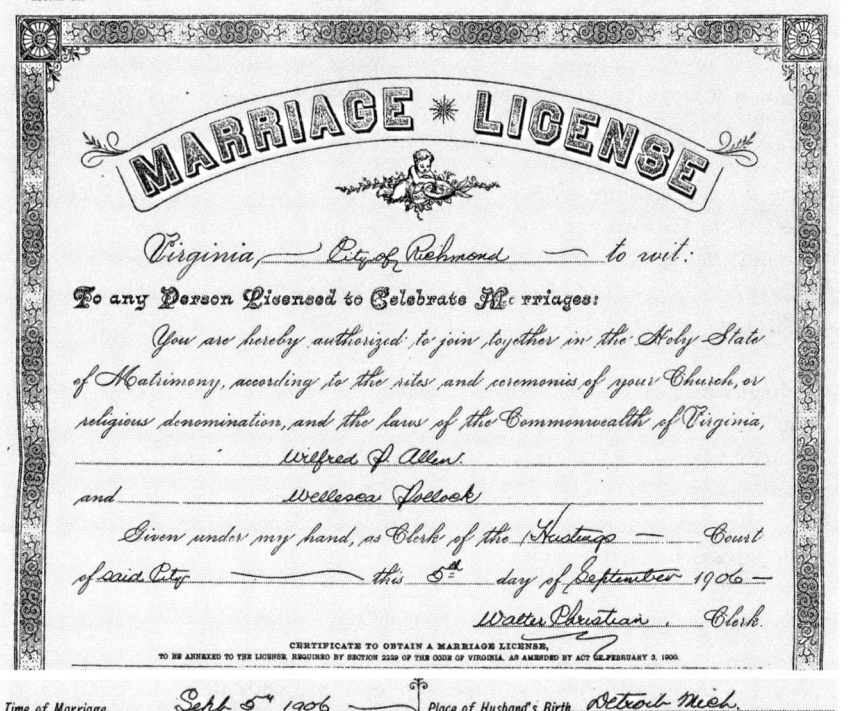

**Figure 19.1** Train conductor Wilfred P. Allen and Wellesca Pollock's license for a marriage that took place in Richmond, Virginia, on September 6, 1906 (filled out by clerk Walter Christian). Mr. Allen's signature was on another document said to match that of H.G. Dyar, Jr., and used as evidence for his dismissal under the employ of the U.S. Department of Agriculture.

Howard, who surmised that Busck contributed to the buzz about Dyar's status, forwarded Clark's letter and his reply to Busck to quell the rumor.[9] Prior to hearing from Howard, Clark wrote Schaus of Dyar's supposed dismissal: "Dr. Barnes of Decatur was with me last week, and had it from Mr. Busck, so it must be true I think." To this Schaus responded:

> Your delightful & interesting letter ... duly reached me ... I am very glad Dyar has left the Museum & I hope your ... suggestion to Dr. Howard will bear fruit, it would be the least he could do, & I really believe I could do more to help the Museum than any one I know, in fact there is no one free who would put as much heart to it as I would.[10]

Schaus's disdain for Dyar was a change from when Dyar and Zella visited him in England (1905) and when he visited the Dyar residence. Perhaps Schaus, who had corresponded with Zella, had ill feelings toward Dyar over the divorce. One indication that Schaus was unhappy with Dyar occurred when Schaus was in a Guatemalan hospital for an appendectomy. He wrote to Howard criticizing Dyar's Lepidoptera curation, observing:

> The types are quite safe ... all arrangements made to have them sent to Museum in case anything happened to me, but so long as I am alive ... I have little confidence in the careless handling heretofore displayed with my specimens. I don't know whether I have told you that I rediscovered many missing types [originally described specimens] amongst material carelessly placed among duplicate North American specimens. I don't complain for it is so typically American to do things in a haphazard way. I am expecting to go north this summer unless war with Germany might make it risky to take my Guatemalan Collections North. I don't think about losing my life, but I do not want my collections to be torpedoed.[11]

On Dyar's return to the museum in 1917 Howard showed him the Schaus letter, and Dyar replied,

> Schaus' specimens were not carelessly handled. He found 3 or 4 types among some drawers of unassorted material, not 'duplicates,' which material had been placed under the North American heading because the contents were proportionately North American. My method of handling large masses of material differs from Schaus', but it is not in any sense careless or haphazard. I prefer my method and will gladly explain it, if you have any lingering suspicions that Schaus is in any way justified in his strictures.[12]

It appears the rift was primarily Schaus's at the time, as Dyar soon favored a position for Schaus under his direction. Never compensated for his massive collection, Dyar supported giving Schaus a salary due to the poor financial status of his assets in Europe after the war.[13]

August Busck (1870–1944) (see Figure 20.1a), a Dane who started at the USNM in 1896, developed fractured relations with Dyar even though Busck became an authority on "micros" under Dyar tutelage. Like other struggling scientists he accepted a loan from Dyar to purchase a home. After returning from a Panama voyage in 1912, Busck was able to repay Dyar and wrote, "my ship has come in . . . if you could be good enough to dig up the little note of mine and let one know just how much it amounts to cover the interest to date. I shall be very glad indeed to send check for the full amount with many thanks for the loan."[14]

By fall 1917 Busck no longer felt beholden to Dyar, perhaps because of his growing allegiance to William Barnes, who provided extra income, and because Dyar's absences increased his number of work-related moth identifications. Busck's change in attitude can be seen in his response to Dyar's request for a specimen preparator.

Dyar had followed Howard's advice to "[t]alk the matter over with Mr. Busck,"[15] but Howard would have been better off if he had handled the situation himself. Busck had a major conflagration with Dyar over his request for Miss Pratt, the assistant he wanted. "Mr. Busck imagined that I was in sufficient trouble so that it would be safe for him to browbeat me openly," Dyar told Howard. "Therefore he accused me in a loud tone of voice, with much bravado and in the presence of all who happened to be in the room, of underhand work and he cried out: 'I advise you to study the ten commandments.'" After Dyar asked him what he meant, Busck repeated it. Dyar deduced he meant the seventh commandment: "Thou shalt not commit adultery." Dyar denied such "false and malicious gossip . . . made against me and which the Secretary was foolish enough to accept, and is the cause of my dismissal."[16]

Howard smoothed things over, and Dyar noted that Busck "offered me 'his girl' for two weeks while he was gone." Dyar at first declined but accepted Pratt in the end, while the USDA sought to hire someone with "Lepidoptera mounting skills."[17]

♧

A celebrated work by his adversary William Barnes, and his professional assistant J.H. McDunnough, greeted Dyar shortly after his return to Washington, D.C. "Check List of the Lepidoptera of Boreal America," published in February 1917, would become one of the best-known American Lepidoptera publications of the twentieth century. The stated purpose was for students "not only to arrange their collections but to have before them in a concise form the latest views on the classification of this order of insects."[18] Dyar's Shakespeareanesque review of Barnes and McDunnough's "Check List" began with a signature soliloquy: "The gentlemen from Illinois have published again. For personas already under criticism, this is nothing short of an 'overt act,' and the temptation is strong once more to obtain their *Caprae hirci*." Once again from his own *Insecutor*, Dyar bristled at the absence of literature citations: "A list like this, without references, is useful inversely in proportion to the attainments of the student. This list, therefore, will be of especial value to beginners and amateurs." Dyar was indignant because the checklist benefited from his 1903 work without attribution.[19]

At the time of his review, Dyar, perhaps realizing his weakened position after his dismissal from the USDA, offered Barnes and McDunnough assistance on a new catalogue of Lepidoptera that they were preparing and that "should be, the standard work on the subject for the next fifteen years," committing the resources of the museum and placing his "personal efforts . . . [for] the benefit of any information

I may possess." Dyar's past hostility was due to "an action" Barnes could "remedy at any time" and his criticisms were more severe because of it. It's unclear to what Dyar was referring, but it may have been Barnes's accusation that Dyar had refused to assist him with his Lepidoptera research. Dyar concluded, "I wish you to study this collection and to accept what assistance I can give for the benefit of your catalogue.... Afterward hostilities may be resumed according to circumstances."[20] J.H. McDunnough would write his own updated checklists of North American Lepidoptera in the late 1930s, but no catalogue as proposed by Barnes, nor any other, has ever replaced Dyar's "A list of the North American Lepidoptera and key to the literature of this order of insects" of 1903.

## RECOVERING FROM BREAKDOWN IN CONNECTICUT

Six months after returning to Washington, Dyar suffered a nervous breakdown and spent late December through January 1918 in Norwalk, Connecticut, at the Hotel Mahackemo (Figure 19.2a). Dyar began attempts at reinstatement to government service by writing the USDA administration, accusing Mrs. Leech of perjuring herself in testimony about his relations with Mrs. Allen. This did not persuade Secretary of Agriculture D.F. Houston, who replied that "the department sees no good reason why your case should be reopened and . . . cannot give favorable consideration to your request for reinstatement."[21] Dyar would renew his efforts near the end of the year.

From Connecticut Dyar answered T.D.A. Cockerell's complaints (sent from the University of Colorado) about the mishandling of his specimens by James Crawford of the USNM's Division of Insects. He was concerned excellent taxonomists such as Cockerell (Figure 19.2b) would no longer identify or donate specimens to the National Museum due to its disorganization but showed compassion and even fondness toward Crawford, who married Wellesca's niece Emily Baker. Dyar observed,

**Figure 19.2** Hotel Mahackemo in Norwalk, Connecticut, where Dyar convalesced in 1918 (*a*); T.D.A. Cockerell, with whom Dyar corresponded with about bureaucratic inefficiencies at the USNM while in Norwalk (*b*).

"his charming wife has recently presented him with their first-born.... Where a man's interest are, his activities are likely to be also, and Crawford is a very agreeable fellow personally. I write this lightly because I think [his] shortcomings a very minor matter. I want to reform the vicious system."[22]

In a lengthy postscript to Cockerell, Dyar wrote about the bureaucratic chain necessary to ship insect specimens from the USNM, observing, "The large and active National Museum is thus run by an undersecretary of the Smithsonian. We have outgrown that. We outgrew it at least 20 years ago."[23] This point of view had its merits and was borne out in the future. At the time of the letter, an assistant to the Secretary of the Smithsonian had always run the museum from an office located in the Smithsonian Castle.

## DYAR'S PARTNERSHIP WITH FREDERICK KNAB

Frederick Knab (1865–1918) (Figure 19.3a) was one of the few colleagues Dyar got along with,[24] by the 1910s. During Dyar's lengthy absences Knab managed *Insecutor* and checked the proofs of the last volume of "The Mosquitoes of North and Central America." In July 1914 Knab wrote Dyar about a variety of topics relating to the journal, including C.H.Tyler Townsend's "valise full of mss." He kept Dyar up to date:

> [N]ot made much progress with the proof, because I have not felt up to the mark. I presume it is the weather—at least I hope it is nothing more! . . . Schaus is talking of going to Rio for a while. . . . Haimbach is here to name up some of his Leps. Busck and I went over to Baltimore the other day to investigate a suburban mosquito problem. . . . all [*Culex*] *pipiens*, breeding in the sewers and cesspools.[25]

Knab was concerned about continuing publication of *Insecutor*. In December 1914, he needled Dyar about the journal's bizarre name, much as entomologists and family members had over the years: "I was beginning to get worried about you and was relieved to see that you are on the mend. . . . Shall I give out the copies to

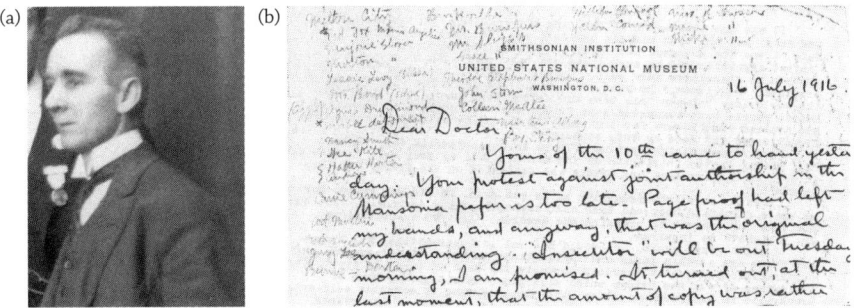

**Figure 19.3** Frederick Knab (c. 1912) (*a*) and letter from Knab to Dyar (1916) (*b*), who was in Reno at the time. Dyar's faint writing on Knab's letter was a list of the characters he used in an unpublished story "A Ladies Man."

your Washington subscribers, in order to establish publication? . . . We are hoping that you will continue to 'menstruate' for another year! I hope this finds you still improving!"[26] The following month Knab tried to lift Dyar's spirits, writing: "I am very sorry to see that you have had a set-back. You sure will have to go slow for a while! But don't get discouraged 'Ischkabibble' [which means "don't worry"]." The following year Knab wrote: "Was glad to hear from you . . . although rather disappointed at the postponement of your return East. But the important thing is to get clear of your domestic complications, of course."[27]

In September 1915 Knab facilitated access to Dyar's "Blue Books" and catalogue of Lepidoptera specimens, writing to Dyar about William Trowbridge Merrifield Forbes (1885–1968) of Cornell University: "Forbes of the Pink Whiskers is here. . . . He is very anxious to see your note-book that has the localities for the numbers on the specimens. We found the notebooks with your biological notes [Blue Books], but not the one for the higher numbers. Do you care to let him see it? And if so, can you tell me where to find it?" Forbes would later write the monumental "Lepidoptera of New York and Neighboring States."[28]

While Dyar was away, Knab's coverage of both mosquito identifications and control campaigns for Howard were among his most taxing duties. In October 1915 he visited West Virginia to advise on the control of the common house mosquito (*Culex pipiens*) (see Figure 21.1b): "The White Sulphur Springs problem was easy. . . . They had installed a sewage disposal plant the year before and now are raising millions of 'em! [*Culex pipiens* larvae love sewage]."

From an insect collector's point of view, Knab teased Dyar about the earthquake in Nevada on October 2, 1915: "I see you have had quite a 'shake' at Reno. Hope it did not jolt you too badly!" Referring to the small points tiny insects are glued onto, Knab feigned concern about the quakes impact on specimens, joking "Nor knock our midges off the card-points!" (see Figure 19.3b for an example of Knab's letters to Dyar).[29] In his role as Honorary Custodian of Diptera at the USNM, Knab also pushed Dyar to collect various biting flies and others as well as mosquitoes while in Nevada: "By all means mount the Chironomidae as far as you can! Take all the biting things. . . . Tabanidae [horse flies] will all be nice from there. In fact, all Diptera!"[30] Knab made sure Dyar received such provisions as insect pins, card points, cyanide bottles, and wooden Schmidt boxes to ship back specimens. Using war imagery, as was his wont, Knab wrote: "I'm quite startled at the rapid expenditure of ammunition, but look forward with pleasure to the final results! . . . More will follow in a day or two. I fear I would hardly hold down a job with the German supply division."[31]

❧

The fourth and final volume of the Carnegie monograph, *Mosquitoes of North and Central America and the West Indies* (1917) by Howard, Dyar, and Knab represented a giant leap forward in our knowledge of these insects. The authors noted that they arrived at "an independent view" of the classification using the larvae alone in 1906, but that "more extensive study" of adults, including their male genitalia "naturally led to some modifications in details."[32] This publication completed a project begun in 1903. Cockerell reviewed the volume for *Science* in June 1917:

"The final part of this great work has at last been issued, amid general rejoicings from those interested in medical entomology, since it contains a full account of the malaria-organism carrier, *Anopheles*. The two parts containing the descriptive matter and synonymy total 1,064 pages . . . . The yellow-fever mosquito alone takes over sixteen pages."[33] The timing of volume 4 was perfect for Cockerell to complete, with help from Dyar and Knab, "The Mosquitoes of Colorado."[34]

⌘

Several months after Dyar returned to the USNM in spring 1917, he worked with Knab on the national mosquito collection (see cover). Knab's frustrations from his health problems—the effects of what is known today as the visceral form of Leishmaniasis, the symptoms of which include headaches, fever, anorexia, and depression. These symptoms were noted by Dyar who wrote in September 1917 to Cockerell, "Poor Knab is in a very bad way, as you probably know, and he only gets in an hour or so a day, arriving late in the afternoon."[35] Dyar's compassion can be seen in several ways in early 1918. As Knab's condition continued to worsen, Dyar wrote Howard: "I know that Knab is very weak, while the trouble having infected his right eye makes it almost impossible for him to do fine work. I want you to consider the mosquitoes that I do [identify/curate] as done by Knab, and credit him with the work, so as to justify his salary as fully as possible."[36]

Knab wrote to Howard from George Washington University Hospital about his imminent demise in February 1918 and died on November 2 that year a short time after Dyar returned from collecting in Canada during the summer. As Knab had recommended to Howard in February, John Merton Aldrich (1866–1934) (see Figure 20.2c) took over his duties as Custodian of Diptera along with becoming Associate Curator of Entomology the USNM in 1919.

Although Frederick Wallace Edwards (1888–1940) and Dyar "contributed most significantly to the broad genus-group concepts that provided the framework on which the 'traditional' classification of the twentieth century was built," Knab also deserves credit for the "complete life stages" classification he pioneered with Dyar, as contrasted with Theobald's artificial groups based on poorly conceived features of only the adult stage.[37]

The influence of how Dyar dealt with death, whether he considered it a celebrated transformation, given beliefs from early exposure to spiritualism remains unclear. It is worth noting, however, that although none may have had closer ties to Dyar than Knab, Dyar left Knab's obituary to others as he had done with Neumoegen, Lintner, Strecker, and Grote—all close associates in entomology.[38] Certainly for Dyar there was no one who could replace his friend and devoted colleague Knab.

# PART IV

## The Final Decade

*Attempts at Reinstatement*

CHAPTER 20

## The National Collection of Lepidoptera, Its Workers, and Their Tiffs: 1920s

*You know Dr Dyar's psychology and his particular fondness for this group of moths* [limacodids], *which he considers as his pet children and you may for this or other reasons decide to return the specimens, but whether you so decide or not, you should know, that it is not at my request or for my sake, that you do so.*[1]

August Busck to William Barnes, October 12, 1920

By 1920 it was clear that Dyar would keep his position as Honorary Custodian of Lepidoptera. He could now, once again, focus on the nation's collection of butterflies and moths. During his tenure with the USNM collection, he had divided it between the "macros" (Macrolepidoptera) and "micros" (Microlepidoptera),[2] ceding more control of the "micros" to Busck (Figure 20.1a) and Carl Heinrich (1880–1955) (see Figure 26.3), who were assigned them by Howard. Dyar handled the "macros" with William Schaus (Figure 20.1b), though he also worked on the snout moths (considered "micros") with Heinrich. While away the only person Dyar thoroughly trusted to take care of the collection was Schaus, largely because of unofficial specimen loans or exchanges involving Busck and Barnes.

Although external pressures to remove Dyar calmed, problems in the confines of the USNM had developed in his relations with Busck and Schaus. However, though he was having difficulties with Busck he was apparently naïve about Schaus.

Some of these divisions between the lepidopterists appeared after Dyar took flight to Reno in June 1915, when the influence of William Barnes (Figure 20.1c) on USNM specialists grew. Barnes gave Busck and Heinrich contracts to work on his collection in Decatur and to Caudell to catalog the food plants of Lepidoptera

**Figure 20.1** Portraits of August Busck (*a*), William Schaus (*b*), and William Barnes (*c*).

caterpillars while in Washington. Dyar had given Caudell and Busck loans or paid their field expenses, but now Barnes helped them financially.[3]

During the late teens or early 1920s Barnes began a questionable practice of pulling identified specimens from the USNM Lepidoptera collection without loan paperwork, "borrowing" them so that his curator in Decatur could compare them with his own to insure proper identification. At the same time Barnes unofficially donated specimens of species not found in the collection, in essence, trading them. These under-the-table exchanges put Dyar in a difficult position, especially because he wished to encourage Barnes to donate his collection to the museum. Likewise, the other USNM lepidopterists wanted to maintain friendly relations with Barnes.

In 1920, after their conflicts over unofficial specimen transactions, Dyar sent Busck a caterpillar to identify from a madrone tree in Seattle. Fulfilling the request, Busck, a giver as well as receiver of the latest gossip, wrote Dyar: "I am glad to hear you are having a good time on the coast. Somebody started the rumor that you had married out there. Is it as good as all that?" This momentary détente, however, was about to come to an abrupt end.[4]

On his return to the USNM, Dyar had a confrontation with Busck after Dyar found some of his limacodids, which he had raised from caterpillars in the 1890s, were missing from the collection. To Howard, he wrote:

> On noting the loss, I endeavored to trace it. I asked Mr. Busck, knowing that Mr. Schaus would never think of touching this material. Busck at first denied, but on being pinned down, admitted that Dr. Wm. Barnes, in my absence, had probably taken these specimens out. Afterward Busck grew angry, insulting and threatening, as is his custom when cornered, and claimed that he was officially in charge in my absence.

"I think it is well known that Busck is only a stool-pigeon for Dr. Barnes," wrote Dyar, "and it was easy for him (without my knowledge) to have himself appointed in charge, evidently with the purpose and intent in advance to rob the collection, and especially *take my pets* [italics mine], for the benefit of said Dr. Barnes, both of these gentlemen well knowing that these could not be had with my consent." Busck admitted he let Barnes go through the collection and take the specimens, for which no records were made. The politically savvy Howard told Busck to ask for a return of the specimens, but Busck refused after Dyar's continuing tirade, purportedly telling Dyar he "did not want any more of . . . [his] gab."[5]

Dyar now asked Howard to get his 10 specimens back from Barnes, pointing out that only Schaus, not Busck, had authority over his custodial duties in the collection when he was away, writing: "While I am not a friend of Dr. Barnes . . . , I have not forbidden him the collection, and I will receive him with politeness." He would allow Barnes to take specimens that could "be spared," and he wished "to treat him well for the reason that he may be induced to leave us his collection."[6] Howard swiftly responded: "I am greatly disturbed by your statements, and have been considering the matter carefully. . . . We will have it distinctly understood that in your absence Mr. Schaus shall have exclusive charge of the Macrolepidoptera."[7]

Dyar looked for more pilfering of the collection and found specimens of tiger moths (*Haploa*) missing. He noted to Howard that he "was shocked to see vacancies" in the series that he had arranged to show the wing pattern variation "from fully marked to pure white" and "with great pains, arranged . . . from one extreme to the other, and right in the middle, one of the key or bridge specimens had been taken out . . . I have not dared look any further for fear I would have a stroke of apoplexy or a nervous collapse."[8]

Venting his spleen, Dyar wrote Howard: "When is a custodian not a custodian' When he is made a monkey of by a d— interloper" and stated Busck had given away specimens without a record. According to Dyar, this was more flagrant than a case in which "A man was dismissed for breaking types [type specimens]." To close, the angry Dyar wrote: "If MR. Busck should resign or be dismissed, I know of a perfectly capable substitute"[9] and soon mentioned that Heinrich was "worth two Buscks already."

The next day, Dyar continued to rant, imagining how administrators would treat his charges against Busck: "I am called 'custodian,' but it seems a purely fancy title. Another man steps in my absence and removes my most cherished treasures? If he did it in my private collection I would have him arrested and get some satisfaction. As it is, Aldrich refers it to Stejneger, and he to Bartsch, and back to you, and you do nothing." He then asked Howard to request the specimens to be returned and Busck to be dismissed.[10]

Howard asked Barnes to return the specimens—calling Busck on the carpet for his unofficial specimen exchanges—while an unrepentant Busck wrote Barnes: "I gave you those specimens, as an officer of the Museum, acting and in charge during Dyar's absence and with full right to do so . . . it was but a very slight return for the valuable material you had just donated to the Museum. . . . The specimens are therefore rightfully yours."[11] Clearly Howard could not tolerate this situation and in November 1920 he wrote Busck:

> You took an unwarranted liberty in giving Doctor Barnes specimens from the portion of the collection in which Doctor Dyar is particularly interested, without his knowledge or permission . . . I think further that you were entirely in the wrong in not listing the material taken by Doctor Barnes. This is from the standpoint of

Museum administration.... I suggest that it will be wise if you will inform Doctor Dyar (by note if you prefer) that you are sorry you lost your temper. Nothing further, I am sure, will be needed to reestablish conditions.[12]

Given these collection management problems, Howard considered dividing up the duties between the resident lepidopterists by moth groups—an idea vehemently opposed by Dyar: "Imagine trying to put this foolish plan into execution in my office," Dyar observed. "Heinrich will want to work on the Pyralidae and you would make him custodian of Pyralidae.... It is one of my pet groups and I would not give it up. Under the present arrangement he can do all the work he likes on these or any other family." Likewise his interests overlapped with Schaus:

> He and I work over exactly the same groups with a minimum of friction and most of the time in brotherly love.... You would start trouble immediately, for I would not give up any [families], and poor Schaus would be left out in the cold. Now he considers himself in co-charge with me (I am delighted) and he can give himself all the airs he likes and nobody cares.

Dyar said Busck was the only problem with the status quo: "Busck ... is not a gentleman. This is not the fault of the system. Do not change the system; change the man.... Let him understand that I am in charge, and he has no authority over the collection (Tineids included) except as delegated by me. When I am absent, no authority falls upon him automatically or otherwise."[13] In the postscript, Dyar expressed the problem of Busck receiving specimens from Barnes in Busck's area of interest (e.g., tortricids) and then exchanging them for specimens in Dyar's groups.[14]

## SCHAUS AND THE PROBLEM WITH HONORARY TITLES

Dyar supported Schaus as his backup when he was away, but one year after Schaus's initial appointment as Assistant Custodian of Lepidoptera in July 1918, the USNM administration, with the support of Leonard Stejneger, the Head Curator of Biology who oversaw the Division of Insects, gave Schaus the new title of Honorary Assistant Curator. This position had more prestige than Dyar's as Custodian.[15] Dyar, who learned about the change in the museum's annual report, wrote Howard:

> William Schaus is "Honorary Assistant Curator" while Harrison G. Dyar occupies the humble position of 'custodian, Section of Lepidoptera.' What I desired and what I understood you to do was to appoint Mr. Schaus Assistant custodian, Section of Lepidoptera, namely assistant to me, and such was always my understanding and I think Mr. Schaus' also.... I would request that if matters are correctly stated in the report that you make a transfer, making me the Honorary Assistant Curator and Mr. Schaus Custodian of Lepidoptera, and ... kindly call Dr. Stejneger's attention to the error in order that he may have it correctly stated in any subsequent publication.[16]

**Plate 1** Caterpillars of the Limacodidae, Dyar's favorite moth family, whose life histories he described from the 1895 to 1914 for the New York slug caterpillar series: Skiff moth (*Prolimacodes badia*) (*a*); crown slug (*Isa textula*) (*b*); saddleback (*Acharia stimulea*) (*top*) and spiny oak (*Euclea delphinii*) (*bottom*) (*c*); pin-striped slug (*Monoleuca semifascia*) (*d*); smaller Parasa (*Parasa chloris*) (*e*); rose slug (*Parasa indetermina*) larva and adult moth (*inset*) (*f*) hag moth (*Phobetron pithecium*) (*g*); red-cross slug (*Tortricidia pallida*) (*h*).

**Plate 2** Dyar's field notebooks describing life histories and illustrations of caterpillars were called "Blue Books" based on the earliest one that he began in 1882: The small, original "Blue Book" (*a*); "Blue Book" and some of the other Dyar rearing notebooks that followed, written into the early twentieth century (*b*); Dyar's unpublished illustrations of variation among skiff moths (*Prolimacodes badia*) (*c*); life history notes and drawings of the mature and first instar caterpillar of the smaller Parasa (*Parasa chloris*) from 1896 and later published in *Journal of the New York Entomological Society* (*d* and *e*) (note: red check mark indicates the descriptions were published); unpublished drawings by Dyar of the variably marked red-cross slug (*Tortricidia flexuosa*) found on American Chestnut in 1896 in Fort Lee, New Jersey.

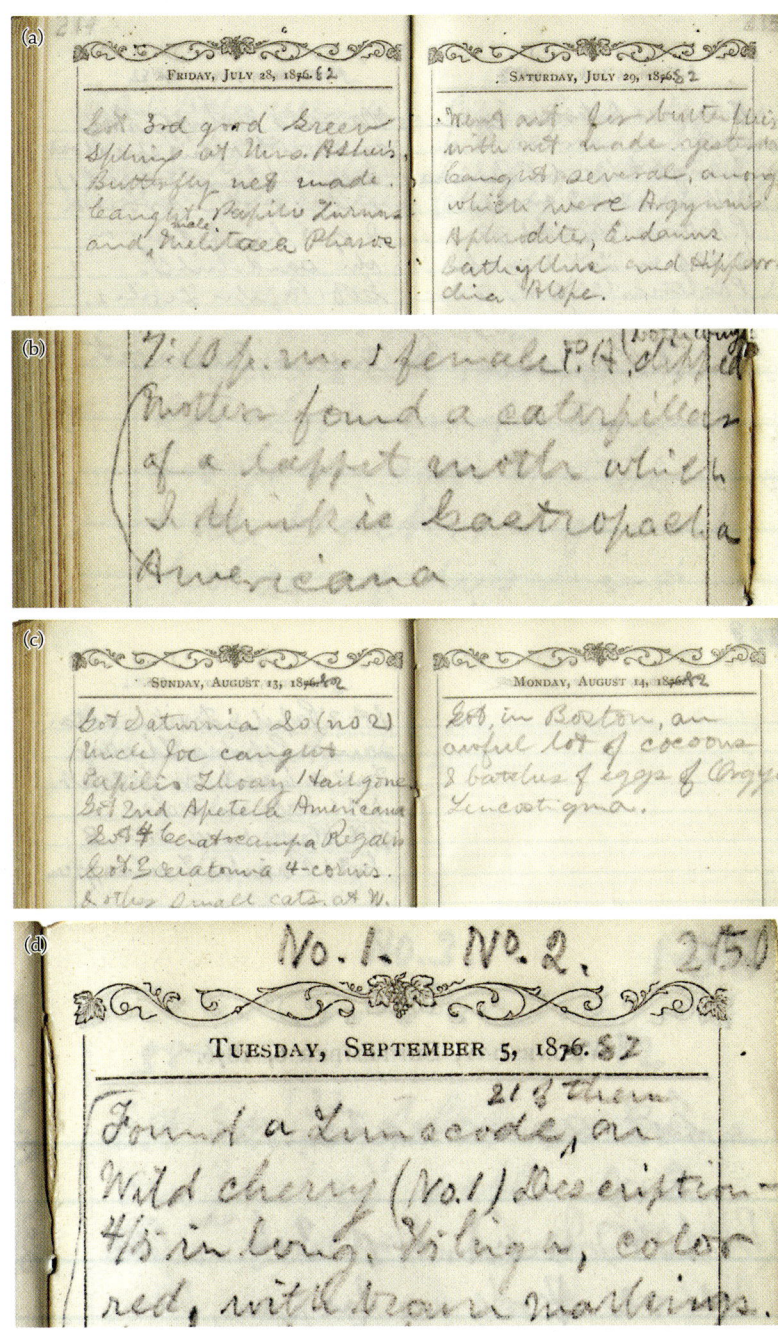

**Plate 3** Entries by sixteen-year-old Dyar found in the original "Blue Book," from the summer of 1882: Collects at Mrs. Asher's with new butterfly net (July 28) and continues collecting butterflies the next day, including aphrodite fritillary (*a*); lappet-moth caterpillar found by his mother (August 3) (*b*); found an io moth caterpillar and Uncle Joe collects a giant swallowtail with him (incorrectly identified as *Papilio thoas*; August 13) (Dyar goes to Boston the next day, right, where he found "an awful lot of coccoons & batches of eggs of Orgyia leucostigma" (the white-marked tussock moth)) (*c*); "Limacode" caterpillar of *Parasa chloris* on wild cherry (September 5) (*d*).

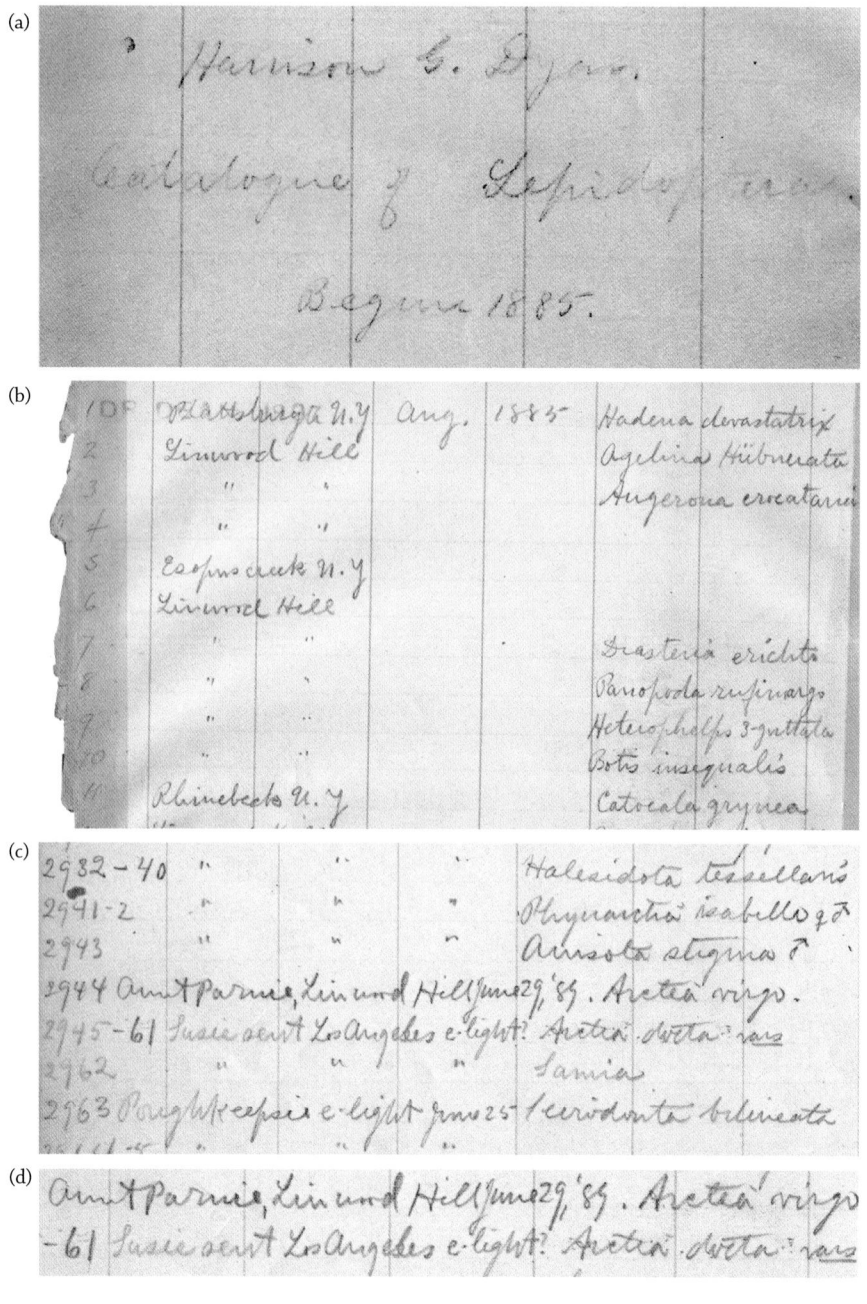

**Plate 4** Dyar's "Catalogue of Lepidoptera," an account of 39,950 specimens he collected, reared, traded, or purchased between 1885 and 1913 (only 69 specimens from 1905 to 1913): Title page (*a*); first page of entries (*b*); page of entries containing Record 2944, a virgin tiger moth "*Arctia* [now *Grammia*] *virgo*" found by his Aunt Parthenia "Parnie" Hannum Colburn at Linwood Hill in 1889 (*c*); detail of "Parnie's moth record" and another collected by "Susie" in Los Angeles (*d*).

**Plate 5** Examples of owlet moth caterpillars reared by H. G. Dyar between 1882 and 1903: Paddle moth caterpillar and adult (*Acronicta funeralis*) (*a* and *b*); smeared dagger (*Acronicta oblinata*) (*c*); green marvel (*Agriopodes fallax*) (*d*); eight-spotted forester (*Alypia octomaculata*) (*e*); laugher (*Charandra deridens*) (*f*); herald (*Scoliopteryx libatrix*) (*g*); camphorweed paint (*Cucullia laetifica*) (*h*).

**Plate 6** Some of the notodontid moth caterpillars and their adults reared by Dyar in the 1880s and 1890s: Lace-capped caterpillar (*Oligocentra lignicolor*) (*a* and *b*); saddled prominent (*Heterocampa guttivitta*) (*c* and *d*); Drexel's Datana (*Datana drexelii*) (*e* and *f*); black-etched prominent (*Cerura scitiscripta*) (*g* and *h*).

**Plate 7** Moths from assorted families reared by Dyar in the late nineteenth and early twentieth centuries: Lappet (*Phyllodesma americana*) (*a*); promethea (*Callosamia promethea*) (*b*); laurel sphinx (*Sphinx kalmiae*) (*c*); faithful beauty (*Composia fidelissima*) (*d*); definite tussock (*Orgyia definita*) (*e*); common aspen leaf miner (*Phyllocnystis populiella*) (*f*); wavy-lined emerald (*Synchlora aerata*; note camouflaged caterpillar on left) and adult (*g* and *h*); curved toothed geometer caterpillar and adult (*Eutrapela clemataria*) (*i* and *j*).

**Plate 8** Dyar's mentors and early correspondents: Charles V. Riley (*a*); George H. Hudson (*b*); Joseph A. Lintner (*c*); Alpheus S. Packard (*d*); Charles H. Fernald (*e*); Augustus R. Grote (*f*); F. H. Herman Strecker (*g* and *h*).

**Plate 9** Entomologists of the U.S. National Museum in their younger days: August Busck in the 1890s and, on the right, digging up Brood X periodical cicadas on the National Mall in 1902 (*a* and *b*); William Schaus, on the right, in the early 1900s and in Mexico or Central America with Jack Barnes (*c*.1906) (*c* and *d*); Andrew N. Caudell and while studying his grasshoppers (*c*.1905) (*e* and *f*).

**Plate 10** Dyar family photos along the northeastern seaboard in the early 1900s: H. G. Dyar and son Otis (*a*); Zella (*b*); Zella, Dorothy, and Otis (*c* and *d*); Otis with uncle (perhaps John M. Holland) (*e*).

(a)

(b)

**Plate 11** (*a*) H. G. Dyar with his children Dorothy and Otis on mules in the western United States (*c*.1912); (*b*) mother-in-law Harriet M. Peabody helping develop commerce for the Navajo through the sale of their rugs in Bluff, Utah (1901).

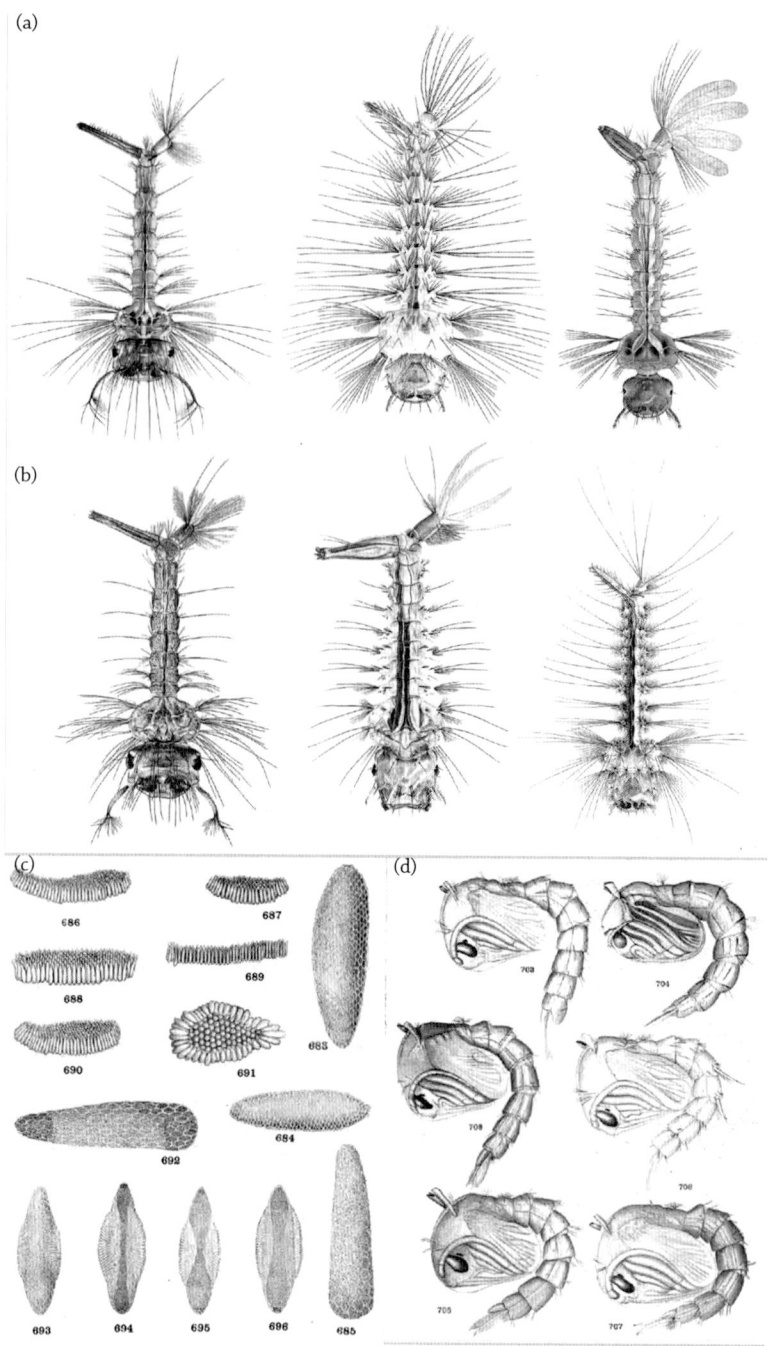

**Plate 12** Illustrations of immature stages of mosquitoes from "The Mosquitoes of North and Central America and the West Indies," by Howard, Dyar, and Knab: Larvae known as "wrigglers" (note: illustrations are flipped upside-down from the original, which places their breathing tubes on the surface of the water as occurs in nature) (*a–b*); eggs in groups (known as rafts) and individually (*c*); pupae (*d*).

**Plate 13** Entomological workers of the U.S. National Museum, Smithsonian Institution, May 21, 1925. *Left to right:* (standing) Arthur Burton Gahan, Charles Tull Greene, William Schaus, Adam Giede Böving, Andrew Nelson Caudell, William M. Mann, Henry Ellsworth Ewing, Harrison Gray Dyar, Eugene Amadeus Schwartz, Sievart Allan Rohwer, Leland Ossian Howard, Raymond Corbett Shannon, W. Samuel Fisher, Herbert Spencer Barber, Virginia (Jennie) Locke, Robert Asa Cushman, Mrs. A. C. Willis; (kneeling) Mildred (Shields) Everhart, Eleanor Armstrong, Janice Kyser, Frances (Kaufmann) Appleby, Carol, unidentified, Mrs. Yates; (seated) Miss Sellins, Mathilde M. Carpenter, Eunice Myers, Nettie Grochek.

(a)
## Old Tunnel Here Believed To Have Been Used By Teuton War Spies and Bootleggers

**Labyrinth Extending Hundreds of Feet Under Exclusive P Street Section Revealed When Truck Sinks Into Entrance Back of Pelham Courts. Walls Made of Enameled Brick.**

Extending underground at least 500 feet, a labyrinth of subterranean passages, long since forgotten, were uncovered yesterday in the rear of Pelham Courts apartments, 2115 P street northwest. In the

The bringing to light of the subterranean passages, which run through the earth for a distance of two blocks at least, was made yesterday by Bishop Hill, of the real estate firm of Moore & Hill, 730

(b)
### 'MYSTERY' TUNNELS BUILT BY SCIENTIST 'MERELY AS PASTIME'

Exercise Sole Motive in 10-Year Work, H. G. Dyar Says.

**SUBTERRANEAN MAZE UNDER HIS PROPERTY**

German Spy and Rum-Cache Theories Continue to Enthrall Public.

Harrison G. Dyar, an entomologist in the Smithsonian institution, last night admitted to a Post reporter that he constructed the labyrinth of tunnels in the rear of Pelham Courts which, since they were uncovered two days ago, have caused the wildest speculations as to their origin and use.

Mr. Dyar made the admission while taking his dinner at his home, 804 B street northwest. His story, which clears the baffling tunnel mystery, was told as a matter of fact way in the professor of "a wife and son." It was briefly and modestly told.

Throughout the narration one could sense the coolness of it—this much, mild-mannered scientist devoting his time and study to butterflies and moths in a government office in the day and secretly dig-

### DIGGER OF "MYSTERY" TUNNELS

HARRISON G. DYAR, Entomologist, who dug the underground passage in the rear of Pelham Courts for exercise.

(c)
**INVESTIGATOR EXPLORING TUNNEL**

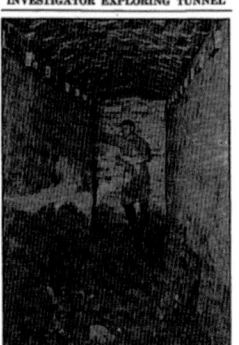

Beneath Washington's exclusive residential section German spies, bootleggers and whatnot are believed to have found their retreat. Here is pictured a turn in the tunnel that is believed to have extended from a lot adjacent to the Pelham Courts to Rock Creek. One of the exploring party is at the turn.

(d)
That Mysterious Tunnel in Washington

(e)
### DYAR'S TUNNEL HOBBY MAY BE CURTAILED

**Scientist Must Get Permits in Future; Officials Investigate Possible Danger.**

Startled by the extent of the secret burrowing of Harrison G. Dyar, scientist of the Smithsonian Institute, Maj. John W. Oehmann, District building inspector, today will inspect the tunnels dug by the scientist and lay before the District commissioners a report of his findings.

Mr. Dyar's hobby of tunnel building was exposed when discovery of an underground labyrinth in the rear of 1510 Twenty-first street northwest, adjacent to Pelham Courts, a large apartment house, was followed by a second discovery that he has extensively tunneled under and around his present home at 804 B street southwest.

(f)

(g)
## "Mystery" Tunnels Digger Burrows New Double Maze

**More Elaborate Subterranean Labyrinth Opens from Cellar of Scientist's Home, Ending in Square Well 24 Feet Deep.**

Born of the same "hobby" that caused Harrison G. Dyar, entomologist of the Smithsonian Institution to dig for ten years constructing a maze of underground passageways

ly within the bounds of Mr. Dyar's property. They are being built, he explained, simply as a means of affording him exercise and diversion from his daily scientific grind. Mr.

(h)
### DUG MAZE OF TUNNELS FOR FUN

**Plate 14** Hullabaloo surrounding the tunnel collapse in Dupont Circle, Washington, D.C., and the revelation of another tunnel on B Street in September 1924: Theories behind who dug the mystery tunnel (*a*); Dyar admits to being the digger (a fact revealed seven years earlier during a similar collapse (*b*); investigator inside the Dupont Circle tunnel (*c*); political cartoon of "Fighting Bob LaFollette" digging his way into the White House (*d*); concern about safety brought an inspector to the B Street tunnel (*e*); photos and headlines about Dyar discussing his hobby and his newer multilevel excavations on B Street, across from the National Mall (*f–h*).

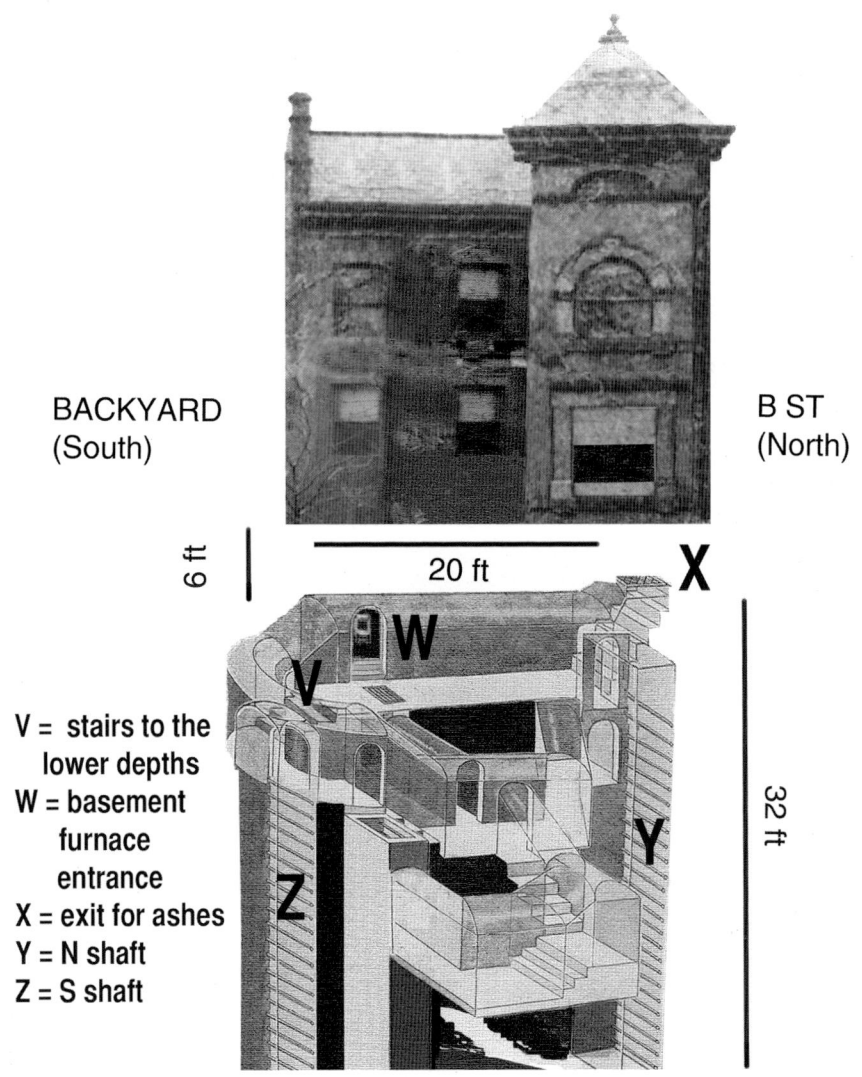

**Plate 15** Diagrams of Dyar's home and labyrinth on the east side of 804 B Street SW (Independence Avenue today). The arch at the basement entrance (W) is where *Facilis decensus averni* was inscribed. Above the V is where Dyar chiseled "February 14, 1923" on his fifty-seventh birthday.

**Plate 16** Dupont Circle tunnel and its collapses behind the Dyar home: Collapse in alleyway, 1924 (*a*); Curd children in the Dupont Circle tunnel after 1958 collapse (one adult child from each of Dyar's families visited the site at the time) (*b*); outside of the blocked entrance in the 1990s (*c*).

Howard replied that it was Stejneger who suggested Schaus's title, to

> give him a better standing in his dealings with outside parties than the title of Custodian [and] ... Schaus functions as Associate Custodian of Lepidoptera, ... the wording of his title [is not] ... a matter of any great importance, nor ... that it was Doctor Stejneger's intention to elevate him over you, as you express it. I do not think that it would be good policy to make any point on this matter now.

The "outside parties" Howard was referring to were private collectors such as B. Preston Clark, now listed as "collaborator" in the same annual report, and William Barnes. Schaus was known to raise money to purchase collections from his wealthy friends. For example, in 1925 he raised the sum of $50,000 for the purchase of the Dognin collection and library.[17]

Howard's response led Dyar to recall Schaus's initial appointment in 1918: "[D]octor Stejneger used the word Curator when he took me aside and solicitously enquired if I had any objection.... No one explained to me that any radical change had been made in the proposition, or if they did it did not get through my skull. Dr. Stejneger's manner is soft and insinuating, may I say reptilian?" Dyar's mention of Stejneger's manner related to the curator's background as a herpetologist (reptiles and amphibians) and the new revelation about Schaus explained why "outside parties like Dr. Britton and Mr. Pierce addressed Schaus instead of me on matters of general information."

Dyar's complaints appeared to have been heard. The 1922 annual report listed Schaus under Dyar as Associate Custodian of Lepidoptera, but the following year Schaus was again listed as Honorary Assistant Curator. Stejneger wrote Dyar, telling him an error had been made in 1922, not 1921.[18] Howard persistently attempted to keep Dyar's spirits up, reassuring him that Schaus understood he was Dyar's assistant:

> I imagine that it will be perfectly harmonious for you to work away—you with the title which gives you charge of the whole collection, and he working on those branches which you may mutually agree he is best fitted to handle.... If you insist upon it, I will see if I can get Doctor Walcott to let you also have the title of Honorary Assistant Curator, but I think that it would be altogether better not to push the matter.

Dyar's response was:

> Your judgment, as usual, is the best, and I will conform to it. The irruption, however, was not without its uses, at least to satisfy temporary annoyance. Mr. Schaus will be Mr. Schaus, especially at times. 'Nuff sed. I have worked for the Museum 27 years, most of the time for nothing. I got Mr. Schaus here—and they set him above me. When I remonstrate—'Sir: You are informed, etc.' I would like to kick them into the sewer."[19]

This situation continued to frustrate Dyar, and the following year he called to Howard's attention a piece in *Entomological News* about Schaus being granted the degree of "Doctor." The article mentioned that Schaus was "in charge of Lepidoptera" at the National Museum. He then asked Howard to get Schaus to publish a retraction of this error. Howard counseled restraint.[20]

Dyar's friendly association with Schaus ended by February 1924 when he grew tired of Schaus's pipe smoke and confronted him. Dyar told Howard of two alternative notes he had considered writing to Schaus. The first was "to say in effect, 'I am a poor weak invalid who cannot stand the rich aroma that emanates from the pipe of a real man. Will you in magnanimous compassion spare my weakness and forbear to smoke?'" The second: "'The vile smell of your pipe is offensive to any man of taste. Kindly cut it out.'" Dyar then wrote: "No. 1, I considered, did not fit the case nor my inclination, and so I chose No. 2. There have been times when I have hit Schaus as hard as this without eliciting resentment, but this was evidently not one of them. However, I am content. I infinitely prefer a row to humiliation."[21]

Howard's response suggested other alternatives: "I note your two available 'attitudes of mind.' The politic and proper attitude would have been a compromise between the two. Self-humiliation was not necessary; neither was the opposite at all necessary. You would save yourself lots of trouble and get along beautifully with everybody if you could see this."[22] Having moved a safe distance from the smoke across the hall, Dyar understood Schaus was smoking less now that "Miss Armstrong has exchanged desks with me." He now spoke neither to Schaus nor Jack Barnes.

Reflecting further about the incident, Dyar replied to Howard, telling him:

> The advice, of course, is excellent, provided one desired hard enough to get along with everybody. For myself, I am satisfied if I get along with a few. . . . Your suggestion that I compromise between the two possibilities of approaching Schaus, reminds me of the man who wished to commit suicide; but having placed himself upon the track and hearing the train approaching, was beset by uncertainty. He had two possibilities before him, one to lie still and let the train destroy him, and the other to get off the track and save his life. Being unable to decide the question he determined to compromise between the two alternatives (as you advise me), and so he got half way off the track. The train cut off both legs and he died miserably in a hospital. I am afraid that compromise would not have been any more successful in my case.[23]

Dyar's inability to compromise and his low tolerance for other people carried beyond Schaus to his friend John "Jack" Barnes. In January 1919, prior to the pipe-smoking incident, Dyar complained to Aldrich: "Will you be so good as to request that an official request be made of Mr. Wm. Schaus, my assistant custodian, not to have his friend John Barnes come to my office any more? The gentleman is not personally agreeable to me, and has offered me an insult. I spoke to Mr. Schaus about it personally, but he insists on ignoring my wishes."[24]

Jack Barnes had an association with Schaus dating back to the 1890s. He was a field companion to Schaus on trips in Mexico, Costa Rica, Guatemala, Puerto Rico, Brazil, French Guiana, and Ecuador.[25] Schaus often referred to Barnes as his "pal," and it was rumored the two men had more than just a friendship.[26]

Even if there were those in the community who took exception to Schaus's lifestyle, it behooved them to maintain a code of silence because Schaus was deeply respected among lepidopterists for his knowledge and connections. Dyar's reference "the tail goes with the hide" could be viewed as a "breach" of this code—perhaps an intolerance of homosexuals was beneath the surface all along. Alternatively, it could have also reflected Dyar's general intolerance of people and of Schaus in particular, given his frustration of being supportive of him coming to the museum and having him receive the title of a superior.

## TIFFS BETWEEN DYAR AND SCHAUS WITH THEIR HELPERS

Both Dyar (Figure 20.2a) and Schaus had direct conflicts with support-staff.

In late January 1927 Schaus (see Figure 20.2b), for instance, bawled out Eleanor Armstrong (Figure 20.2d) for "talking loud" about drawings she was doing for a mosquito monograph. Schaus went up to her and said, "'You'll have to be quiet or get out of this room.' She replied, 'Why, Dr. Schaus, I am not used to being talked to like that.' 'Shut up,' he replied." Schaus also yelled at illustrator Mary Foley [Benson] (Figure 20.2e), who according to Dyar was "as quiet as a mouse."[27] The following month Dyar had a tirade with the staff librarian, Mathilde "Tilly" M. Carpenter (Figures 20.2d) when two visitors from the National Geographic Society inquired about scales on butterfly wings for an article in the *Magazine*.[28] He wanted to show his visitors Scudder's work on microscopic butterfly wing scales in the library and found "Miss Carpenter engaged in a lively personal telephone call." When Dyar found only the index volume was on the shelf, he "went to her desk," but she remained on the phone. "Thinking I had kept my visitors waiting long enough, I replaced the book on the shelf and started to leave the room. Immediately Miss Carpenter yelled at me: "What did you want? What did you want? . . . When you use the telephone you take as long as you want. . . . She said some other unpleasant things, ending with "What are you going to do about it? What are you going to do about it?" yelled at the top of her voice after I had returned to my room."[29]

Carpenter said she did not know the call was personal until she answered and was unaware of Dyar's presence.

> While I was talking over the phone he came to my desk and interrupted my conversation . . . [and] before I had finished Dr. Dyar slammed a book he held in his hand onto the shelf with a terrific bang and started to leave the room. I hung up the receiver and asked him what he wanted. Instead of telling me . . . he berated me in a most offensive manner about receiving personal calls. . . . "You are not paid for receiving personal 'phone calls'["] . . . I have endured with great patience and forbearance other outbreaks of rage and rudeness from Dr. Dyar. His anger rises at

**Figure 20.2** Dyar (*a*); Schaus (*b*); and John M. Aldrich (*c*); illustrator Eleanor Armstrong (*second from left above*) and museum librarian Mathilde Carpenter (*bottom right*) (*d*); illustrator and future aviator Mary Foley (Benson) painting tussock moths (*e*).

the slightest provocation. Once in a rage because I did not have a medicine dropper in stock, he made some rude remarks about the way the supply room was kept and banged the door of that room with such force as to crack the glass in it.[30]

Dyar sent his account to the librarian's supervisor, J.M. Aldrich (Figure 20.2c), who became Associate Curator of Insects of the USNM after Crawford left in 1918. Aldrich asked Carpenter for her version, infuriating Dyar, who expected only his side of the encounter to be placed on file. "You certainly acted in an unnecessary and irritating manner," he wrote Aldrich. "I did not put the matter before you as a judge or advocate.... Does the policeman take the affidavit of the thief as to his side of the case? No, this is for the court.... It would be advisable, if possible, to have Miss Carpenter's statement withdrawn."[31]

Aldrich replied that his superiors, Doctors Wetmore and Stejneger, approved his actions. Not mollified, Dyar wrote: "your action was irregular and ... calculated to transform my complaint from a dignified record into a personal altercation. It is what we call in the vernacular the 'double cross.'" Dyar had the last word: "She is a person entirely unfitted for the position of librarian, which is one of service and courtesy. This boisterous, screaming plebian is out of her element here, and the sooner she is also out of the job the better it will be for the peace and efficiency of this section."[32]

Carpenter continued to work at the museum for many years and produced bibliographies of USNM entomologists after they died. Apparently Tilly Carpenter had the last word after all: She did not produce a list of Dyar's publications post mortem.

◊

Dyar had other conflicts with Aldrich as the Associate Curator of Insects and as Custodian of Diptera (the slot Frederick Knab wished for him). One in particular was in November 1926 when after Dyar identified Lepidoptera for Charles Johnson of the Boston Society of Natural History, Aldrich hand delivered the specimens to for return to Johnson. This Dyar felt was a breach of the policy that specimens could be kept in exchange for identifications.

After Johnson wrote that he had no authority from the Boston Society to allow specimens to be kept by the USNM, Dyar replied:

> I will not name the specimens for Dr. Aldrich after the discourtesy he practiced.... [Retaining specimens is] the only way we get any return for the expenditure of time and brain power; but the good Doctor did not play the game according to Hoyle, and hence the rumpus. It is far better to always send things officially through the regular channels (Museum or Bureau) and then the chance of personal friction is lessened.[33]

Other accounts suggest that it was not only Dyar and Schaus with anger problems among the scientists. Indeed, F. Martin Brown (1903–1993), who visited the USNM in the 1920s, described his fellow lepidopterists as a "disagreeable bunch," who kept themselves in their closed-up offices.[34]

In spite of these dark episodes, however, there was some levity. For example, at some point between 1927 and 1928 Louise M. Russell (1905–2009), working for the USDA as a scientific aide for Harold Morrison, recalled some merriment among her cohorts at Dyar's expense. They giggled at the observation that Dyar's handlebar mustache made him look quite similar to the mosquito larvae he studied: complete with mouth bristles.[35]

## CHAPTER 21

## Mosquitoes and Dyar's Pursuit of Reinstatement

*I have hopes of being back at the Museum by Feb. 1 and I will work up all the accumulated mosquitoes promptly. I want to stay here till I have finished with this doctor, as he seems to be doing me good. It will not be a long delay, so the more mosquitoes there are on hand the better, as I am very fond of them.*[1]

H.G. Dyar to L.O. Howard from South Norwalk, CT, January 14, 1918

On Christmas Eve 1918, Dyar began his second ledger titled "Matters referred from Bur Ent: Incoming mosquitoes and lepidopterans." It followed a similar one written while a government worker, documenting that he did the same valuable work as he had done previously, but without pay, important to justify rehire, and continued until October 1925, well before his reinstatement was seriously considered.[2]

Dyar's first attempt to receive financial support following his dismissal came before he left on a four-month trip to British Columbia, the Yukon, and Alaska in April 1919. He wrote Howard to request that the Bureau of Entomology (USDA) cover his field expenses related to collecting "injurious Lepidoptera on spruce" in the Northwest. Dyar wanted "a lump sum to avoid" the scrutiny of the Secretary of Agriculture because he "was dropped from the rolls" and the complexities of officially approved travel in Canada.[3]

Howard rejected the request because Dyar's dismissal made it politically impossible for the Bureau to financially support him. Insulted, Dyar accused Howard of "suffering from a case of unnecessary fright" about the Secretary of Agriculture, found his letter "both illiberal and unnecessary," and threatened to discontinue his mosquito "observations . . . [for USDA] gratis."[4] Exasperated by more requests from Dyar about the trip, Howard replied: "I have your two notes . . . I did not enter into detailed explanation of my statement . . . because I did not think it would be

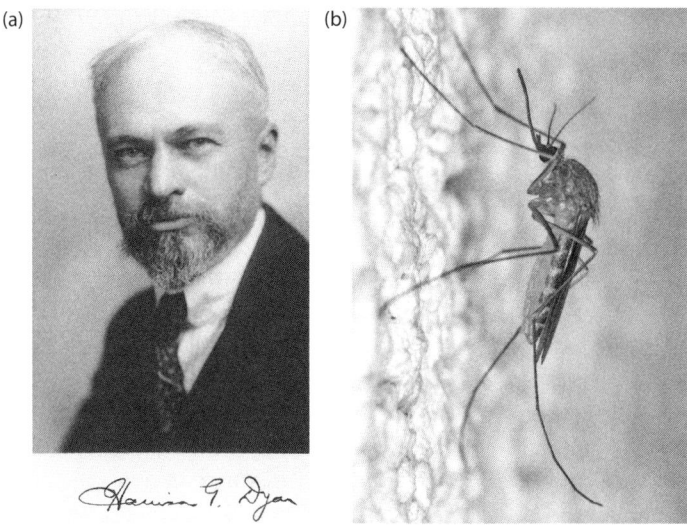

**Figure 21.1** Portrait of H.G. Dyar (c. 1920) (*a*); specimen of *Culex pipiens* mosquito (*b*).

necessary. The Secretary would certainly not approve any such recommendation from this office. I hope you will have a successful trip."[5]

The ever-determined Dyar replied:

> I ... still think, that the Secretary should be satisfied with dismissing a man without blacklisting him in addition. A collaborator as I understand it, is anyone competent outside of the Department, and I think my rights as an outsider should be respected ... What is particularly annoying ... is that my dismissal is due to the fact that the intelligent (?) detectives in the Secretary's office ... could not judge evidence, and accepted as gospel truth what everybody else in the case ... knew to be lies. I am not done with the matter and expect to return to the charge next fall.[6]

Dyar remained dedicated to his work, as was evidenced by his concern over the accuracy of mosquito identifications. He thus instructed Howard to hold specimens for his return, writing: "hope not to find material from my pet correspondents (like Mrs. Bonne-Webster) again in the hands of Mr. [Charles Tull] Greene, as was the case last year. Not that Greene isn't a good fellow, for he is—but you know." Howard clarified that Greene was "to consult Doctor Aldrich and both of them to see you ... and have a perfect understanding with you about just exactly what mosquito work should be done in your absence."[7]

∞

In November 1920 Dyar perceived a breach in protocol in his role as a mosquito authority at the museum when he overheard August Busck—whom Dyar was angry with for allowing Barnes to take moth specimens from the USNM—dictating a letter to the stenographer in response to an inquiry from Dr. E. Martini, "a tall lanky German" mosquito worker who visited the USNM in 1914.[8] This Busck incident powerfully demonstrated to Dyar his need to defend his "turf" as the lead worker on mosquitoes.

An incredulous Dyar wrote Howard that Martini

> ought to have written to you or me, as Knab's successor; but I hear [about it] from conversation floating over the intervening furniture . . . and of course [Busck] did not communicate with me [about Martini] . . . I know he used the word Culicidae [mosquitoes], . . . but what Busck does not know about the Culicidae would fill several volumes . . . I suggest that you write to Dr. Martini and say we wish [to] have a copy of his paper.[9]

Howard soon revealed to Dyar that "the letter was written by Dr. Martini to me, and not to Mr. Busck. . . . and Busck translated it" and had sent "Martini . . . two volumes of the [Carnegie] Monograph . . . [,] told him of your great activity, and urged an exchange." Howard also mentioned that "Busck may have written him a personal letter, since he knew him and was with him and Knab a good deal when he was here."[10] To this Dyar answered: "My conclusions were drawn from scraps of conversation and admittedly indefinite. This is of no consequence; but I am relieved to find that Dr. Martini addressed you."[11] While his misunderstanding about the Martini letter may have been of "no consequence," Dyar's poor rapport with many around him did not help his cause.

In May 1921 Dyar wrote Howard of his latest rebuff by the Secretary of Agriculture in his attempts at being rehired: "I do not care, except that I do not like that record which he has against my name. I suppose it really amounts to little and I might as well forget it." This was exacerbated by an edict to take away authority from those holding honorary custodial positions by William de Chastignier Ravenel, administrative assistant to the Smithsonian Secretary—in charge of the National Museum.[12]

In the early 1920s Dyar's expertise on mosquitoes was clearly his best hope of compensation for scientific work. In the fall of 1921 Howard encouraged him to write "The Mosquitoes of the United States." Because he had recently published "The Mosquitoes of Canada," Dyar wrote: "To include all our species would be simply to reprint the Canadian fauna proper . . . The reason I did the Canadian paper at all was because I had explored Canada pretty thoroughly personally, which is not the case (as yet!) with the southern United States." However, Dyar decided to go forward with this project and would include records from his recent western trips. The tome was published the following year.[13]

## DR. CLARA S. LUDLOW AND THE ARMY MEDICAL MUSEUM

The idea of Dyar seeking a position from the U.S. Army to identify mosquitoes had its origins in 1918 when he suggested to Howard to advocate for an anatomist position held by an officer at the Army Medical Museum.

**Figure 21.2** Dr. Clara S. Ludlow (*a*); view of the Army Medical Museum that Dyar had from his front porch at 804 B Street SW (*b*).

Three years later Dyar suggested to Howard the need for mosquito work at the National Museum, which could be funded through the Army Medical Corps. Howard concurred that "Army work on mosquitoes ... [should] surely be done in the National Museum, and I am only waiting for an opportunity to push the matter. Some of the best men in the Medical Corps quite agree with me."[14]

Clara Southmayd Ludlow (1852–1924) (Figure 21.2a) worked on mosquito taxonomy at the Army Medical Museum (Figure 21.2b) from 1916 until her death at age seventy-one. Dyar and Knab wrote critical reviews of her mosquito work and Ludlow's responses are legendary. Dyar and Knab accused Ludlow of mixing a mosquito from the Philippines with one from Pennsylvania, which she named *Anopheles perplexens* in a 1908 article in *The Canadian Entomologist*. Because Ludlow stored specimens inside pillboxes with the data written on the covers, they believed she mixed covers, giving the specimens incorrect locality information. In her response in *The Canadian Entomologist* she gave an account of bringing specimens to Dyar and Knab at the National Museum and admitted it was impossible to "keep track of all the lids" on the pillboxes, but destroyed the specimens that could be confused, rather than risking inaccurate label information associated with them. She also recounted Dyar's advice to publish the species in question, *perplexens*, as new to science. Curiously, Ludlow did state an error was possible and the mosquito may have been from the Philippines, but not by switching the lids. Today we find, the species is indeed native to Pennsylvania.[15]

Ludlow, referred to Dyar and Knab's rebuttal as "Dr. Dyar's recent article." Because it was a coauthored piece with Knab, this infuriated Dyar, who wrote: "[T]he responsibility [of the article] is jointly shared, and Dr. Ludlow should have addressed us both." Ludlow, in turn, wrote in her response: "The senior author is responsible for what appears under his name, whether he wrote it or not." She then referred only to Dr. Dyar in her next point:

> It would probably have simplified matters if it had occurred to me to state definitely that, while the specimens are shipped to me in boxes, the collection has

never been kept in them ... moreover, my method of keeping my collection, even if it were "unfortunate" as Dr. Dyar persistently insinuates, is strictly a personal matter, and lies quite outside Dr. Dyar's province.[16]

Dyar's suggestion of becoming Dr. Ludlow's co-worker in October 1918 was surprising, given their combative relationship. His proposal, however, was to completely avoid direct contact with her. The plan was to earn a salary at the National Museum and for her to remain in the Army Museum across the National Mall. To receive his help with identification, she was to submit "specimens by mail or messenger when desired, and not be always obliged to come personally, which is often a hardship to her in her bad state of health." The last part of Dyar's work plan was that if Ludlow desired "to consult our collections" she would rely on his "personal judgment and the making of numerous [dissection] slides."

Given his poor finances it's not surprising Dyar would propose such a plan. Apparently to show good faith, he indicated that if he returned to the Bureau of Entomology, he would keep an honorary post with the Army.[17] The idea may not have been seriously considered but if it had been, Dyar's critiques of Ludlow could not have been helpful.

By March 1920 Dyar's relations with Ludlow were back to normal, but he was beginning to show concern about the effect of his critiques on his livelihood, writing to Howard: "The methods of our esteemed [?] contemporary, Dr. C.S. Ludlow come to notice unpleasantly again in an article just appearing in Psyche, where she describes two new species from Panama. ... I am writing these to you to vent my feelings, lest I publish something not according to Hoyle and offend the ARMY."[18] Dyar and Ludlow went on to coauthor three papers during 1921 and 1922. Perhaps he became more cooperative to gain a position.

After Ludlow died in 1924, Dyar got to look at her specimens. His letters to Howard about doing so show little respect for her. The first was about type specimens Dyar couldn't find, but on Friday the thirteenth of February 1925, he wrote Howard: "Bad day for the stock market, but good for skeets. Major Callender sent word that due to the scare caused by you butting in, the Ludlow executors 'coughed up' all the rest of the mosquito material, and it is now in the Army Medical Museum."[19]

Howard replied, "I am extremely glad that so much of the Ludlow material has been exhumed."[20] Continuing to look through Ludlow's specimens, Dyar wrote: "How she must be turning in her grave to think of ME pawing over her treasures and noting all her mistakes and blunders. She was 'mad' at me for the last year of her life, you know. Well, anyway we'll get the Philippine species straightened out. ... I shall go over every specimen finally, if my life is long enough."[21]

Before Ludlow's death, during his trip to Yosemite in June 1924, Dyar finally gained some employment from the Army, receiving pay from a position in the Sanitary Corps. In a jocular mood, buoyed by his new position, something he dearly needed with his increasingly sagging finances, Dyar wrote Howard,

I just got word that I am a captain in the Army. I salute the Bureau and its amiable Chief on behalf of the Service

Sincerely yours,

Capt. Harrison G. Dyar,
Sn.–O.R.C., A.M., Ph.D.[22]

## TRAVELS FOR MOSQUITOES: 1916–1924

Dyar's fieldwork from 1916 to 1924 was mostly mosquito related, primarily from the Tahoe area and the Pacific Northwest, including Alaska and the Yukon. He also traveled to Colorado and Panama in 1923. Although Wellesca's name does not appear on the collecting records or in the literature for these trips, she was traveling with Dyar on most if not all of them.[23]

While waiting for his divorce trial in Reno and on the government rolls, Dyar wrote Howard about his collecting efforts in San Diego that he would: "continue to look for *Aedes calapus* according to your wishes. I plan to stay here continuously 5 or 6 weeks or until the collecting season begins in the mountains." According to a report in *Insecutor*, Dyar found this species, a yellow fever mosquito of concern, and others. His general observation was that there were unfavorable conditions in San Diego for mosquitoes because the "main residential part of the city lies high and is continually swept by cold sea breezes."[24]

Dyar went on to Yosemite in May 1916 and then to the Tahoe region, where he remained from late that month until July. While on the California side of Lake Tahoe at the Fallen Leaf Lodge, Dyar had field assistance from its proprietor William W. Price. Normally discovery of a new mosquito variety at your lodge would not be particularly good for the tourist business. However, Price and his wife were Stanford graduates interested in developing nature study walks and nature games for their guests. It was said that Price "had the ability to infect others with his enthusiasm and love of nature" and had built "an alter to the spirit of the open air." He collected mosquito larvae in a marsh at Tallac for his visiting expert and to show his appreciation, Dyar named a variety of *Aedes palustris*, var. *pricei*, for him.[25] At the end of the western trip Dyar made a short collecting stop at Roseville, California in August.[26]

In May 1917 Dyar collected in Washington State at Hoquiam and Bellingham and returned to Longmire Springs in June, where he had visited several years earlier with Caudell. That month he also visited such places as Glacier, Seattle, Ashford, and Lake Cushman. Through mid July his collecting continued in Sandpoint, Idaho; Montana, including Missoula and nine other sites; and Spokane, Washington.[27]

Dyar began his fourth visit to Canada in late June 1918 at White River, Nipigan, Dryden, and Kenora, Ontario, and Winnipeg Beach, Manitoba in early July. He also spent a month in Alberta, including collecting on Lake Louise, and on his return through Saskatchewan at Saskatoon and Prince Albert during August.[28]

 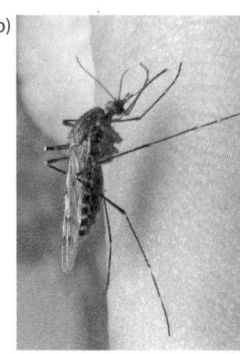

**Figure 21.3** Dyar went through Ketchikan, Alaska (*a*) in 1919; one of the mosquitoes Dyar encountered during the Yukon portion of the trip was *Culiseta alaskaensis* (Ludlow)(*b*).

In 1919 Dyar began a collecting trip to Alaska and the Yukon after five weeks in Alberta and British Columbia in late spring. Departing from Prince Rupert, British Columbia in June, the couple took the inside passage ferry route of the Alaska panhandle to Ketchikan (Figure 21.3a), Cape Fanshaw, Juneau, and Skagway. At the end of June the couple began three weeks of collecting in the Yukon at White Horse and went as far north as Dawson, where they found *Culiseta alaskaensis* (Figure 21.3b). Dyar described swarms of male mosquitoes (always nonbiting) at White Horse while "walking toward the bluff [on] any still evening," seeing species "first . . . in the tops of the small pines," followed by another "over open spaces between pines and willows, then, on reaching the high spruce trees" and at Dawson found a swarm "over willow bushes on the hillside any time after 4 p.m. that the sun went behind a cloud."[29]

During his time in the Yukon he collected at Ft. Selkirk ("Selkirk"), the Tahkini ("Tahkeena") River and Carcross until July 21, in British Columbia at Lake Atlin and Bennett, and back through the inside passage in Alaska. After returning to Prince Rupert, they collected in British Columbia along the train route at Hazelton and Prince George, and through Saskatchewan and Manitoba through mid-September.[30] In total, the 1919 trip yielded over 11,000 specimens including many new records and new species of mosquitoes, black flies, and midges.[31]

Dyar made two separate western trips for mosquitoes in 1920, and although it's unclear why, after the first month-long trip to Garfield and Salt Lake, Utah in April and May, he returned to Watkins Glen, New York, where the Allen children generally stayed for the summer. Dyar stopped in Washington, D.C. in early June before returning to Nevada, where he collected at Carlin and Tahoe City, California. He remained in the Tahoe area until July 1 and then visited Washington State, British Columbia, and Oregon.[32] In August Dyar collected at San Juan Capistrano, California, and at Kerrville and San Benito, Dyar's only collecting records for Texas.[33]

In late July during his 1920 travels, Dyar collected Lepidoptera "on the crags about Crater Lake." His account of the tiny Muir's crambid moth (*Gyros muirii*) reads

"The little red things were ... flying in sunshine and alighted for a few seconds on rocks or sand. With much trouble (and many failures) I secured a small series and took pleasure in anticipation in filling this gap in the Museum collection."[34]

Dyar collected at Truckee, California, in 1921, four days before his reported marriage to Wellesca in Reno on April 26, and as "newlyweds" near Lake Tahoe and along the little Truckee River, which flows through Reno, until May 11. After arriving in San Francisco in mid-May he headed up the coast to Orr's Hot Springs in Mendocino County, California, and to Medford, Oregon, for three weeks of mosquito collecting including visits to Prospect and Engineer's Camp, above Whiskey Creek. Through mid-July, Dyar collected a month in Glacier National Park, Montana followed by Grand Forks, North Dakota, and East Grand Forks and Moorhead, Minnesota, in late July.[35] Dyar spent time in August at Rehoboth Beach, Delaware, where he collected the eastern saltmarsh mosquito *Aedes sollicitans*.[36]

Dyar and his new wife's 1922 summer trip to the Midwest involved health-related matters. He developed goiter, so following Winnipeg, Manitoba, and Thief River Falls and Warroad, Minnesota, the couple made a detour to the Mayo Clinic in Rochester, Minnesota. Not favoring surgery, the Dyars went to Long Beach, California, for alternative medicine,[37] stopping in Yellowstone National Park, where they collected from late June through mid-July, shelving a visit to the Colorado Rockies because the mosquitoes had an early flight season. Housed by the Yellowstone superintendent in exchange for providing mosquito-control advice, the Dyars collected their specimens along with black flies at sites including Old Faithful, Yellowstone Canyon, and Mammoth Hot Springs. For pleasure he spent an "evening at the 'bear dump' watching the antics of these animals," whereas mosquito-wise Dyar, the entomologist, noted that the hot springs "were not a factor in [mosquito] breeding conditions."[38]

❦

Fieldwork in 1923 more than compensated for negligible collecting the previous year. The Dyars took two back-to-back mosquito trips, the first being on the Western slope of Colorado around Grand Lake (May 23–June 26), followed by mosquito collecting in Panama (Figure 21.4a)—Dyar's only truly tropical collecting trip—where the couple met up with Raymond Shannon (Figure 21.4b), for whom a trail on Barro Colorado Island, Panama, is named.[39]

When Dyar left notice about being away he told Howard that "Schaus [could] handle the butterflies and the mosquitoes will remain fatherless."[40] To this the Bureau Chief responded:

> I am sorry that from April 1st to September 1st the mosquitoes of most of the United States will be in confusion. No one—not even the mosquitoes themselves will know their names; and this may react disastrously on the public. However, I am glad that you are going to have a good trip, and hope that you will take the best of care of yourself while in Panama.[41]

**Figure 21.4** Passport photograph of Dr. and Mrs. [Wellesca] Dyar used for their trip to Panama (1923) (*a*); Raymond C. Shannon (1925) (*b*), who met the Dyars for a mosquito survey of the Panama Canal Zone in July 1923.

Dyar wrote Shannon from Colorado that they expected to board the *SS Carrillo* on July 4, 1923, and reach Cristobal, Panama, a week later. His plans on arrival were to,

> establish headquarters in Ancon at once. You can get a good laboratory there with assistance of [James] Zetek. We need plenty of table room, a lot of wide-mouthed jars and some pans. Of course all the isolation tubes we can get, and I will get some in Washington if time and possibility permit.... Mrs. Dyar [will] watch the cultures while you and I get new material.

Dyar disagreed with Shannon's suggestion to "build a house on an island in Gatun Lake" because he believed they would "waste much time there waiting for the material to develop," while limiting collecting "to the local fauna" of species from "a drier faunal region." His preference was to have a central headquarters and investigate more the wet region by "making trips to the Atlantic watershed bringing home the material in bulk" where the specimens could "be studied at leisure" rather than "be isolated" in one location.[42]

Dyar and Shannon collected on Barro Colorado Island (July 17–21) and revisited the site at least five times in August, often on the same day as other locations. Among 16 other collecting sites in late July to August 20 were Frijoles, Gamboa, Mount Hope, Gatun, Close's, Cano Saddle, and Fort San Lorenzo. The Dyars departed from Cristobal, Canal Zone, on the *SS Toloa*, arriving in New York City on August 26.[43] Following the Panama trip, Dyar and Shannon noted the 20,000 specimens of larval mosquitoes at the USNM represented 87.6% of the species known to occur in the United States and 69.4% of those from Panama. Dyar went on to publish his "Mosquitoes of Panama" in 1925.[44]

In 1924 Dyar went to Bremerton and Hoodsport in Washington State in early May, followed by his final trip to Yosemite in June and July. The only specimens reported from the Pacific Northwest were one species of midge (*Chaoborus trivittatus*), larger than many mosquitoes but considered to be a mosquito at the time (midges are usually tiny, nonbiting flies), and in Yosemite they found *Aedes ventrovittis* mosquitoes, although one-time resident *Aedes increpitus* had been exterminated by oiling.[45]

After his return to Washington, D.C., in July, Dyar remained at home. The fall was to bring up some bizarre revelations, to say the least, about his activities in the District of Columbia.

CHAPTER 22

⚭

# Dyar and His Tunnels: Dupont Circle & B Street

*We have kept the secret passage and the trap doors and false backs to the book-cases all just as they were and as soon as little John and little Edith are big enough they shall play secret castle in them, just as their mother did. Are we afraid of the ghosts? Oh, dear, no. If the spirits of dear old grandpa or kind old Uncle Jim should meet any of us even in the dark old secret passage, we'd be so glad to see them and thank them for coming clear from Heaven just because they remember that they loved us, wouldn't we, children? And we'll always take care of the place where their poor old bodies rest, that they don't need any more.*[1]

"The Ghost Makers" by Harrison G. Dyar, December 19, 1913 (unpublished)

Although there are many bizarre tales in the Dyar legend, none compare with those surrounding his underground labyrinths. On September 26, 1924, the District of Columbia was abuzz with news of mysterious tunnels discovered when an alley collapsed after a truck backed into it near the Pelham Courts apartments in Dupont Circle (Plate 16a, see Figure 22.1b for map location). Horse stables owned by Edward B. McLean, publisher of the *Washington Post*, abutted the alley. The front-page of the *Post* proclaimed (Plate 14a):

> Old Tunnel Here Believed To Have Been Used by Teuton War Spies and Bootleggers!
> Descending through the opening ... the searchers stood in a passageway high enough and broad enough for a man to walk with ease ... One of the most astounding features.... was the.... carefully, even artistically formed [walls] of white enameled brick, pronounced valuable by builders.... On the ceiling were pasted numerous copies of German newspapers dated during the summer of 1917 and in 1918.... Electric wires with bulbs attached which were found in some sections of the tunnel would point to its modern construction, but these devices are believed to have been introduced by recent occupants, probably bootleggers....

(184)

**Figure 22.1** Dupont Circle home in 1990s (*a*). Investigators in the Twenty-first Street tunnel following the 1924 collapse (*b*). Map (*c*) where the collapses and other tunnel-related events occurred in relation to the Dyar residence "A, B": "C" is perhaps where Dyar began digging in 1906; "D" is the alleyway where 1924 collapse occurred; "E" is the site of the 1917 collapse off P Street prior to construction of Pelham Court apartments; and "F" is the site of the 1958 collapse; a satellite view (*d*) of the residences "A, B" and neighborhood.

One of the most authentic stories concerning the tunnels which could be gathered traces their origin to the Civil War, when they may have been used for the protection of Confederate soldiers hiding in Washington. Subsequently, it is stated, the labyrinth came into the possession of Dr. Otto von Golph, a German chemist, who embellished and furnished them. The chemist then is believed to have employed the underground chambers as laboratories for his scientific experiments![2]

The story was featured in newspapers throughout the United States (Figure 22.1c photograph was in many of the articles). The public flocked to see the mysterious underground maze and reporters searched for explanations. What they didn't expect was that they were excavated by a slight and "a bit stooped . . . government scientist." Dyar admitted to his strange pastime (Plate 14b and frontispiece), recalling the excavation began in 1905 or 1906:

> Mrs. [Zella] Dyar wanted a bed of hollyhocks, and a little garden for vegetables. . . . Well, I volunteered to dig the garden. When I was down perhaps six or seven feet,

surrounded only by the damp brown walls of old Mother Earth, I was seized with an undeniable fancy to keep on going.

Dyar kept going, excavating a labyrinth of underground tunnels, six feet high, and extending some two hundred feet, with walls of brick and plaster. There were four entrances, one with a little house concealing it. There was also a place below most of the passageway in which he "constructed a ladder of pipes set in concrete."[3]

A *Washington Evening Star* interview with Dyar dispelled images of a Frankensteinian lair. Complimented for his construction skills, the naturalist smiled, "Well, I never was taught engineering or how to lay bricks. I've spent my life chasing bugs." The German language newspapers, he explained, dated from before the war. However, the dates of 1917 or 1918 on the newspapers indicated they would have been placed in the tunnels after he had moved from the neighborhood.[4]

The tunnels were such a sensation they even made it into the political cartoon arena for "Fighting Bob" La Follette's presidential campaign (Plate 14d). The cartoon by W.A. Rogers had La Follette with a pickax digging his way to the White House, c/o William Jennings Bryan's brother Charles's "Secret Tunnel."[5] There were also various humorous references, such as George Rothwell Brown's Washington "Post-Scripts column": "*'What man dost thou dig it for? For no man, sir.'* A Post reporter does some digging on his own account, and unearths Harrison G. Dyar, of the Smithsonian, as the deep student who dug the famous P street tunnels: but here's no mystery about them—the professor is an entomologist, and digging tunnels is his bug."[6]

Dyar continued work on the tunnels until 1915. He had purchased additional land behind the Twenty-first Street home in Dupont Circle in 1906, purportedly for an alley and more room for a garden,[7] but it provided a larger area for tunneling. Dyar said he had avoided digging beneath the alley to the west in the direction of the stable (lot 40), but curiously, this is near where the collapse occurred in 1924 (Plate 16a, Figure 22.1b).

The two large lots Dyar purchased on P Street for the garden was the future site of Pelham Courts apartments and a place to where the tunnels reached hundreds of feet away (see vacant lot location of "E" and surrounding lots in Figure 22.1b). These lots abutted the 1510 and 1512 Twenty-first Street properties and the stable. The lots on P Street would come to serve another function besides providing space for the labyrinths beneath: "The dirt from the tunnels in the rear of my house on Twenty-first street I dumped on the vacant lot where Pelham Courts now stand [on P Street]." Soil from the tunnel was also used in the gardens.[8]

The tunnel in Dupont Circle was an underground playground for his children and others in the neighborhood. "My son, Otis Dyar, ... used to play in the tunnels ... other boys played in the tunnels and while they didn't annoy me they were somewhat of a nuisance to some of the neighbors," Dyar said. Dorothy had a Halloween party in the tunnels "with all her friends, decorated with jack-o-lanterns." He also recalled the police visited several times, either in response to complaints about children or simply to investigate.[9]

Although the tunnels provided Dyar with a diversion and a place for children to play, excavating beneath the earth had its dangers, as he narrowly escaped being buried alive during a tunnel collapse while digging. After this calamity he was to have "come into the house shaken, covered with dirt—head to foot, but he went back into the tunnels."[10]

## THE B STREET TUNNEL

A few days after Dyar admitted to digging the Twenty-first Street tunnel in September 1924, he confessed to creating a multileveled subterranean maze, still under construction, at his residence at 804 B Street. The tunnel was under his property to the east and between his house and Saint Mark's Lutheran Church (see Figures 22.2a, b, 23.2b; Plate 15). It was "[a]lmost a quarter of a mile . . . lined with concrete. The deepest passage . . . extends 32 feet down. . . . The catacomb [is] constructed in three levels, with steps and iron pipe ladders leading between different tiers."[11] The tunnels could be reached both from the southeast corner of the basement of the house

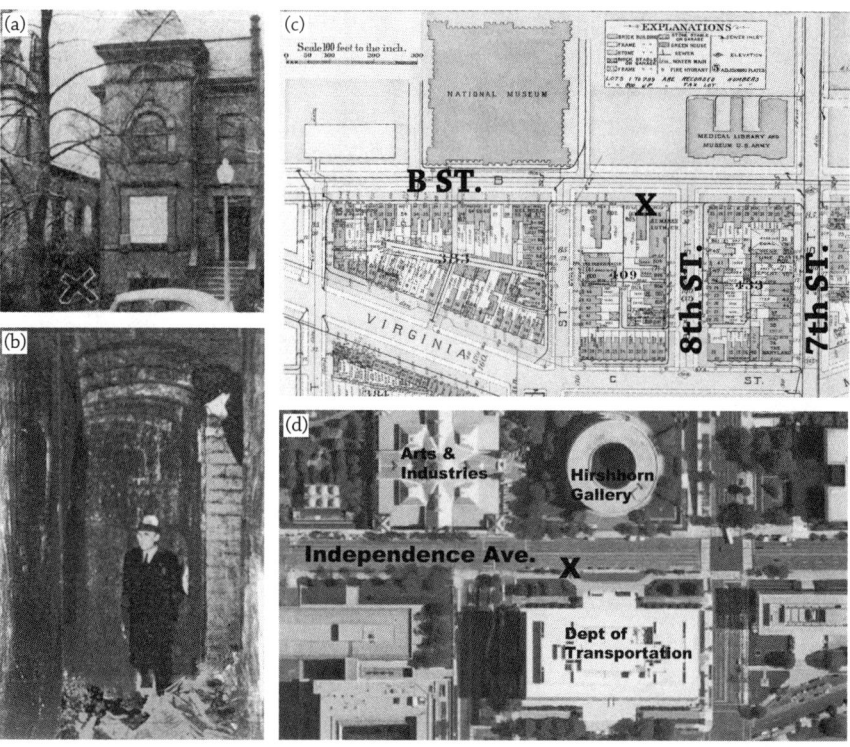

**Figure 22.2** Dyar home at 804 B Street and Saint Marks Church (*left*) (X is the location above the entrance to the tunnel on the east side of the house) (*a*); inspector standing in upper level of the tunnel (8 ft. tall) as another person (*right*) looks from basement entrance to the tunnel (1942) (*b*); map of the B Street home (X at front near tunnel entrance) and neighborhood (1920s) (*c*); satellite view of location of where the razed B Street home, church, and Eighth Street would be today in relation to Independence Avenue and the Department of Transportation (*d*).

or the front yard, and the dimensions were six feet tall, or as tall as eight feet near the basement entrance,[12] by three-and-half to four feet wide with dirt floors. "The dirt from these tunnels is piled in a mound in my back yard," Dyar said.[13]

Dyar decided to build this set of tunnels to make "an underground entrance to his furnace cellar"[14] (see Plate 15), his son Wallace made clear, although he, and Dyar as well, failed to see that digging tunnels took much more effort than carrying ash:

> He was having his heart problems, and he was finding it very difficult to bring the ashes up from the furnace in the basement.... So he conceived a thought of starting a tunnel right beside the furnace, which he did.... He could carry ashes out. Instead of having twenty-five or thirty steps, he only had about twelve or fourteen. So that started it.[15]

However, what began as a simple idea, such as digging a garden bed or ash removal, became something entirely different. To enter the tunnels through the basement, there was a small gothic-shaped door in the southeast corner: On the arch was inscribed the Latin motto *Facilis decensus averni* from Virgil's *Aeneid*, "The way down to Hell is easy," and the next sentence, though it was not on the tunnel entrance, was *sed revocare gradum superasque evadere ad auras, hoc opus, hic labor est*: "The path out of Hell is hard."[16]

The B Street tunnel was illuminated electrically and on the walls and archways were sculpted heads of persons and animals. Wallace recalled his father using a long extension cord to provide light as he dug and suggested the sculptures were placed on the walls, not carved into them. On the first level, "the tunnel ran straight toward B Street for the length of the house. The floor of the 'first level' is of hard packed clay, the walls concrete and the ceiling brick." On one archway near the cellar entrance was an inscription "H.G. DYAR FEB 14 1923" (his fifty-seventh birthday). Dyar would inscribe dates of completion on various parts of the elaborate tunnel but had not completed the project by September 1924.[17]

To get bricks for the archways and ceiling, the family would go with Dyar (who the children called "Dak") by car—mother would drive. Wallace recalled one trip to Alexandria, Virginia, observing that it

> was kind of funny, because [Dyar] went in there with authority, and he went into the brickyard, and he started picking around the brick. This laborer or watchman ... came around and wanted to know what his business was. So he said he was trying to pick out some brick.... He was very particular about the kind of brick he put into his tunnels. I think this was some particular archway he wanted to push some color into it.... Well, the man gave him a real hard time. He said, "You get out of here.... You've got to have permission to come in here." Well, they had quite an argument.... Dak lost the argument. The man was bigger and heavier than he was.[18]

As Wallace recalled, "We loved the tunnels. We'd play down there. And they were quite safe. We couldn't get the cats to go down there. They were scared to death of it ... I had to carry them." Wallace also remembered telling "ghost stories down there sometimes, because it was perfectly dark, like the Luray Caverns or any cave. You could scare the wits out of yourself if you wanted to, because there was no daylight there."[19]

## THE 1917 COLLAPSE OF THE DUPONT CIRCLE TUNNEL

Given the hullabaloo about the tunnel's origin when the story broke in 1924, it is astonishing that Dyar had already been identified as the digger and architect after a similar collapse seven years earlier. A plausible explanation for the memory lapse was that tunnel was quietly reported on May 19, 1917, during the month following U.S. entry into the World War. The account in the *Washington Times* referred to Dyar as "an entomologist, employed by the Department of Agriculture," reporting:

> A mysterious subterranean tunnel, ten feet below the surface of the ground, with secret entrances, and dark, ghostly looking caches, or vaults, suggestive of ancient castles in the day of knighthood, connecting two houses in the most fashionable of Washington with a small, obscure brick structure to the rear, has been unearthed by a steam shovel crew in excavating for an apartment to be located on the north side of P street between Twenty-first and Twenty-second streets.[20]

The collapse occurred soon after Dyar sold his land to developer Harry Wardman, who began digging the foundation for of Pelham Court Apartments, "when the huge steel jaws [of a large steam shovel] brought up a mouthful of bricks ... the surprised operator saw through the hole a cross section of a walled up tunnel. The remaining bricks were pulled out, revealing a secret entrance, which had been subsequently walled up." The hole led "back into the bowels of the earth" and could be easily seen from the street by "hundreds of passersby" (see "E" in Figure 22.1b for approximate location of collapse). Further north, another branch of the tunnel went between the home on Twenty-first Street and a small house in the back of the lot, which Dyar had built.[21]

Dyar's "immediate neighbors" said "he worked early and late for several years on this tunnel." They also noted, "Frequently the owner of the property would be seen shoveling dirt and spreading it over the surface of the ground, and from time to time loads of brick and other material would be dumped there ... but [only] Mr. Dyar was ever seen to work there."

## WHY DYAR DUG THE TUNNELS

Dyar dug seemingly because when he started something, he would relentlessly "keep on going." The compulsion required him to continue digging and to construct a lengthy, brick-lined underground marvel is exemplified in his scientific research and in other areas of his life. Certainly his personal fortune enabled him to spend large blocks of time on his predilections, but compulsiveness appears to have been ever present throughout his life.

Dyar's compulsiveness to "keep on going" shows again in his travels. For instance, the extent of his sojourn in 1891–1892 took close correspondents like Lintner, and perhaps even Zella, by surprise.[22] Rather than simply a whim to stay on the road, Dyar's success at finding new caterpillars fed his desire to seek more.

Examples in his entomology research show how Dyar went from unmatched productivity in one area and then shifted into another. For example, after years of rearing and measuring Lepidoptera caterpillars, Dyar described the life histories of all sawfly caterpillars he could find over several years. What is particularly amazing was he continued productive work on Lepidoptera at the same time. Toward the end of his sawfly rearing he took the plunge into classification of mosquitoes based largely on the larval stages for the remainder of his life. It's not hard to imagine a seminal moment that moved Dyar passionately in a new direction. His father was said to be similar: "whatever the subject ... [he] never left what he entered upon until he believed he had mastered it to this fullest ability ... arranged his results, and passed on to another problem or topic."[23]

Dyar's compulsiveness is also evidenced in the volume of short stories he wrote, whose origins appear to have been from hearing ghost stories or attending séances as a youth. A collection of anecdotes written to his sister grew into hundreds of stories, including a couple of novellas. As Dyar said in 1924 about digging tunnels: "Ah well! Great oaks from little acorns grow."[24]

Why Dyar dug long horizontal tunnels in Dupont Circle and deep, multi-level tunnels on B Street has perhaps a simple explanation. The scientist had about 50 feet until he hit bedrock and a long stretch of land at Dupont Circle, whereas at B Street he had about 200 feet to hit rock and a relatively small lot (see Map 6).

Dyar justified his tunneling to himself by saying he did it for exercise and "attributed his good health" to the hobby. His numerous health issues in the teens and 1920s (tachycardia, influenza, chronic insomnia, nerves, etc.) evidenced that he seems to have been deceiving himself. The physical exertion had other benefits, bringing him a sense of calm from his stormy intellect. Perhaps Dyar became the "meek, mild-mannered scientist" portrayed in the 1924 newspaper exposé of Dyar, but only after hours of digging. Given his critical nature, one can only wonder what Dyar would have been like without exercise,[25] but at some point the tachycardia brought his prolific digging to a close, after dodging the issue of needing permits.[26]

For Harrison Dyar, being in the tunnels may have represented connections to his childhood, parents (including his father's invention for underground dirt removal),[27] other relatives, and perhaps even their spirits (see quote at beginning of chapter).[28] Dyar's inscription of Virgil's quote from the *Aeneid* on the archway leading to the tunnels has religious undertones ("The path to Hell [sin] is easy"), especially when the adjoining line "The path out of Hell is hard" is added to the quote.[29] On the surface Dyar looked for vindication from charges against him, but underneath he sought redemption.

CHAPTER 23

# Personal Life and Bahá'ís in the 1920s

After living with Dyar's first family for decades, Harriet M. Peabody, now nearly eighty moved back East from California in the early 1920s. At the time she wrote Dyar's brother-in-law Dr. Knopf in New York seeking medical advice. In regard to Zella and the children, Mrs. Peabody relayed the latest news that the family in Berkeley had reached a relative calm after the turbulent and lengthy divorce. She also took a jab at Dyar and Wellesca, writing: "I do not know anything about the family on B. St. as I have not yet seen any of their church people and I do not care to see or hear much about them."[1]

In her salutation to Knopf Mrs. Peabody's letter shows a bond of friendship, quite opposite of what she had with Dyar, referring to Knopf as "My Dearest and only Son."[2] Their affinity formed through years she and Knopf had been warriors aiding tuberculosis victims and his giving medical advice to the family. The following year she died of stomach cancer in the town of Mexico in Oxford County, Maine, on December 9, 1922, home of her brother John M. Holland. Perhaps Zella was present at her mother's bedside.[3]

After Zella settled in Berkeley, the Dyar children became active in the First Unitarian Church. Dorothy (Figure 23.1a) became a secretary on the board of trustees. Otis took the secretary position at the Channing Club—the young people's society—and later became president. In March 1918 Otis played a policeman, who arrested the hero, in the club and the church's Women's Alliance production of a farce called "Luck?"[4] Dorothy made the honors list at Berkeley as a junior and registered to vote along with her grandmother, as women had suffrage in California by 1911.[5]

In her church-related work, Dorothy wrote an article called "Youth's Challenge to the Church" for her father's *Reality* magazine while studying religion at Columbia University in 1922 and 1923. Zella moved to New York to live with Dorothy, apparently staying at 16 West Ninety-fifth Street, Perle and Dr. Knopf's residence. Dorothy received her B.D. degree and was ordained in November 1927. Soon after, she began her career as a minister in Seattle at University Unitarian Church and served as Dean of the Tuckerman School in Boston.[6]

**Figure 23.1** Children from Dyar's first family: Dorothy Dyar in college yearbook (1918) (*a*); Otis Dyar, as a boxer, while at University of California, Berkeley (*b*); Otis with his Model T in southern California (*c*).

Otis graduated from Berkeley in engineering, while also distinguishing himself as a boxer (Figure 23.1b, c). In May 1920 there were indications he forgave his father for leaving the family in this letter written by Wallace McLean to Dyar: "I had heard only in an indefinite way about Otis' attitude, and it [is] extremely gratifying to know that some one has brought about a change of feeling." Otis later became an electrician, collector of arrowheads and various other objects, an outdoor enthusiast, and was known as "Uncle Oat." Otis and his wife Catherine were unable to have children and settled in Arcadia, California, where her family revered him.[7]

Some work-related trips Dyar had attempted with his grown children, albeit no-goes, show that he cared for them. In the summer of 1921 Wellesca wrote, "Dr. Dyar would go with a party to Alaska who were to be sent by the Carnegi [sic] fund, or he was thinking of taking his young lady daughter or young man son. Thus I would not be his entomological assistant this year."[8] Dyar also pushed to have Otis accompany him and Shannon to Colombia in 1924 for a mosquito-related expedition, as he wrote Howard: "I have a son, aged 24, whose work is bad for his health. If you can use him in the proposed mosquito exploration, keep me advised. I cannot furnish him gratis, but on same conditions as you pay others. He could join your Dept. if desired."[9]

## LIFE ON B STREET

*We played hide and seek around the dinosaurs. . . . Just loved it. We explored the old National Museum and the new National Museum. I guess that's about one of the best educations I ever got. We'd go up there through the Indian exhibits and the exhibits of the whales and fish.*[10]

Wallace J. Dyar, on living across the street from
the Smithsonian and the National Mall

The dynamics of Dyar's second family in the early 1920s was different from that of his first. The children believed they had a father none could remember. Wellesca had attempted to convince them Wilfred Allen was their father: They suspected otherwise, having known Dyar since birth, either as a visitor to the house or a

**Figure 23.2** Children from Dyar's second family (*left to right*): Wallace Joshan, Harrison Golshan, and Roshan (*a*); National Mall (1933–1937), across from their home on B Street SW (*right*), Saint Marks Lutheran Church (behind the fire wagon; note Eighth Street SW to left of Saint Marks) (*b*); Wallace J. Dyar's high school yearbook photo (1930) (*c*).

boarder. Their name for him, Dak, may have come from their combining the names they had heard: Dad, Dick, or Doc. Wellesca had affectionately called Dyar by the nickname of Dickie, while he referred to her as Kit or Kittie. Earlier, he referred to her as "mama" around the children, so that they understood Wellesca was their mother because they had been with Mrs. Leech for such a lengthy period and would call Wellesca "Mrs. Allen."[11]

In April 1922 the District Supreme Court granted Dyar's petition to adopt Wellesca's children, the three Allen boys (ages thirteen, ten, and eight), and legally change their names to Roshan Allen Dyar, Harrison Golshan Dyar, and Wallace Joshan Dyar (Figure 23.2a). The notice in the paper read: "Mrs. Dyar's former husband, according to the petition, left this city to Philadelphia and has not been heard from since. Mrs. Dyar was married the second time at Reno, Nev., in 1921."[12]

When Dyar was home on B Street he worked in the first floor den, where he played the grand piano. The kids, Wallace recalled, "weren't to go back in and disturb him. So he conducted most of his work and his hobbies there, with his stamp collecting and his musical and his entomological work."

Wellesca, Wallace continued, "used to proofread with [Dyar]. I don't remember whether she did the reading or he did, but . . . did a lot of editing there [in the den]." They took the "portable typewriter . . . to all the expeditions. . . . He'd make all his notes . . . while he was out, and they'd . . . permanently put it in writing. Then they would edit that for the various publications that he had. . . . The one I remember because of the working was the "secular city metros [probably *Insecutor Inscitiae Menstruus*]."[13]

The children

> had one of these player pianos that you pushed the pedals. You could either play it or push the pedals, and music comes out. . . . We loved to play that one. But [Dyar] played very well. That was another one of his ways of getting away from concentration required in his work. It was sort of relaxation for him. . . . He played almost

exclusively by heart. Of course, he could read most of the music that he played. I don't think he really wanted to work at it. I think he had learned the piano probably in his college days or at some point earlier in life.[14]

A friend gave this account of Dyar as musician at his grand piano:

When Aseyeh took me into the spacious living room . . . we found the Doctor seated at the piano, lost in the strains of one of Rachmaninoff's Symphonies . . . Dyar, the scientist-was now- Harrison Dyar the musician. After an hour spent in a delightful musical treat he became a boy, wrestling and playing with his three fine sons.[15]

The second floor consisted of a parlor and several bedrooms for the children and boarders, including Bahá'ís. Looking down from the top of the stairs you could see the first-floor hallway.

In the late teens and early 1920s, the family had a maid and cook, and Dyar no longer drove but had a chauffeur for his old Pierce Arrow automobile. Later, without the chauffeur, Wellesca drove until she had a serious accident with a streetcar in Georgetown in 1920 or 1921. She was severely injured and hospitalized for some time.[16]

According to her son Wallace Dyar, Wellesca shopped for food at the Central Market, across the street from the new National Museum on the north side of the National Mall and at the wharf on the Potomac, bringing

home fresh vegetables and meat . . . pumpernickel bread, and we just loved it. . . . Mother, being part German, had learned a number of recipes, and even though we had a cook. . . . [She] would supervise the cooking and put her two cents' worth in. Later on, she did all the cooking, but we had a cook many, many years. In fact, that was the last employee we ever let go . . . I remember we went down to the wharf quite often. In fact, we used to love to watch the ships come in with their catch. Yeah, we had fresh fish.[17]

When the children arrived at the table for dinner, Dyar would make sure they had washed their hands: making them throw down their hands, he would say: "Peedle, peedle, peedle, listen to the germs. Go wash your hands." At the table, Dyar would often tell funny stories and quote Shakespeare. The meat carver for the family, he did not like vegetables very much, telling Wellesca "Take that rabbit food away from me. I just want some good dark meat." Wellesca, in contrast, was a vegetarian. Dyar had a taste for shellfish and also was said to be "an experimenter with food," which included raw oysters, a treat that he introduced to Wallace (see Figure 23.2c). Tales of one of his true experiments—as passed down among generations of entomologists—recall that during a meal with Ray Shannon, Dyar put castor oil on his ice cream![18]

The family was fond of pets, especially cats.[19] The three boys, first as Allens, then as Dyars, would go across the street to the National Mall (Figure 23.2b). According to Wallace, the explorations of its museums were "the best education I ever got."

They went over by themselves, even though Dyar "worked there." In fact, Wallace "never saw his office" and "never knew where his office was. But . . . enjoyed going to the museum."[20]

Dyar and Wellesca left their young children with Mary Robbins Meade (1861–1940s) at Watkins Glen, New York, when they traveled. When the children were older they spent more time at Skyland with their uncle George Freeman Pollock. Meade, a writer and spiritualist/theosophist in the "new thought movement," lived in Watkins Glen for most of her life and was the great aunt of writer and composer Paul Bowles, who named a character Nelson Dyar.[21]

In July 1916 Meade wrote Wellesca in Gold Lake Camp, California, from Watkins Glen, New York, to discuss the possibility of the Allen boys staying with her during the winter: "I cannot think of anything that would be better for the children, nor for you, then to have them here in these lovely surroundings with the sweet spiritual life, and the external comforts and care that I could give them."[22] Beginning in 1919, the boys would spend most of the next four or five summers in Watkins Glen, while Dyar and Wellesca were out West.

In 1919, when Dyar and Wellesca were in Alaska, Meade wrote Wellesca about the boys:

> Your letter to Joshan and the package both to be opened July 22nd [his sixth birthday] are here. . . . I have not done anything about teaching the children this summer since school closed. It has been all that I could do to concentrate upon their health to make them grow strong, but Roshan will take his first music lesson Saturday, and I am slowly trying to teach Bun [Golshan] to write. . . . The children all send love. You will enjoy Nippy's [Joshan] letter I am sure, it is so original.[23]

Dyar and Wellesca's last will and testament, written in 1922, shows the trust and faith they held for Meade, séances and all. In the event of their untimely deaths the boys were placed under her care.[24]

## DYAR'S HETERODOX BAHÁ'Í BELIEFS

*Dear Mr. Young: I make it a point never to get angry with children. But, listen, you should not have read our book. This is solid food for the mentally advanced. One does not give meat to sucklings. We purposely did not send it to any of the "believers," knowing their weakness; but now they got hold of it on the sly and their stomachs hurt.*[25]

H.G. Dyar to Edward Young, about the book "Short talks . . ."
by Aseyeh P.A. Dyar and Harrison G. Dyar, March 20, 1923

Dyar's reputation for being "a thoroughgoing materialist"[26] overlooks the emphasis on reincarnation and spirits in his stories and discounts his becoming involved in the growing Bahá'í community. Certainly Dyar's Bahá'í association had much to do with Wellesca, who loved Bahá'í teachings, which included those of prophets of the major religions.

In the Western world 'Abdu'l Bahá chose to emphasize "principle above personality, teaching a scientific religion of universal ideas" rather than the "personal devotion to himself and his father" followed by more orthodox observers in the East. Furthermore, in North America there was a divide between those wanting a more exclusive religion versus a more inclusive one: The latter is referred to as the "Bahá'í Movement" and appealed to Dyar because he opposed Bahá'í as an organized religion.[27] Indeed, Dyar was primarily motivated to get involved with the Bahá'ís to push forward his own philosophical views to the further advancement of humanity, and he cherry picked from 'Abdu'l Bahá's "Twelve Basic Bahá'í Principles." He accepted only those he could wholly embrace from his scientific background, numbers 2 and 5: "Independent investigation of truth" and "Religion must be in accord with science and reason." Others he seemed less inclined to favor: 1) The oneness of mankind; 3) The foundation of all religions is one; 4) Religion must be the cause of unity; 6) Equality between men and women; 7) Prejudice of all kinds must be forgotten; 8) Universal peace; 9) Universal education; 10) Solution of the economic problem; 11) An international auxiliary language; 12) An international tribunal.[28]

Dyar's belief in eugenics and his antisuffragette stance, in contrast to Zella, were clearly opposed to Bahá'í teaching, as well as his acceptance of separate Bahá'í meetings for blacks and whites in Washington. The nation's capital was a segregated city, and his racial views were not unusual. Clearly Dyar was against interracial marriages. Once again, this was a norm in American society, but from a eugenic perspective, less in the mainstream.[29] However, these positions on race placed Dyar and Wellesca outside of views 'Abdu'l Bahá promoted during his U.S. visit in 1912.

Wellesca was a more devoted follower of 'Abdu'l Bahá than Dyar, and although we know little of Harriet M. Peabody and Zella Dyar's involvement among the Washington Bahá'ís, it is safe to say the Dyar scandal hurt her standing in the community. Wellesca's use of 804 B Street as "Bahá'í House" or "Bahá'í Hospice" could also have fueled controversy, as Dyar's money was used to build it while he was married to Zella. According to Wellesca, there were Washington Bahá'ís who shunned her by not going to a meeting "on account of gossip implicating my character." The term "shunned" appears in various letters and relates to "troubles" she had caused the Dyars and in other areas such as being actively opposed to 'Abdu'l Bahá's wishes for racially integrated meetings.[30]

In response to a letter by a Washington Bahá'í community member in 1918 or 1919 regarding her role in Dyar's divorce, Wellesca wrote:

> I had nothing to do with the trouble in the Dyar family, which began, I am informed, in 1902. Mrs. Dyar's statements in her divorce bill were wholly false and unfounded. She voluntarily withdrew the bill two years ago, which is the same effect in law as if it had never been filed. A person is entitled to be considered guiltless until a verdict has been rendered against him and the guilt established. I am therefore wholly innocent of any of these charges. . . . Let me ask you this: Are we inviting souls to the cause [Baha'i] on account of our personally impeccable characters, or on account of the worth of this Glorious Message?[31]

In 1919 a disheartened Wellesca, as Aseyeh, wrote to 'Abdu'l Bahá about her troubles with her fellow Bahá'ís. The following translated reply from her religious leader encouraged the promotion of unity in Washington and her to spread the word "in all regions":

> Three of your letters have been successively received and perused. 'Abdu'l Bahá is so immersed in merciful emotions and in turning to the Kingdom of God that his highest wish, therefore, is that the friends of God shall forget and lay aside every concern, engage in the propagation of divine teachings, act in accordance with the exhortations and the admonitions of His Holiness Baha'o'llah; guide the heedless, illumine the hearts and attract the souls through the fragrances of God. . . .
>
> I have named thee 'Aseyeh'; reflect thou what love I have entertained for thee! The preservation of this station is all-important; it is a supreme favor.[32]

Several months after hearing from 'Abdu'l Bahá, Aseyeh spoke on about the Bahá'í cause with missionary zeal during the summer trip to the Pacific Northwest with Dyar in 1919. Notices of a lecture to be given in Hazelton, British Columbia, in September were handwritten on letter-size paper, some in Dyar's distinct calligraphy. These had the following on them: "Free Lecture, The Great Peace, 8:15 o'clock at Saint Andrews Hall, Speaker Mrs. Aseyeh Allen of Washington, DC. Mrs. Allen has been lecturing in Yukon Territory, Alaska and in Prince Rupert Everybody Welcome."[33]

Dyar also received a tablet from 'Abdu'l Bahá in support of Aseyeh Allen, referring to her difficulties with the Bahá'í community in the District of Columbia.[34]

Soon after the leader's death in November 1921,[35] Aseyeh gave a talk titled "The Teaching of Abdul Baha" at a Washington, D.C. Unitarian House on Columbia Road under the auspices of the Liberal Religious Union. The talk was preceded by other lectures between July and September, first in Seattle to a class of psychology students at Bahá'í Rizwan titled "The Higher Thought in Religion" and "Troubled Waters." The remaining lectures, including those titled "Religious Sects" and "The New Baha'ism," were in Brownsville, Texas (to an audience of two ministers and a layman), at a Bahá'í meeting in Washington, D.C., and the next year, in Bahá'í Assembly Halls in Seattle and Spokane, Washington; Glacier Park, Montana; and to the faculty of the State Agriculture College at Fargo, North Dakota. The last talk in the series, "The New Baha'ism" was written in March 1922, but not presented.[36]

After her last lecture Mrs. Dyar placed her print order at a cost of $100 for the compilation of the lectures named: "The Teaching of Abdul Baha. Short Talks on the Practical Application of the Bahá'í Revelation" (Figure 23.3a). She coauthored them with Dyar and published them the same year.[37] It is unknown whether these talks were intended for publication prior to the death of the Bahá'í leader. However, as the notice of the lecture in the newspaper showed, his death could be viewed as an opportunity for the Dyars to push for more influence shaping Bahá'í to be more independent from the rules of religion by pushing "Independent investigation of truth" and "Religion must be in accord with science and reason."

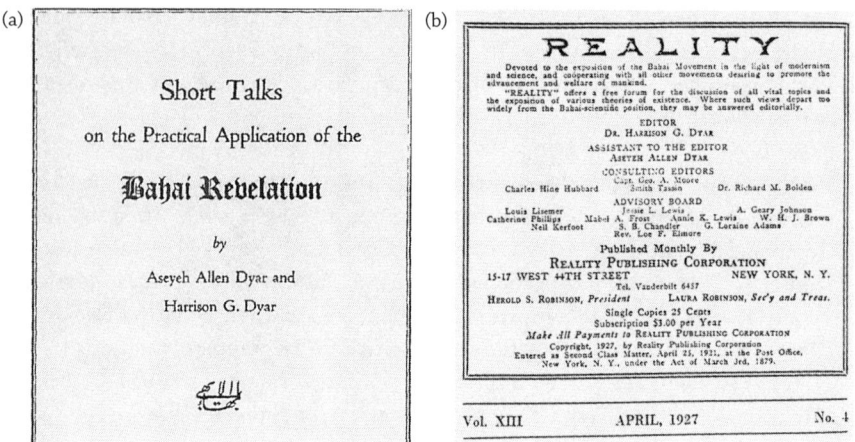

**Figure 23.3** Book of Wellesca's (Aseyeh's) lectures on Bahá'í matters (*a*); cover page of *Reality* edited by H.G. Dyar and assisted by his wife (*b*).

*Reality Magazine* (Figure 23.3b), financed and edited by Eugene and Wandeyne Deuth, began publication in 1919 to represent a group of New York Bahá'ís with an inclusive attitude. Featured in the front of each issue of the magazine were the "Twelve Basic Bahá'í Principles."

The earliest known connection between Dyar and *Reality* was a form letter sent to subscribers of the magazine written about the New York Printers' Strike on November 14, 1919. The letter suggested there was some support for the strike from its readership by stating, "When the Baha'i principles of life will have become a fact and are conscientiously acted upon by all classes, such draw-backs to human progress will not occur. STRIKES WILL BE ABOLISHED."[38] By 1922 there were those—such as editorial board member Horace Holley—who took a contrarian view of *Reality* as an official Bahá'í publication: He wrote an editorial in *Reality* reflecting this view. Herold Sweetser Robinson, brought in by the Deuths to help market *Reality*, purchased a controlling interest in the magazine and installed Mary Hanford Ford as editor and Dyar as supervising editor above her. However, she held the position for only six months after disagreeing with Robinson and Dyar's vision of representing "the independent investigation of truth, rather than furthering the 'Baha'i cause.'" Dyar then became editor (Figure 23.3b), who, as a scientist with his own nonreligious agenda, fit more closely with Robinson's view that the Bahá'í movement was free from formal Bahá'í rules.[39]

Dyar likely bankrolled *Reality* and kept his editorial position for the rest of his life—"persecuting monthly ignorance" as he did with *Insecutor*. Dyar's first editorials and articles communicated his views more than once per issue. In July 1923, for example, he published "Mr. Bryan and the Monkeys," arguing against William Jennings Bryan, who would, in 1925, during the Scopes Monkey trial, advise the West Virginia legislature to pass a prohibition on mentioning that humans are historically related to monkeys or other animals. Dyar argued against Bryan, writing there was nothing about humans being "unrelated" to monkeys in the Bible and that God had written the story of evolution, now visible through scientific evidence:

> He wrote it in the rocks: He wrote it in the skeletons of the animals and in man: He wrote it in the tissues of the plants, in every cell, fibre, nerve and organ of every living thing; He wrote it in the stars; there is no place in this vast universe where the Hand of God has not written.[40]

In June 1924, Dyar wrote of eugenics, as he had done before, in an editorial titled "The Old New England Stock" in which he complained about the "dwindling" birthrate of ancestors of the old English colonists. Proud of his derivation from New England, Dyar advocated the production of large families by such "superior stock in order to accelerate the evolution of the human race." The article was followed by Dyar's biography, replete with his ancestry, borrowed from the genealogy he made decades earlier.[41]

*Reality* served both as a mouthpiece for Dyar's nonreligious agenda among Bahá'ís and an outlet for his short stories; some, such as "The Man Who Did Return" and "An Anecdote of the Law," were actually written the previous decade. In "Anecdote," the protagonist Phil Utterly had been "courting two girls to the onest, the rascal . . . and the thought of possessing one made him sad at the thought o' losing the other." In the end, Utterly manages to marry his two sweethearts at the same time. Perhaps this simultaneous marriage was Dyar's way around 'Abdu'l Bahá's stance of allowing bigamy between a husband and two wives as long as permission was granted by the first: a continuation of previous Bahá'í doctrine.[42] In Utterly's marriages there was no first wife! Given the difficulties Dyar and Wellesca faced in reconnecting with the Bahá'ís in Washington it seems surprising he would even mention bigamy. Furthermore, it had been illegal in the United States since Lincoln, though possible for Bahá'ís in the Middle East.

"A Dream of the Pope," "As a Man Thinketh," and "The Heaven That Bends Above Us" were more direct representations of Dyar's own definition of what it meant to be Bahá'í. "Pope" is about a Dyar dream in which, as editor of *Reality*, he meets the Catholic leader at a "new Vatican" in Denver, Colorado. His pope believes in a church that welcomes all faiths—evidently one of his desires for the Bahá'í movement. "The Sixth Race" is an odd amalgamation of eugenic philosophy, science fiction, and reincarnation, featuring highly evolved souls that are "mostly in the Astral." The astral entities consider founding a new race on earth, given "problems" of the "intermingling" of the "five races." Because there were not enough bodies from the cultivated classes of Europeans to be inhabited, given their lower reproductive rate, the "Great Adept" chose among the spirits who would inhabit bodies to "First reproduce abundantly . . . and second . . . advocate the betterment of the race by purposeful selection."[43]

From the time that Dyar became *Reality* editor the more liberal "Bahá'í Movement" perspective generated opposition from the developing Bahá'í religious organization in the United States. Dyar sought to quell the conflict by presenting a proposal of cooperation with the Bahá'í National Spiritual Assembly in August 1924.

Shoghí Effendí Rabbání, the new leader of the faith, advised he would require the magazine to avoid the "unfriendly" or "harmful" matters, presumably meaning such topics as eugenics. As Dyar continued the status quo, the 1925 Assembly recommended that Bahá'ís cease any involvement in the magazine.[44] Although continuing as editor until his death, Dyar's writing and storytelling were less evident toward the end of the magazine's run (1927–1928).[45]

CHAPTER 24

# Unity in the USNM Lepidoptera Section and Acquiring the Barnes Collection

In the 1920s there were signs of internal cooperation among USNM lepidopterists, in part because they shared friendly relations with outside collaborators, although the decade was replete with feuds among these personalities. Perhaps there was no better example of common ground than their shared desire to obtain the Barnes Collection (Figure 24.1a) and to hire its curator.

Foster Hendrickson Benjamin (1895–1936) (Figure 24.2a), the last of three Barnes curators, played the most important role in unifying the USNM lepidopterists. This came in part because William Barnes needed good relations with them, especially his nemesis Dyar, for the purchase of his collection to go through. In March 1922 Barnes wrote Dyar that Benjamin, a recent Cornell graduate, was coming to Decatur in a few weeks "to take charge of my collection . . . He wants to work over the *Euxoas* [moths] in connection with his Doctor's degree."[1] Benjamin, a specialist on owlet moths, succeeded Arthur Ward Lindsey, who had come to the Barnes Collection after J.H. McDunnough. Lindsey was a specialist on the skippers (a group of day-flying Lepidoptera related to butterflies).

Benjamin's career and fate were closely tied to the Barnes Collection. Although it was clearly in his own self-interest in addition to that of Barnes to promote tranquility with the USNM crowd, he put special effort into attempting to keep peaceful coexistence between old warriors Dyar and Barnes. One difficulty was Busck, Schaus, and Heinrich, all of whom had friendly relations with Barnes, whereas Dyar did not. Another hurdle Benjamin faced was that Dyar and Schaus frequently disagreed on identifications each was trying to verify for Barnes: in other words, the proper genus, species, or even family of a particular moth. Indeed, one legend says that no sooner would Dyar or Schaus put a name on a specimen in the USNM collection than the other would change it.

Strained relations between Dyar and Schaus in the 1920s made it difficult for them to have open discussions. This was compounded when Barnes asked Benjamin

**Figure 24.1** William Barnes in his enormous collection (*a*); letter from his nephew suggesting some of his behind-the-scenes efforts to influence the sale to the U.S. government (*b*).

to consult Schaus rather than Dyar for his taxonomic opinion. From Dyar's prospective, this was a breach in the lepidopterists' hierarchy at the USNM, and Dyar complained to Benjamin in August 1925. He wrote, "It is an annoyance to me to see your articles one after another in which my species are referred to the synonymy on the authority of Dr. Schaus. How much longer will this discourtesy continue? You are aware that I am in charge of the Lepidoptera, and why should you go over my head to have Dr. Schaus pronounce on my species?"[2] Once again Dyar stressed that Schaus was his assistant and that Busck and Heinrich had no official status over the National Collection.

In an uncomfortable situation, Benjamin responded to Dyar, pledging, among other things, to notify him in advance of publishing synonyms of species named by Dyar and giving him the option of publishing them first. He also expressed a desire to send paratypes of species described from the Barnes Collection to the USNM. Benjamin explained he corresponded with Schaus because Dyar refused to do the same for Barnes.[3]

Dyar then told Benjamin that there was nothing personal between them, as he liked Benjamin's ideas on nomenclature and appreciation of Grote. He pointed out that the synonym didn't bother him, and he was free to publish about the error based on "his personal observations on specimens in the National collection." Rather, it was conferring with Schaus that got his dander up. He also mentioned his past problems with Barnes had been over for some time: "I am not in this matter nursing any grudges against Dr. Barnes. I closed that incident, with success as I thought years ago. If Dr. Barnes is still sore, it is a one-sided feud."[4]

In November 1925, Dyar was not the only "persecutor of ignorance." It was Benjamin rather than Schaus who uncovered more synonyms in Dyar's work. The young curator wrote Dyar about the mistake rather than having him find out in print. In a further attempt at full disclosure, Benjamin offered to publish the manuscript in Dyar's journal *Insecutor*. The ultimate olive branch, however, was Benjamin's decision to name a limacodid for Dyar from Texas, which so happened to be in one of Dyar's favorite genera from his youth. As Benjamin wrote, "I have taken the liberty of naming the new *Phobetron dyari* in acknowledgement of your kindness to Dr. Barnes." In response Dyar wrote "Many thanks for giving my name to the new *Phobetron*." Never afraid to give his scientific opinion, Dyar tweaked: "I fancy that the genus *Euphobetron* would be more appropriate."[5]

Interestingly enough by mid December 1925 Dyar had come to accept Schaus as the Honorary Assistant Curator, at least in some respects. He wrote Benjamin about Schaus's title along the lines explained to him by Howard: "to appear to better advantage to foreign correspondents."[6]

Keeping peace between Dyar and Barnes, as always, was clearly a challenge, though now it seemed to fall as much on Benjamin as Howard. During the month prior to Benjamin naming the limacodid for Dyar, Barnes made more "thefts" from the USNM collection. Predictably this brought fireworks. In his complaint to Howard, Dyar wrote:

> Barnes was here, and I gave him access to the collection. In fact he came and began work before I knew he was here. Mrs. Locke informed me of the fact, and I went in and helped him find some things. Now after he has gone, I find that he helped himself to material out of the collection. . . . I asked Heinrich about it, but he said he knew nothing about the matter, and referred me to Busck. . . . Busck . . . has been guilty before and not properly penitent, and is in any case an overbearing presumptuous bully, who deserves kicking into his proper place on general principles. *Lest affairs become too monotonous, I feel obliged to start something occasionally* [italics mine].[7]

Howard wrote a terse reply on the same day: "If at any time I find things getting monotonous, I will surely call on you for excitement!"[8]

Seeing no humor in Howard's retort, Dyar emphasized his point by writing about a drawer in the moth collection as "filled with traces of Dr. Barnes' work [theft]."[9] After receiving a letter from Dyar about the latest fiasco Barnes admitted taking specimens once again and wrote the following: "In each and every case there were a number of specimens in the National museum collection so that even in the extreme event that some railroad accident [on his way home] or other might destroy them, there would be no great harm done." He then wrote about his "generous gifts of specimens to the USNM" that from his perspective more than made up for his pilfering from the national collection.[10]

◊

Clearly Benjamin was liked by all of the USNM lepidopterists. They wanted to bring the talented curator to the USNM along with the Barnes Collection. Benjamin, however, worried about his prospects of landing a job at the National Museum. Busck thus observed in a letter to him, "I know of your anxiety for the future and hasten to send you just a personal line . . . that things shall come out OK. Do not worry. All your friends here—inc. Dyar, Schaus & Rohwer are backing you for a position here—as are of course Heinrich and yours truly."[11]

Sievert A. Rohwer (1887–1951), mentioned in the letter, was an influential supporter of the Barnes Collection purchase. Rohwer, who towered in height over most of his fellow entomologists at the USNM (Plate 13), was an expert on sawflies and directed the taxonomy section of the Bureau formed in 1925, and later the Plant Quarantine Division. His support along with the lepidopterists was crucial in order to gain Benjamin among the ranks. At the close of the letter Busck provided more reassurance with the following: "There will be an exam shortly ... for a position here [for a salary of] $2,400–3,000 right in the Lep[idoptera] room with us all—and you will ... [be notified] about this. ... You can assuredly pass [the] exam easily and then [sic] is no reason to have misgivings about the matter."[12]

Dyar and Benjamin agreed on more than just moving the Barnes Collection to the USNM. Both disparaged entomologists who practiced the "art" of creating names for forms or subspecies simply to have one's name attached after the Latin binomial. This became known as the *mihi* (of me) *itch*, a term Benjamin attributed to Dyar in a letter to a colleague Ottolengui: "A lot of darned idiots will say 'This form I have is just as worthy of a name as that other form which is kept on the Lists, so I guess I'll name it, put my name on the Lists, and get some types in my collection.' This is simply another phase of that mental disease which our friend Dyar amply described when he called it the 'me-he itch.'" Although Benjamin attributed the *mihi itch* to Dyar, it predated him.[13]

One particular example of internal cooperation among the USNM lepidopterists was a working relationship that Dyar had developed with Carl Heinrich, with whom Dyar wrote a joint paper on an economically important group of snout moths.

Other lepidopterists from outside the walls of the museum besides Benjamin worked in concert with the quartet of USNM lepidopterists. These collaborations, in turn, helped them develop more of a unity of purpose. Among the best known of these workers were Frank Morton Jones and W.T.M. Forbes (Figures 24.2b, c). Jones was interested in case-making caterpillars such as the clothes moths (Tineidae) and the case bearers (Psychidae). On his trips from his home in Wilmington, Delaware,

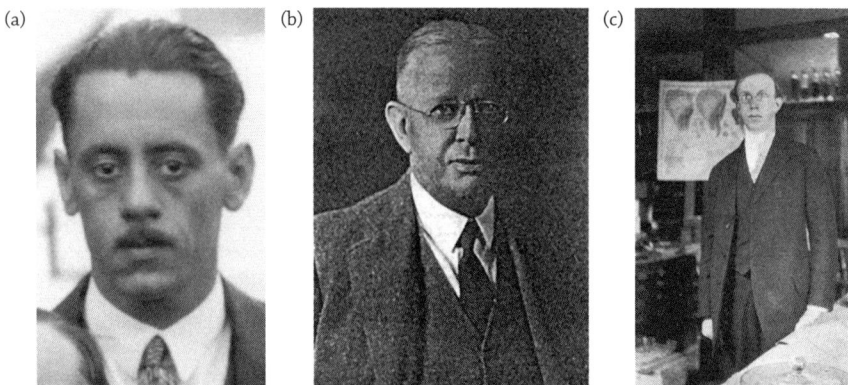

**Figure 24.2** Lepidopterists who collaborated with USNM moth workers in the 1920s: Foster Benjamin (*a*); Frank Morton Jones (1869–1962) (*b*); W.T.M. Forbes (*c*).

Jones would arrange to meet up with Dyar, Busck, and Heinrich. Dyar wrote Jones in November 1922, showing a rather unusual openness to collaborate: "Are you intending to work up the genitalia of the Psychidae? I have made mounts of several of the species. . . . I may write something on the subject myself. Or perhaps we could write a joint paper. What do you think?" In a follow-up letter Dyar wrote "Busck and Heinrich will certainly be interested in the other case-bearers which are not Psychids. I shall anticipate your visit with pleasure."[14]

There were clearly some positive developments in Dyar's relations with his compatriots in the mid to late 1920s. At times they were even on the same side of the battlefield.

## WILLIAM BARNES'S CONCERN OVER DYAR'S INFLUENCE ON THE PURCHASE OF HIS COLLECTION

In 1922 William Barnes attempted to garner an appropriation from the U.S. Congress for government purchase and move the collection to the USNM—by the 1910s and 1920s the largest in North America not housed in a larger natural history museum—growing from 200,000 to a half million specimens. It appears the surgeon had always intended for it to be sold to a large U.S. museum and his earliest known interest in the USNM as repository was in the 1910s. The justification was that the specimens would help the government protect U.S. agriculture. The proceeds, besides compensating Barnes, were to cover construction of a new Macon County Hospital in Decatur, Illinois.[15]

In his efforts to have his collection purchased, Barnes hired his nephew William C. Barnes, a private secretary to congressman Allen Francis Moore, who represented Barnes's legislative district in Illinois (including Decatur and Springfield). Dr. Barnes apparently had his nephew look into removing damaging files at the museum and other places that could affect the sale of his collection (Figure 24.1b).[16] These files appear to have been related to a fair purchase price estimated to be $100,000 by Dyar, Schaus, and Busck rather than the $300,000 price tag, or the correspondence relating to Barnes's habit of taking moth specimens important to Dyar from the USNM collection.

> *Dear Uncle Will: . . . It has also been possible to secure from the files of the Department of Agriculture certain material which I didn't want him to see when he investigated. This was . . . through a friendly young lady in the Department. I am going to try and clean out some other file cases at the Smithsonian and the Library Committee of the House . . . send me a little money . . . wining and dining and getting things out of files and bills assigned where the rules say they shouldn't be a trifle of a drag on my rather depleted bank account . . . nor do I care to put some of the things on my expense account. You understand I presume.*[17]

Barnes wrote Dyar in March 1922 that there had been a promise from the U.S. Congress "to get the bill for my collection out of a committee and put it on the Calendar" for a vote.[18] Several months earlier Rep. A.F. Moore contacted Smithsonian Secretary C.D. Walcott with a request to provide him with an estimate of the Collection's value,[19] although it appears that J.M. Aldrich, USNM Associate Curator of Entomology, had the information from the museum's lepidopterists.[20]

Growing impatient with the passing months, Barnes, a Democrat, wrote Moore, a Republican, asking, "How is the bug bill coming on? . . . I got the impression that he [Barnes's nephew] seemed to think if you pressed the bill very much it might result in losing you some votes in this district. I am very sure if the bill went through, we could off-set this with the Democratic votes we could get for you. . . . It probably would have been better to have someone introduce it from another district entirely."[21]

Besides the problem of gaining approval from Congress, Howard noted in a letter to Dyar in 1924 that the "present secretary" (of Agriculture) would not approve the purchase "at the price mentioned [$300,000]."[22] In spite of concerns expressed by Barnes to Benjamin that Dyar would somehow block the purchase of the Barnes Collection, Dyar was always a staunch advocate, as seen in 1925 when Dyar wrote Barnes that the collection would "be put to the utmost . . . use and service not only to the government departments but to [U.S.] collectors" and it "would be a calamity" if it went to Europe. He also mentioned the disadvantage of having "no congressmen to urge the matter on [in the District of Columbia]." Dyar concluded that he didn't want to undermine efforts to purchase the collection by Smithsonian Secretary Walcott before congress, as he was concerned such efforts would "appear officious or meddling and do more harm than good."[23]

Because it was a tough sell to the U.S. Congress for the USDA to purchase the collection, Jeremiah O'Conner of the Smithsonian was charged with finding private funds. He wrote Barnes a discouraging letter on December 20, 1927. Dr. Walcott, the Smithsonian Secretary who had been behind the purchase, had died and his acting replacements Charles Greeley Abbott and Alexander Wetmore were "opposed to the campaign for the collection." This was because it could interfere with their efforts to raise ten million for the Smithsonian Endowment Fund. In spite of this conflict, O'Conner sent "out an appeal to some of the most hopeful prospects" including John D. Rockefeller, Jr. Unfortunately for Barnes, after investigating the purchase, Rockefeller declined to contribute.[24]

Because private money was not found, there was need for a congressional appropriation. However, it took until 1930 for one to pass in the U.S. Congress and the actual purchase of the collection did not occur until 1931, after the deaths of both Dyar and Barnes. The final transaction was at a greatly reduced sum of $50,000. For 473,000 specimens this rounds out to a dime per specimen—"a bargain for the largest Lepidoptera collection ever purchased by the government."[25]

## CHAPTER 25

☙

# Final Travels, Projects, and Days as Custodian of Lepidoptera

*You know there is nothing I like less than killing mosquitoes. The mosquitoes are the subject of my interest, not their absence, and so I feel that all anti-mosquito work is directly detrimental to my special interest. I love to see the mosquitoes in vast swarms, and if I had my way, all the oil would be poured over the human exterminators. However, as a matter of what is expected of an individual by the community at large, I give the benefit of my experience when asked for.[1]*

H.G. Dyar to L.O. Howard, January 18, 1925

Dyar and Wellesca made it to Glacier National Park for three months, beginning in April 1926: It was his last western collecting trip. Unable to perform the rigors of fieldwork, Dyar was assisted by ranger Paul Schoenberger of the National Park Service, who went to the head of Swiftcurrent Pass and to the foot of the Grinnell Glacier (Figure 25.1) to find mosquitoes for him. He found *Aedes nearcticus*, a species Dyar had named earlier, "breeding in large numbers." Among the few insects the doctor reported from the trip other than mosquitoes was a new species of drain fly in late July.[2]

The couple encountered forest fires in Montana, and Wallace McLean, replying to the news from Wellesca, observed, "If I were [there] you would find me . . . far more distant than six miles from a forest fire, so you had better . . . get well beyond the range."[3]

In mosquito-related correspondence, Dyar told Howard "I wish Mexico belonged to the United States. We'd show these incompetent Spaniards how things are done; and also get some collecting done in that unknown country." Dyar was also in contact with Altus Lacy Quaintance, the acting Bureau Chief in Howard's absence to complain about entomology in Mexico, sarcastically noting he received nonbiting crane and drain flies that were supposed to be mosquito samples from a professor in Mexico.[4]

**Figure 25.1** Grinnell Glacier at Glacier National Park, Montana, a site Dyar could only see from a distance in 1926, as he required others to go collect mosquitoes for him at these elevations.

Although there were no reports of Lepidoptera on the trip, Dyar took the opportunity to name a new genus and species of owlet moth sent to him by J.H. Horton, *Hortonius euenemus*. He let the collector know that *Hortonius* was named for him and that he planned to retain it "in the Natl. Coll." Dyar's decision to name a new moth away from the National Collection was curious, and after Dyar published his description of the moth, Foster Benjamin informed him that both genus and species had already been described under other names (*Euscirrhopterus gloveri*). Although Dyar acknowledged his error to Benjamin, such mistakes were common for the overextended scientist.[5]

## DYAR AND HOWARD ON ISSUES RELATED TO MOSQUITO CONTROL

Dyar sorely needed income in the 1920s but had misgivings about profiting financially from mosquito control, in essence destroying creatures he so admired, particularly if they posed no threat to humans. Overzealous control efforts could result from inaccurate information about how mosquitoes could spread disease, as shown in a 1922 *Science News* article claiming that disease-carrying mosquito larvae could be transported into homes through the water supply (it is adult mosquitoes that are disease vectors). Dyar forwarded the piece to Howard with the following note: "[The] article by Dr. W.H. Ballou on mosquitoes . . . is about as bad as could be.

I read some to Mrs. [Wellesca] Dyar till she asked me to stop, as she did not want to hear more of such entomological heresies."[6]

The mosquito expert detested the control methods of the day, especially oiling ponds to kill larvae at America's hallowed national parks Yosemite and Yellowstone. Playing the role of environmentalist/conservationist, Dyar reported to Howard in 1924 about the need to protect Yosemite's mosquitoes. He did not feel morally obligated to help in mosquito control efforts to destroy the harmless species "unless asked." For example, he didn't share knowledge with controllers from his own published work that larval *Aedes ventrovittis* lived in pools from melting snow earlier in the season than when they treated with oil.[7] He also complained that oilers in Yosemite National Park had "exterminated the large isolated colony of [*Aedes*] *increpitus* ... and also all the Chironomidae, dragon-flies and frogs. I am ... for a closed season to protect the mosquitoes, though I fear *increpitus* will never come back.... A few mosquito bites are good for hikers and lend zest to the fisherman's waiting."[8]

Later commiserating with Dyar, in January 1927, Howard asked him if oiling would stop if he were to tell sanitary engineer Harry B. Hommon, the person in charge of control efforts at Yosemite, about the ineffectiveness of "oiling of pools that could not contain mosquito larvae" based on "cases reported to me by competent entomologists." Dyar replied the next day: "Hommon is the very fellow I showed up for an ignoramus (skeetologically) in the Yellowstone several years ago ... At Old Faithful the people were complaining of the dreadful 'mosquitoes' and itching themselves. Maybe they complained to Mr. Hommon. But all you had to do was to glance at the bites—*Simulium* [black fly] bites, every one. Next thing they will be digging, tearing up and destroying the pretty rivers so the tourists won't be bitten. Why not do it right? Blast the whole place down with dynamite and steam shovels and plant it smoothly to lawn grass?"[9]

Dyar's attitude about insect-control measures carried over into Lepidoptera as well. Encouraging Mary J. Bronn of the University of Oklahoma in her study of owlet moths, Dyar complained of the problem: "If you would work up the life histories and give good descriptions of the larvae of half a dozen Noctuidae that were not worked up before ... it would be a distinct contribution to knowledge and a better thesis than any amount of general work. The Bureau of Entomology has no one working on this subject, and has not had since [Albert] Koebele's day. All they want to do is to kill bugs, and they make no contributions to knowledge whatever."[10]

Late in 1927, after Howard stepped down as Bureau Chief, he requested Dyar's "critical comments" for an article related to mosquito control, titled *Mosquito Work Throughout the World*. Dyar included Howard among the mosquito-killing conspirators and his rapid reply was:

> Of course there is no criticism except the philosophical one that your attitude is too homo-centric. You slash about, and don't care what becomes of the rest of creation so long as somebody can make a subdivision on a salt-marsh. You say 'all anti-mosquito work is sanitary work.' There I differ. Destruction of innocuous or only slightly annoying species is a waste of time from the human stand-point,

and from the general stand-point is a piece of egotistic presumption. Do you realize that out of the five hundred species I am listing [In "The Mosquitoes of the Americas"], less than fifty are of any affect on man?[11]

Howard requested a paragraph from Dyar about his recent activities relating to worldwide mosquitoes before the New Jersey Mosquito Exterminators Association in January 1928. Two days later, Dyar replied that he "did no fieldwork in 1927" and had spent his time on "the mosquito book." This was a marked change for Dyar. He had been devoting his time to fieldwork since childhood. Continuing to work gratis, Dyar also identified recently collected specimens by several South American workers, including Drs. M. Nunez Tovar and F.M. Root from Venezuela.[12]

While preparing his final mosquito monograph in 1927, "The Mosquitoes of the Americas," Dyar's dealings with his protégé Raymond Shannon became tense. Although Shannon spent time with Dyar and Wellesca in Panama in 1923, his letters were cordial rather than affectionate. Prior to the war and the Panama expedition, Shannon lived with and worked alongside Knab between the last half of 1914 to the fall of 1916, when he went to Cornell. Perhaps Shannon resented Dyar for the demands he made on Knab, who was like family to him, as he was becoming progressively ill.

Shannon, now around thirty years old, may also have been trying to establish his own professional identity, seeking a certain amount of independence from the domineering Dyar. Furthermore, Dyar's poor attitude about extending an invitation to Eugene Amadeus Schwarz (Plate 13) (the elderly patron of the Washington entomological community) for a return trip to Panama in 1924 probably hurt their relationship.[13] Dyar, now frail himself, was concerned that Shannon's lending assistance to Schwarz would interfere with his own goals; however, it was natural for Shannon to help his benefactor, who had paid for him to attend Cornell. The trip was not destined to be, but Shannon, nonetheless, may have harbored resentment about Dyar's attitude.

In December 1927 Dyar expressed his concerns to Howard about Shannon's working independently on the mosquito faunas of Chile and Argentina. He railed about not seeing the specimens Shannon named with del Ponte. When reviewing their paper before publication, Dyar believed the new species to be synonyms, whereas the authors failed to discover those that were actually new species. Dyar's frustration was understandable because his lack of access to Shannon's specimens from southern South America would make completing his own "Mosquitoes of the Americas" more difficult.

Dyar thus wrote Howard: "I would not look at his [Shannon's] Chilean things now on a silver salver, being properly 'mad' and through with begging and coaxing." According to Dyar, when requesting he see medically important *Culex*, Shannon did "not want to permit that." To Dyar, this revealed Shannon's true motives, as he put it, "[S]o here came Miss Pussy out of the bag and pricked her whiskers—he doesn't want me to study his material." But Dyar wanted such comments kept private, concluding, "This is just between us."[14]

After Dyar had completed "The Mosquitoes of the Americas" in May 1928 Howard wrote, "It looks like a very good job, and I am sure that you have done a wonderful bit of work and in a marvelously short time. . . . It contains an extraordinary amount of information, and I am more than surprised to see how many of the larvae are known to you. I shall keep the book by my side."[15]

Howard requested help from Dyar to write his review on "The Mosquitoes of the Americas." Dyar thought the review was "a little lacking in pep" and suggested something be included about the stable classification that resulted from the use of larvae in both his and Edwards's work. However, Dyar's formula for adding "pep" included denouncing Frederick V. Theobald for criticizing Dyar and Knab's work, while mentioning that Lord Walsingham let Theobald go at the British Museum. Howard would not comply with such requests, wanting to avoid controversy.[16]

## DYAR REVISITS HIS "PET" GROUP, THE LIMACODIDAE

The fondness Dyar felt toward the Limacodidae never waned. As he finished the manuscript of the complete New World limacodid fauna in 1927 for inclusion in Adalbert Seitz's *The Macrolepidoptera of the World* Dyar hand printed the title page. His calligraphy (Figure 25.2c) is reminiscent of the style seen on many of his short stories, posters for Wellesca's campground Bahá'í talks, and the carved letters he used on entryways to his tunnels.[17]

Dyar's involvement in the Seitz limacodid volume began with a note from Prof. Max Draudt, written on April 28, 1927. Draudt, serving in Seitz's stead and acting on a suggestion from William Schaus, wrote to determine if Dyar was interested in authoring the comprehensive work.[18] Dyar worked with E. Martin Hering and Walter Hopp, each a prominent lepidopterist at the Museum für Naturkunde in Berlin, to make a more complete manuscript. Hering needed help in identifying South American specimens, especially of *Sibine* (now *Acharia*), a very difficult genus and a Dyar favorite. This was a mutually beneficial proposition given the limacodid specimens present, including types, in each of their museums. Hering sent Dyar photographs of types in the Berlin collection and specimens of *Sibine* for him to identify with the help of Carl Heinrich's dissections.[19]

At around the same time he was corresponding with Hering, Dyar was identifying limacodids and related families from the American tropics for Cornell's W.T.M. Forbes (Figure 24.2c). Dyar was impressed with the number of species in the parcel and told him to send "[a]nything more that turns up in the next twenty years I will be glad to see—provided I am still on the job then."[20]

On November 12, 1927, the *Journal of the Washington Academy of Sciences* received a manuscript from Dyar containing brief descriptions of thirty-eight new species of Limacodidae and nine Dalceridae, a closely related moth family. More details about the new species would appear in Seitz's *Macrolepidoptera of the World*, but Dyar wished "to validate the names as soon as possible." Twenty-two of the species Dyar named in the 1927 article were in the genus *Sibine*, a group known as the

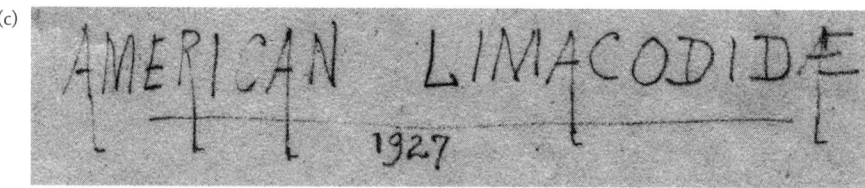

**Figure 25.2** Dr. and Mrs. [Zella] Dyar, probably on the steps near his sister Perle's brownstone by West Central Park (c. 1925 or 1926), and labeled in his own hand (*a*); type specimen of limacodid species *Sibine* (now *Acharia*) *zellans*, named by Dyar in 1927 (*b*); handwritten title on Dyar's manuscript for "The Macrolepidoptera of the World" (*c*).

saddlebacks because of striking oval patterns on the backs of the these caterpillars (Plate 1c). Dyar gave women's names to nearly all of these species but didn't make specific comments on those honored. However, *zellans* (Figure 25.2b), *norans*, *berthans*, and *gertrudens* are of particular notoriety, as they appear to be named for his former wife, mother, and daughter (Bertha was Dorothy's nickname), and cousin Gertrude Dyar. Another, *ximenans*, was likely named for the comely young lady he had mentioned to Zella in 1915 from Truckee, California: Ximena McGlashan.[21]

Because his mother used the nickname Nora, it appears he named the species for her rather than his sister Perle Nora. Given Perle's involvement with illustrating and collecting for Dyar, it's surprising that he did not attribute one for her. At around the same time, in a short separate article, Dyar apparently named species for two of his mother's sisters that doted on him: one a limacodid and the other a related megalopygid. These were *Venadicodia ruthaea* (Aunt Ruth Marble) and *Megalopyge partheniata* (Aunt Parthenia "Parnie" Colburn).[22]

The 1927 paper was not the first time Dyar had dubbed a moth for Zella. He had named tiger moths for her, including the species in his "List" of 1903 (see Figure 9.2b) and the other a genus *Zellatilla* in 1914. For Dyar to name a species for his ex-wife Zella, seven years after the divorce, was perhaps a form of "Dyarian" absolution for the pain he caused her. It may have also been that Dyar felt a "Zella" species deserved a place along with Wellesca's among his treasured moth group, the Limacodidae. After all, Wellesca received the honor in 1900. It may have also been an expression of Dyar's interest in genealogy, as he could not deny Zella was the mother of Dorothy and Otis. Although Dyar named species for relatives and acquaintances simply to honor them, he named moths for the collectors, including his son Otis and mother-in-law Harriet in 1903. Around the same time Dyar christened the species for Zella, he was photographed with her in New York near the Knopf's brownstone in 1925 or 1926 (Figure 25.2a).[23]

## FINAL DAYS AS CUSTODIAN OF THE USNM COLLECTION OF LEPIDOPTERA

Considering Dyar's financial difficulties, it is remarkable he continued as the unpaid Honorary Custodian of Lepidoptera. He found it impossible to give up the only status he possessed at the United States National Museum. Part of his duties over the years was to locate specimens on request in the growing collection. In June 1928, W.J. Holland (director of the Carnegie Museum in Pittsburgh) requested label information for a rare butterfly, the Nokomis fritillary. Holland was a famous lepidopterist, known for his two popular books, *The Butterfly Book* and *The Moth Book*.

Unable to find the specimen, Dyar's reply to Holland gives a bird's eye view of how the collection was arranged toward the end of his tenure. It also shows a difference in Dyar's curatorial philosophy from Schaus's, who evidently like to dispose of tattered specimens: "I searched the collection for a specimen of *Argynnis* [*Speyeria*] *nokomis* from the Bitter root Mts.... There are no old specimens in the collection, so thinking of the weeding process to which Friend Schaus likes to subject things. I looked in the duplicates also, but without result. We do not have any specimens antedating the Riley collection."

Dyar then referred to the fire in the Smithsonian Castle of 1865 as a possible culprit: "I think that everything from the 'Smithsonian Collection' including [Townend] Glover's specimens, were destroyed." His final hypothesis implicated Schaus, who had more interest in butterflies, and could have removed the specimen: "The remote chance remains that it was removed to the "duplicates" and then given away or destroyed." Dyar appreciated the importance of specimens, regardless of whether they were in perfect condition. He expressed his desire to no longer have specimens of lesser quality in a separate collection: "As soon as I can get the space I intend to abolish 'duplicates' and transfer everything to the regular series that has any history with it, irrespective of poor condition."[24]

In the fall of 1928 William Schaus received an inquiry from Mr. John F. Gates Clarke of Bellingham, Washington, to identify some of his moth specimens. The letter ended up in Dyar's box to answer and in October Clarke received this reply:

> There are two methods in use for naming material for correspondents. I. Material sent in bulk from which the museum keeps what is desirable for the collection and returns the rest named. II. The correspondent sends numbered specimens [number on the pin] which he has compared with others in his collection. The names are sent by the numbers and all the material remains in the National Collection. In either case the specimens should be sent mounted and fully labelled . . . we have no preparators to attend to this extra work and the material should be ready to go into the collections—that part of it which we keep.
>
> Very truly yours, Harrison G. Dyar, Custodian of Lepidoptera.[25]

Little did either man know, but the torch had been passed to the next generation of lepidopterists of the National Museum of Natural History. "Jack" Clarke would join the USDA staff at the Bureau of Entomology in 1936 and later became chairman and builder of the Smithsonian's Department of Entomology.

## CHAPTER 26

## Financial Collapse and Final Push for Reinstatement: 1925–1929

*I have been in communication with Dr. Stockberger, Director of Personnel, Dept. Agriculture as to the possibility of my getting back into the Bureau, and I have received what I regard as a most encouraging reply. He ignores the charges and the dismissal, although I discussed these in my letter . . . The implication is plain that if there were such a position open, there would be no objection to my making application for it."*[1]

H.G. Dyar to L.O. Howard, December 15, 1927

Dyar's mounting financial losses during the teens and 1920s were due to risky investments on major luxury apartments and mining.[2] Furthermore, throughout these decades Dyar was unable to get the rental prices he desired in New York and Washington.[3]

For his mosquito work Dyar received some income from the sanitary corps in the 1920s, but that money, along with revenue from tenants, was inadequate to cover promissory notes that were frequently coming due. Thus it became necessary for him to take out second or third loans to cover these notes. Other losses resulted from financing *Insecutor*, alimony payments, various Bahá'í-related activities, and his collecting trips.

Dyar, nonetheless, acquired a luxury apartment in February 1925 in an apparent attempt to achieve solvency. The purchase price was a staggering $2,300,000 (30.5 million in today's dollars) for "The Sixteenth Street Mansions," alternatively known as The Chastleton (Figure 26.1), and brought him once again into the national spotlight.[4] This was the last of three notable apartment complexes built by the famous architect and real estate developer Harry Wardman (1872–1938) that Dyar owned. He purchased the other two on Seventeenth Street in 1917.[5]

The purchase price of The Sixteenth Street Mansions was thought to be a bargain, and Dyar secured a loan of $200,000 to make the purchase through a deed

**Figure 26.1** The Sixteenth Street Mansions, known today as The Chastleton, in northwest Washington, D.C. (1921), before Dyar's purchase in 1925 for $2,300,000.

of trust on another property.[6] His goal was to rent out only 60 of the 200 vacant apartments and run them "in a manner fitting the building and location—that is, first class" and the tenants "should have everything for their comfort to which they are entitled." His realtors Higbie and Richardson were to manage the property.[7]

In this era of rapid exchanges of real estate large and small, Dyar unloaded the Sixteenth Street Mansions a month later. The purchase price by Frank C. Wolfe "was not made public." Wolfe, president of Marlborough Hotel Co., had plans to "remodel the 980-room apartment house into a hotel." The following November Dyar sued Wolfe to recover $2,085 due on a note, presumably from the sale.[8]

Major financial calamity hit Dyar in 1927. According to his son, Wallace: "the crash came of 1927—not the '29 crash . . . he lost his money, his holdings." However, there was no 1927 crash and if anything there was a real estate bubble, which suggests other reasons for Dyar's financial decline, such as loans coming due from purchases of apartments exceeding his rental income. Furthermore, according to Wallace Dyar "entrusted his investments to others, and they had several of them. One . . . was a gold mine investment, then there was a lot of investment in real estate in the Washington area."[9] The gold mine was most likely the "Eureka Secret Canyon Mine," which Wellesca had encouraged others to invest in by sending out circulars.[10]

The Eureka investment was of great concern to Herold Robinson, publisher of *Reality*, because Wellesca and others among the Bahá'ís were counting on big payouts from the mine to help fund the magazine, but the mining district failed to produce much of anything after the 1890s.¹¹ Robinson, who in fact, tried to caution her about this investment in the "Eureka Mine," wrote:

> Last November [1926, when] you mentioned this mining proposition to me . . . I made my own careful investigations . . . [with] people who are supposed to know . . . that the promises . . . [of] big dividends . . . was all nonsense, hot air as they called it, and that when the time arrives, they would give their friends and stock-holders a very plausible and good excuse and ask for further money.¹²

Such creditors as C.H. Pope (Munsey Trust Co.) and James J. Lampton were also after Dyar in 1927. Pope wrote Dyar about three payments of $1,600 that were past due on "second trust notes secured on St. Albans, 2310 Connecticut Avenue." He was glad, he said, that Dyar was "gradually working out" financial "difficulties and at an early date . . . will be able to build up your account and reduce the indebtedness." Later in the letter Pope indicated that Dyar had other issues resulting from the death of his business partner, Alfred Higbie, in December 1926. He wondered "what arrangement" Dyar was "able to make with Mr. Higbie's Estate [about] the note of $25,000", which was being held "as collateral to [another Dyar] loan of $12,500."¹³

The previous July, Higbie was thinking of retiring to the country and wrote Dyar, teasing him about his mosquitoes and about some more serious business: "I expect to go down to the farm tomorrow and make a thorough inspection and then decide what is best to do. It is admirably located and equipped to raise chickens and hogs *together with mosquitoes on the side* [italics mine]. I wish you would let us have this deed back at your earliest convenience as we want to close this not later than August 1st."¹⁴

Dyar's financial situation made it impossible for him to continue publishing *Insecutor*: In the Valedictory of *Insecutor* Dyar wrote: "[W]e hope [to] have presented matter of permanent scientific value. Each year there has been a financial deficit, and as year by year the costs of publication have mounted, this deficit has increased in size, although not relatively so, as our subscribers have all been loyal; but this is not the principal reason that has decided us to close."¹⁵

At the time of *Insecutor*'s demise, Dyar continued editing *Reality*. However, there was a notable absence of his short stories after "Anecdote of the Law" appeared in April 1927, and by September 1928, the editorials were all being written by William E. Mann.

Dyar did continue to write and edit his stories, including some of his longest. There were around 70 written between 1923 and 1928, though many were never published, and none outside of *Reality* other than "Diamonds Going and Coming" and "Annabel Lee," which the Stratford Co. in Boston published in January 1926. A favorable review in a Richmond, Virginia paper reported "Diamonds" (a novella) was "Told in a racy, entertaining manner" and "a clever detective story of a series of baffling diamond robberies" and "Annabel," which made reference to the poem by Edgar Allen Poe, was a "murder mystery forming the climax of an otherwise pretty little romance."¹⁶

## THE LAST PUSH FOR REINSTATEMENT: JANUARY 1929

Dyar clearly needed funds to support his family at the end of 1927. He took out a second mortgage on the B Street home, and Roshan faced the prospect of having to leave Purdue University. Wellesca's heart disease kept her either in the hospital or confined to the second floor of their house. Dyar's health was failing as well (note that in Figure 26.2a Dyar appears frail).[17] It was time to pursue a salaried position through reinstatement to his old government job. What was once optional had now become a necessity.

Dyar's push for reinstatement was evident on December 15, 1927, when he wrote Howard about a reply from Warner W. Stockberger, the Director of Personnel at the USDA.[18] Dyar was encouraged that Stockberger did not mention his dismissal as problematic. Now was the time, he believed, for Howard (or other supporters) to step in and justify his hire because he was, in essence, already doing the job gratis by making all the mosquito identifications and many on Lepidoptera. Howard replied cautiously, "While I am not sure that you have drawn the right inference from the wording of your reply from doctor Stockberger, I nevertheless hope that you are right. Your work is so valuable to the government that I think you ought to be paid for it, and I will do what I can; but you know, of course, that I no longer have any executive functions."[19]

By March 1928, L.O. Howard, now *emeritus* to Bureau Chief Charles L. Marlatt, was helping get Dyar back on the rolls. He wrote Dyar: "I have just received your note which encloses the data about the mosquito work. I am glad to have them. I have sent the memorandum about possible reinstatement to Doctor Marlatt." Perhaps another reason for optimism was that the Secretary of Agriculture and senior administration of the USDA had changed since Dyar's dismissal. Clearly both Marlatt and Howard were behind the effort to appoint Dyar as mosquito expert.

Months went by and the situation had not been resolved, and on August 31, 1928, Dyar pleaded with Marlatt to make the Secretary of Agriculture and the Civil Service understand he had worked for the Bureau of Entomology for over ten years without pay, and they should reinstate him.[20] In October Stockberger expressed doubts to Marlatt about the attempt:

> In addition to the difficulty in obtaining approval from the Commission for appointment under this exception of the rules, there stands the further obstacle of the record of his removal from the service on charges. Such appts. are not subject to the retirement act and may be terminated at any time. Frankly, I anticipate that there would likely be considerable difficulty in obtaining approval of the civil Service commission for Dr. Dyar's reemployment in the Department.[21]

Dyar was beginning to wonder if he would fall victim to administrative red tape. Indeed, he had expressed doubts about whether Marlatt would help him and wrote to Howard, "I believe I can be useful to the Bureau and I know the salary will be useful to me, and this gratis work that I am doing cannot continue.... [I]f Dr. Marlatt

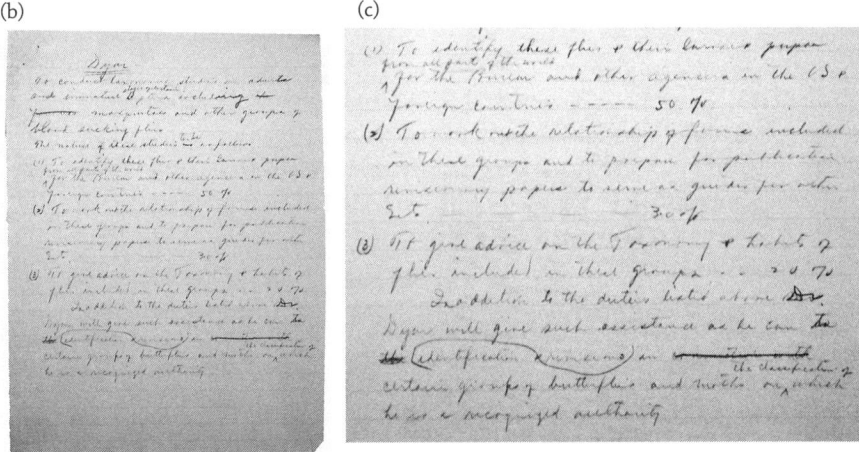

**Figure 26.2** One of the last known photographs of H.G. Dyar (1928?) (*a*); job description to reinstate Dyar to government service (date unknown) (*b*,*c*).

was interested to try he could get around the technicalities, and I wonder if a word from you, if you felt like it, might not serve to stimulate his interest."²²

Howard showed Dyar's letter to Marlatt the same day and wrote the following on his behalf:

> He has done a lot of identification work in the Lepidoptera and had handled all of our taxonomic work with mosquitoes. I don't know what we should have done without him in the important mosquito work.... If, as he states, there remain 'only a few technicalities in the way,' I think that a strong effort should be made to remove these barriers.... It may interest you to know that Dyar and Edwards of the British Museum are the only two persons alive in whose taxonomic work on mosquitoes I have implicit confidence.... [T]here is no one in this country who possesses a tithe of Dyar's knowledge, and only one aside from Shannon (namely, S.B. Freeborn of California) who could be trusted to begin this work here at all authoritatively.²³

Buoyed by Howard's plea to Marlatt to employ him, Dyar in gallows humor wrote to his retired boss: "Many thanks for the letter you wrote to Dr. Marlatt. I am glad that I lost my money to hear such splendid appreciation." Dyar also excitedly mentioned that he had received from F.W. Edwards in London "a box full of five [mosquito] species from the Belgian Congo, mostly new to our collection."²⁴

Marlatt assured Dyar in October 1928 he had every intention of rehiring Dyar, writing:

> You have quite radically misinterpreted my attitude, which, from the beginning, has been one of strong desire to see you back in the service, realizing your long period of useful work and particularly the fact that you have been working now for years without salary.... During the last week I have been working on the case, getting information which may be to your interest, and just now am ready to take it up with the authorities in the Department who will be responsible for your reinstatement or reemployment.²⁵

Dyar thanked Marlatt and in November wrote, "I hope that my reinstatement is progressing rapidly. Finances are becoming pressing and unless I can get some funds before Christmas I shall have to take my boy out of college."²⁶

A report from Stockberger and the Civil Service Commission had still not come, and it appeared that "a request from the Secretary [of Agriculture]" was required, but he would not be back in Washington for a week. Marlatt suggested bringing Dyar in as "a unique specialist." He also suggested an appointment under "the classification of 'Agent,' and this phase is just now also being looked into." Marlatt also referred to a letter Dyar wrote in August about having "papers establishing ... [his] innocence of the old charges." He requested copies of the papers to help "in taking up the case with the Civil Service and the Secretary."²⁷ Presumably the papers gave evidence of blackmail by Leech.

By November 19, 1928 Dyar's case was still in the hands of the administration and Stockberger, the Director of Personnel. A year had passed since he began his latest reinstatement attempt, which now included strong letters of endorsement from Marlatt and Howard. Dyar's letter focused on his need of money:

> My finances have become desperate and I cannot hold on much longer. My wife is in Gallinger Hospital with heart disease; I have two boys attending High School here and another at College at Purdue University. He is about to finish his first half year. I am obliged to hire a woman.... I can little afford and there are taxes and heating. We have two mortgages on the house, which must be kept up and if we were thrown on the street, I have no funds to rent a place. It is impossible for me to wait around while indefinite negotiations go on and I certainly beg that quick action may be taken.... If I can get the salary soon, it will save the situation. Doctor Howard earnestly desires that I be with the Bureau, and if it is to be done finally, cannot a little speed be shown, when time is so important and vital to me?[28]

After receiving Dyar's letter, Stockberger wrote Marlatt and gave him a copy, writing: "He [Dyar] has been advised that the initial steps looking forward to appointments are necessarily taken in the bureau in which the employee is to work." Dyar, having seen the letter, wrote to Marlatt about Stockberger's advisory, with the question, "Is it possible that after all this time no initial steps have been taken toward my employment? I cannot understand why this should be when I have taken pains to explain the urgency of my situation."[29]

Dyar's financial may have been desperate but he was still making real estate deals, and on November 13, 1928, the following announcement was made: "Transfer of the Parkside Apartment Hotel, 1336 I street northwest, yesterday, by Frank C. Heigel [sic] to a group of local investors headed by Harrison G. Dyar for about $400,000 was announced last night by Vernon B. Lowrey, attorney for the buyers.[30] It is likely this purchase, much like others made by Dyar and Wellesca over the years, was made through exchanges of property and accruing more debt. Perhaps the rents from the purchase would provide liquidity to pay Dyar's mortgage and basic expenses. However, it may not have been enough to cover his accrued debts, including one to A. Joseph Hower, who won a lawsuit on November 23, 1928, in which Dyar was ordered to pay $3,500 for default of a loan. On the other side of the ledger, George Freeman Pollock defaulted on considerable loans from Dyar.[31]

✧

In spite of the gloomy outlook, Dyar's reinstatement was almost certain to take place. Marlatt or another administrator wrote a detailed job description as a prelude to bringing Dyar back to work, presumably during the last months of 1928 (Figure 26.2b, c). His involvement in additional work on Lepidoptera would have exceeded 100% of his time:

> Dyar –
> To conduct taxonomic studies on adults and immature stages of certain Diptera including mosquitoes and other groups of blood sucking flies

The nature of these studies to be as follows

(1) To identify these flies & their larvae & pupae from all parts of the world for the Bureau and other agencies in the US & foreign countries _____ 50%

(2) To work out the relationship of forms included in these groups and to prepare for publication revisionary papers to serve as guides for other Ents _____ 30%

(3) To give advice on the taxonomy & habits of flies included in these groups _____ 20%

In addition to the duties listed above Dyar will give such assistance as he can to revisions & identification in certain groups of butterflies and moths on the classification of which he is a recognized authority.[32]

## JANUARY 1929

The cold and damp winter chill in Washington, D.C. during January makes it a dreary place, and Dyar could no longer travel to warmer ones. On Friday January 4, 1929, lepidopterists of the National Museum and Bureau of Entomology gathered with Howard for a photograph (Figure 26.3). Dyar appears contented, unlike in other photos of him. Perhaps he knew he was about to be reappointed by the USDA.

On Saturday morning, January 12 at 11 a.m., Dyar gave a lecture on mosquitoes to Howard University medical students. He had given similar lectures the previous two years at 3:30 p.m. at the end of his workday during the week.[33] Four days later Dyar wrote what may have been his last letter, one to Frank Haimbach of Philadelphia:

> There is no list of the Walsingham numbers other than those published in the paper to which you refer. Busck suggests that other numbers were not identified or perhaps were not Tineids etc. Or these may be the only ones reported on. Anyway we have no list and do not know of the existence of any.
>
> Very truly yours, Harrison G. Dyar.
>
> P.S. I am very sorry that bad health made me go home without looking at your material the last time you were here. Better luck next time. HGD[34]

On Saturday, January 19, 1929, Wallace recalled:

> One of those realtors was trying to get him to sign over something so he could sell one of these properties ... the papers were at the museum ... later in life, when he couldn't make it to the museum, he'd go over there in the morning for a few hours, and then he'd finish what work he was doing at home, because he couldn't make that long trip walking. So anyway, he had already been over [at the museum] in the morning [came home and had to get documents to sign at the museum] ... so he went over there to do that, the second time that day. That's when he had his stroke at the museum.... They called. Of course, Mother could [not] answer the phone, so I answered the phone, but being sixteen years old, I was, "What do I do?"

Figure 26.3 Last known photograph of H.G. Dyar with (*left to right*) Schaus, Howard, Heinrich, Busck, and artist Francis Noyes taken in front of cabinets in the USNM collection (January 4, 1929).

I was in a panic, so to speak. I knew it would distress Mother, but I had to tell her. They asked me to come over. They had to have somebody in the family to tell him what to do with him, because he was paralyzed. I remember very well, he looked at me more or less in desperation, very sad, because I didn't know what to do.... So finally we realized he had to go to the Emergency Hospital.... Then they brought him home, and I remember Mother leaning over the banister. She called, "Dickie?" He couldn't talk to her. So I remember they took him to the hospital, and he died there. So it was very sad.... but at least she saw him, and he her.[35]

Dyar died on Monday afternoon, January 21, at Garfield Hospital, less than a month shy of his sixty-third birthday. Wellesca later suggested that although Dyar, "like Moses ... did not get to the 'promised land' [his reinstatement], he had the satisfaction of knowing it was to be his in a short time."[36]

The funeral services were held three days later at Lee's Chapel on 322 Pennsylvania Avenue. The body was cremated and later buried at Glenwood Cemetery in Washington, D.C.[37]

## THE TUNNELS AFTER DYAR'S DEATH

The tunnels had a life all their own after Dyar's death in 1929. Stories about the Twenty-first Street tunnel in Dupont Circle resulted from a collapse, much as in 1917 and 1924. A cave-in on Twenty-second Street in 1950 was related to a cavern below and although it had nothing to do with Dyar's tunnels, it brought back stories of the Smithsonian scientist.[38]

The B Street tunnel achieved separate notoriety with a detailed illustration in *Modern Mechanix* in 1932.[39] In the 1930s the home and tunnel were sold to the government, which helped Wellesca and the children financially. According to Wallace Dyar, the USDA had considered using the tunnel for experiments in growing mushrooms.[40] During World War II, another use was suggested: The executor of Dyar's literary estate proposed the tunnel as an air-raid shelter.[41] On March 4, 1942, articles on the B Street tunnel appeared in two Washington newspapers after the tunnel was inspected by the District related to widening Independence Avenue (formerly B Street).[42] It was determined that the multilayered tunnel was too narrow to protect District residents from an attack.

The Twenty-first Street tunnel behind the home collapsed once more in 1958, which made the Washington, D.C. newspapers.[43] Children of the owners, the Curd family, were shown in the opened tunnel with the brickwork visible (Plate 16b).[44] This brought out Dyar's daughter Dorothy Dyar Hill, her daughter Roberta along with Roshan Dyar's son Lowell, who was a cousin and a friend.[45] At another time on the same day, Wallace Dyar and his son also visited the tunnel.[46]

Perhaps the tunnels did have a purpose other than exercise after all: They brought Dyar's two families together.

# Epilogue

## WELLESCA'S ADMISSION TO THE FICTITIOUS MARRIAGE

There is little doubt that Wellesca Pollock married Harrison G. Dyar under the assumed name of Wilfred P. Allen on September 5, 1906, in Richmond, Virginia. Their children were never told, according to Wallace Dyar.[1] But in the weeks following Dyar's death, Wellesca admitted to the deceit after she received an Alumni Federation of Columbia University questionnaire, dated the twenty-fourth of January, evidently in response to the obituary in the *New York Times*. The date that the widow filled out the questionnaire was February 14, 1929: her husband's sixty-third birthday (see Epilogue Figure 1a).[2]

The alumni card on file today (Epilogue Figure 1b), which has a large stamp that reads "DEAD," has the dates of Dyar's two graduate degrees at Columbia, names of relatives, including the two wives, etc. The date of marriage to Mrs. Wellesca Pollock given as the usual "April 26, 1921" from an earlier questionnaire was crossed out and changed to "Sept. 5—1906" (see Epilogue Figure 1d), the date of Wellesca's marriage to Wilfred Allen, in response to her note at the bottom of the form she had received (see Epilogue Figure 1c).[3] Wellesca, after so many years, admitted, albeit to alumni files, that Allen and Dyar were the same person. Indeed, no extant evidence verifies that Harrison and Wellesca were married on April 26, 1921, so it would seem that her only real wedding to Dyar was in 1906.

## IT TOOK TWO ENTOMOLOGISTS TO REPLACE DYAR

Not long after Dyar's death the Bureau of Entomology brought in both a dipterist and lepidopterist to their taxonomic staff: Alan Stone and Foster Benjamin. These actions illustrate the large hole in services that had been provided by Dyar at the USNM. Among Stone's first tasks was to go through the entire mosquito collection and organize or identify "a large quantity of material" that Dyar was unable to process before he died. Benjamin organized the "Macrolepidoptera" that was received

**Epilogue Figure 1** Columbia University alumni questionnaire sent to Wellesca following the death of her husband (*a*); alumni card stamped "DEAD" (*b*); detail of alumni form showing that Wellesca wrote Sept. 5, 1906 as the date she married H.G. Dyar (*c*); detail of alumni card showing the previously reported marriage date of April 26, 1921, between Dyar and Wellesca is crossed out and changed to the September 1906 date Wellesca reported in 1929 (*d*).

via purchase of the Barnes Collection, as desired by Dyar and the other USNM lepidopterists.[4] Barnes died in 1930.

## FINAL DAYS OF ZELLA P. DYAR AND WELLESCA P.A. DYAR

On August 14, 1938, Zella was in an accident while on a visit to Long Beach, California. She had been living with her son and daughter-in-law in nearby Arcadia. As reported in the *Los Angeles Times*:

> INJURED WOMAN FOUND IN HOTEL ... Found unconscious in a hotel hallway in Long Beach yesterday, Mrs. Zella P. Dyer [sic], 69 years of age, of 515 Herman [sic—Otis at 515 Norman] street, Artesia, is being treated for

a possible skull fracture and a fractured right leg ... Authorities are puzzled as to the exact cause of the woman's injuries, police said. Earlier in the day a trailer had struck her as she sat on a bench bordering a downwalk near the Long Beach Municipal Pier. The trailer was attached to an automobile driven by Buell V. Howard, a young resident of Long Beach. Treated only for leg lacerations, her only apparent injury, she returned to her hotel where she was found later lying in the hall.

She would die four months later on December 8, 1938.[5]

Wellesca died on June 21, 1940, at the same age as Zella—sixty-nine years. In her last years, Wellesca preferred being called by her Bahá'í name, Aseyeh. She was said to be "a very old lady, friendly and affable and little reserved. Gray haired." She liked sitting out on an old dark-green lawn chair behind the house and located by a patch of trees near the edge of Piney Branch Park.[6] *The Page News and Courier*, of Page County, where Skyland, Virginia, is located, had the following announcement about Wellesca's death. It reflects that George, who survived her, was a resident and known to the locals:

> Sister of G. Freeman Pollock dies of heart ailment. Baha'in services were held Saturday night. Unitarian services were held at 2 o'clock Sunday afternoon at Gawler's Chapel, 1756 Pennsylvania Ave., N.W., with burial in Glenwood Cemetery, Washington, D.C ... died Friday [21st] in Emergency Hospital of a heart ailment after a week's illness. Her home was at 3441 17th St. N.W., Washington, D.C.[7]

The two memorial services reflected her life-long involvement in the two religious communities.

## FINAL RESTING PLACES

Wellesca buried the ashes of her husband, Harrison G. Dyar, Jr., at Glenwood Cemetery in Washington, D.C., in 1936 along with the remains of her family, which included her parents, George H. and Louisa[e] Pollock, her siblings Uila P. Bradway and Amia L. Baker, and her husband John Baker. Wellesca's brother George Freeman Pollock purchased the cemetery plot, although following his death in 1949, Pollock's ashes were scattered at Skyland. Most surprising of all, however, was that after Wellesca's ashes were interred at the cemetery, the last burial in the plot was James C. Crawford (in 1952)—the entomologist whom the USNM hired instead of Dyar in 1908 and who married Wellesca's niece, Emily Baker[8] (See Epilogue Figure 2a). Easily seen from the markers is the grave of entomologist Charles Valentine Riley, a mere 15 yards away (see Epilogue Figure 2b).[9]

Zella was interred at Forest Lawn Cemetery in Los Angeles.[10]

**Epilogue Figure 2** Final resting place of Harrison G. Dyar, Jr., and Wellesca (Aseyeh) Pollock Allen Dyar in the Pollock plot at Glenwood Cemetery in Washington, D.C. It is marked by one large oval stone with "Pollock" carved in it and four smaller stones, one on each side and two others (*bottom of photo*) to mark the plot boundaries (*a*); burial plot of preeminent entomologist Charles Valentine Riley of the U.S. Department of Agriculture and U.S. National Museum, in close proximity to Dyar's plot (*b*).

## GRANDCHILDREN FIND EACH OTHER AT THE SAME SCHOOL

Perhaps the best known of the apocryphal stories about Harrison Dyar, Jr., was that the children from each family met at school, learning that each had a father at the Smithsonian that studied butterflies. Although the story is false, it contains a nugget of truth, albeit one generation removed. Roberta Hill, the granddaughter of Zella, and Lowell Dyar, the grandson of Wellesca, were each born in 1935 and met in the same class at the Friends School. Both suspected that there was social engineering that brought this about. In other words, half siblings Dorothy and Roshan had conspired to bring them together. Each was an unusual child who loved to collect various things, including streetcar license plates.[11]

# CHRONOLOGY OF H. G. DYAR, JR., AND FAMILY

| Event Title (in Washington, D.C., unless mentioned otherwise) | Event Date |
|---|---|
| Joseph DYER born, father of Joseph DYAR: Boston, MA | 1687 |
| Joseph Dyar born, father of the "AR" Dyars: Boston, MA | Feb. 7, 1719 |
| Jeremiah Dyar born, grandfather of H.G. Dyar, Jr.: Boston, MA | 1771 |
| Aaron Cushman Hannum born, maternal grandfather of H.G. Dyar, Jr.: Williamsburg, MA | Oct. 24, 1798 |
| Harrison Gray Dyar, Sr., born: Harvard, MA | Mar. 1, 1805 |
| Aaron C. Hannum and Parthena Robinson marry, maternal grandparents of H.G. Dyar, Jr.: Williamsburg, MA | Dec. 12, 1819 |
| Dyar, Sr., operates telegraph line, several miles long on race track: Long Island, NY | 1826 |
| Dyar, Sr., patent application for subterranean excavations: Panton Square, England | Aug. 1831 |
| S.F.B. MORSE GIVES FIRST PUBLIC DEMONSTRATION OF HIS ELECTRIC TELEGRAPH: MORRISTOWN, NJ | Jan. 1838 |
| Dyar, Sr., and Heming patent process for making soda ash (sodium carbonate): Cavendish, England | Sept. 1838 |
| Harriet Holland [Peabody] born, mother of Zella P. Dyar: Canton, ME | Sept. 12, 1841 |
| Eleonora Rosella Hannum [Dyar] born, mother of Harrison G. Dyar, Jr.: Adams, Berkshire, MA | Oct. 25, 1842 |
| Dyar, Sr., president of New York and New Haven Railroad | 1849 |
| Dyar, Sr., lives and owns 331 Fifth Avenue (kitty corner to future site of Empire State Blg): New York, NY | 1863–1866 |
| Nettie Colburn and Parthenia R. Hannum (aunt of Dyar, Jr.) conduct a séance for President and Mrs. Lincoln: Washington, D.C. | Feb. 1863 |
| Captain Josiah Cushman Hannum (uncle of Dyar, Jr.) with U.S. troops that quell New York City draft riots: suffers partial hearing loss | Jul. 13–16, 1863 |
| LINCOLN ASSASSINATED AFTER LEE SURRENDER AT APPOMATOX: WASHINGTON, D.C. | Apr. 15, 1865 |
| Dyar, Sr., and Eleonora Rosella Hannum (parents of Dyar, Jr.) marry: New York, NY | May 9, 1865 |
| Dyar, Jr., born: New York, NY | Feb. 14, 1866 |

| Event Title (in Washington, D.C., unless mentioned otherwise) | Event Date |
|---|---|
| Dyar, Sr., purchases property on the Hudson River from Dr. Federal Vanderburgh's (d. 1/21/1868) estate: Linwood Hill nr. Rhinebeck, NY | 1868–1869 |
| Nora Perle Dyar Knopf born, sister of Dyar, Jr.: New York [prob. city] | Jan. 2, 1868 |
| Zella Peabody born, first wife of Dyar, Jr.: Canton, ME | Apr. 3, 1869 |
| Wellesca Pollock born, second wife of Dyar, Jr.: Weston, MA | Feb. 12, 1871 |
| Pollock family move to Washington, D.C. | 1874 |
| H.G. Dyar, Sr., dies: Linwood Hill, NY | Jan. 31, 1875 |
| C.V. RILEY HEADS FEDERAL COMMISSION FOR ROCKY MOUNTAIN LOCUST AND APPOINTED USDA ENTOMOLOGIST | 1876–1878 |
| PRESIDENT JAMES GARFIELD SHOT BY GUITEAU AND DIES IN WASHINGTON, D.C. | Jul.–Sept. 1881 |
| Dyar, Jr., graduates from DeGarmo Institute secondary school: Rhinebeck, NY | May–Jun. 1882 |
| Earliest date in Blue Book series: Dutchess Co., NY | Jun. 17, 1882 |
| Begins Roxbury Latin School, lives with mother on Mt. Pleasant near school: Boston, MA | Sept. 1882 |
| Zella Peabody at Everett Girl's School and Mrs. Harriet M. Peabody matron at children's mission: Boston, MA | 1882–1883 |
| Eleonora Rosella Dyar is trance medium for the First Spiritualist Church of Boston | 1884–1888 |
| Dyar's unpublished "Catalogue of Lepidoptera": First entry Plattsburgh, NY | Aug. 1885 |
| J.B. Smith appointed as assistant to C.V. Riley (USDA): Washington, D.C. | Aug. 1885 |
| Attends M.I.T., lives with mother at 170 West Chester Park (she dies in 1888): Boston, MA | 1885–1889 |
| Fieldwork in New Hampshire | Jun.–Aug. 1886 |
| Zella and mother move to Pasadena, CA | 1887 |
| Eleonora Rosella Dyar marries Dr. Jacob N. M. Clough: Boston, MA | May 22, 1887 |
| G.H. Hudson collects moths at electric lights for Dyar in Plattsburgh, NY | Jul.–Aug. 1887 |
| Requests publications and identifications of "Microlepidoptera" from C.H. Fernald | Nov. 1887 |
| Perle Dyar collects butterflies and moths: Altamont, FL | Feb.–Apr. 1888 |
| GREAT BLIZZARD OF 1888: THE WHITE HURRICANE | Mar. 12–13, 1888 |
| Eleonora Rosella Dyar Clough dies from heart-valve disease: Boston, MA | May 18, 1888 |
| Zella Peabody collects butterflies and caterpillars: Pasadena, CA | Jul. 7–8, 1888 |
| Dyar, Jr., begins scientific correspondence with H. Strecker: Boston, MA | Oct. 7, 1888 |
| Graduates M.I.T. in Chemistry | May 28, 1889 |
| Aunt Parnie Hannum collects *Arctia* for nephew: Linwood Hill, NY | Jun. 29, 1889 |

| Event Title (in Washington, D.C., unless mentioned otherwise) | Event Date |
|---|---|
| Begins 8-month trip collecting at electric light at Albany Hotel and Union Station: Denver, CO | Jul. 23–24, 1889 |
| Collects at Glacier Point and Yosemite Valley, CA | Aug.14–20, 1889 |
| Dyar, Jr., marries Zella M. Peabody: Los Angeles, CA | Oct. 14, 1889 |
| Perle N. Dyar marries Dr. S. Adolphus Knopf: Los Angeles, CA | Oct. 19, 1889 |
| Honeymoon/collecting trip: AZ, LA, FL | Nov. 3, 1889–March 13, 1890 |
| Final measurements of caterpillar head capsules for "Dyar's Law": Dutchess and Ulster Cos., NY | –Sept. 21, 1890 |
| Moves to 400 West Fifty-seventh St, New York, NY | Sept. 1890 |
| "Dyar's Law of Geometric Growth" published (*Psyche*) | Nov.–Dec. 1890 |
| Begins fifteen-month trip: Denver, CO | Apr. 1891 |
| Yosemite and Santa Barbara, CA | May–Dec. 1891 |
| Oahu and Hawaii | Jan.–Mar. 1892 |
| Portland, OR | Apr. 1892 |
| Sitka, AK, and Yellowstone Park, WY | Jun.–Jul. 1892 |
| Returns to M.I.T. (natural history coursework): lives in Roxbury, MA | Sept.1892–May 1893 |
| Chosen secretary of Cambridge Entomological Club: Cambridge, MA | Mar. 1893 |
| COLLAPSE OF NEW YORK STOCK EXCHANGE: BEGINNING OF FOUR-YEAR DEPRESSION | May 1893 |
| Takes embryology course, does intensive fieldwork on sawflies: Woods Hole, MA, and Plattsburgh, NY | Jul.–Sept. 1893 |
| Begins graduate school at Columbia College, New York, NY | Sept. 1893 |
| Neumoegen names genus *Dyaria* after "his faithful co-labourer and friend" | Sept. 1893 |
| L.O. Howard appointed chief of USDA's Division of Entomology after Riley resigns | Jun. 1894 |
| Fieldwork: Plattsburgh and Keene Valley, NY, and Dixfield, ME | Jun.–Sept. 1894 |
| Neumoegen and Dyar complete final part of "Revision of Bombyces" | Nov. 1894 |
| Mrs. Harriet M. Peabody starts New England Peabody Home for Crippled Children: Weston, MA | 1895 |
| Neumoegen dies of tuberculosis | Jan. 21, 1895 |
| Reads Ph.D. dissertation "On Certain Bacteria from the Air of New York City" | Feb. 18, 1895 |
| Fieldwork: Jefferson Highlands, NH | Jun.–Aug. 1895 |
| Death of C.V. Riley from bicycle accident: Washington, D.C. | Sept. 15, 1895 |
| Corresponds with L.O. Howard, Chief of Division of Entomology, USDA | Nov. 4, 1895 |
| Leases property at 599 Broadway to United Society of Shakers of New-Lebanon, for sixty-three years: New York, NY | Jan. 1, 1896 |
| Dorothy Dyar born (first child): New York, NY (names moth *Macrurocampa dorothea*) | Apr. 10, 1896 |
| Speaks about earliest larval stages of limacodids (NY Entomological Society) | Oct. 20, 1896 |

(232) Chronology

| Event Title (in Washington, D.C., unless mentioned otherwise) | Event Date |
|---|---|
| Henry Skinner writes Dyar, admonishing him for caustic remarks in print | Jan. 29, 1897 |
| Expresses interest to L.O. Howard about working at the U.S. National Museum (USNM) | Feb. 9, 1897 |
| Martin Larsson Linell, USNM assistant, dies | May 3, 1897 |
| Receives position as "Honorary Custodian" of the Section of Lepidoptera, Division of Insects, USNM | Nov. 12, 1897 |
| Fieldwork: southern FL and Bahamas | Jan.–Mar. 1898 |
| Purchases home for family at 1512 Twenty-first Street NW: Washington, D.C. | 1898 |
| SPANISH AMERICAN WAR | Apr.–Aug. 1898 |
| Fieldwork: Washington, D.C., and Bellport, NY | Apr.–Sept. 1899 |
| Entomological Society of Washington meeting: Dyar home | Jan. 4, 1900 |
| Collecting trip with family: Palm Beach, FL | Jan. 11–Mar. 6, 1900 |
| Otis Peabody Dyar born (second child): Washington, D.C. | May 13, 1900 |
| Dyar family meets Wellesca Pollock: Stony Man, Skyland, VA | Jul.–Aug. 1900 |
| Names limacodid moth for Wellesca: *Parasa wellesca* | Nov. 1900 |
| Begins term as president of Entomological Society of Washington: E.A. Schwarz home | Jan. 3, 1901 |
| Lepidoptera collecting with A.N. Caudell: foothills near Denver, CO | May–Aug. 1901 |
| Dyar writes short story "Various Things": Denver, CO | May–Jul. 1901 |
| First mosquito collecting: Bellport and Plattsburgh, NY | Aug. 9–27, 1901 |
| PRESIDENT WILLIAM MCKINLEY ASSASSINATED IN BUFFALO, NY | Sept. 6, 1901 |
| First mosquito publication | Dec. 1901 |
| Gives presidential address for Entomological Society of Washington: Dyar home | Jan. 9, 1902 |
| Mosquito and moth collecting: Center Harbor, NH | Jun.–Aug. 1902 |
| "List of North American Lepidoptera" published; includes new species named for Zella | 1903 |
| Field expedition: Kaslo, BC | May–Sept. 1903 |
| BEGINNING OF U.S. CONSTRUCTION OF PANAMA CANAL (APPROVED BY U.S. SENATE IN 1902 BY NARROW MARGIN) | May 4, 1904 |
| Mosquito fieldwork: Weekapaug, RI, and Lake Champlain, NY | Jul. 2, 1904 |
| Mosquito fieldwork with Caudell: Jacksonville, FL | Mar. 1905 |
| Mosquito fieldwork with Knab: West Springfield, MA | Apr. 1905 |
| Trip with Zella to England: British Museum and private collections | May 13–Jul. 7, 1905 |
| Fieldwork with family: Tupper Lake, NY | Aug. 1905 |
| Zella, Harriet Peabody and children leave for over six months in California | Nov. 1905 |
| Probable onset of the Dupont Circle tunnels (while family away) | Mar.–Apr. 1906 |
| SAN FRANCISCO EARTHQUAKE | Apr. 18, 1906 |
| Secret rendezvous with Wellesca: Grand Canyon, AZ | May 7, 1906 |

| Event Title (in Washington, D.C., unless mentioned otherwise) | Event Date |
| --- | --- |
| Visits family (Dorothy requires surgery) and Mrs. Peabody runs into Wellesca unexpectedly: Santa Monica, CA | May 8–30, 1906 |
| Mosquito fieldwork with Caudell: Southern California, Pacific Northwest and western Canada | Jun.–Aug. 1906 |
| Marriage of Wellesca Pollock to Wilfred P. Allen: Richmond, VA | Sept. 6, 1906 |
| Buys lot behind Twenty-first Street NW home | Nov. 22, 1906 |
| Featured in magazine article about wealthy Americans that work gratis for the government | Jan. 1907 |
| Wellesca departs to visit 'Abdu'l Bahá in Acca, Palestine | Apr. 2, 1907 |
| Attends 7th Intl' Congress of Zoology as USNM representative: Boston, MA | Aug. 19–24, 1907 |
| Evelyn G. Mitchell sues Dyar for libel over review of her mosquito book: $35,000 | Apr. 27, 1908 |
| James C. Crawford, rather than Dyar, appointed Assistant Curator at USNM | Jul. 1908 |
| Wilfred "Roshan" Allen born (Wellesca's first child) | Dec. 29, 1908 |
| Staff begin move into new USNM building | Aug. 11, 1909 |
| Wilfred Allen lives at 1228 B Street: 1910 Census | Apr. 16, 1910 |
| Fieldwork (perhaps last trip with Zella and children): Weld, ME | Jul. 21–Aug. 12, 1910 |
| Wellesca writes Zella about how Dyar came to build for her the home at 804 B Street SW | May 12, 1911 |
| Otis Dyar visits USNM but father is at B Street home | Sept. 1911 |
| Harrison "Golshan" Allen born (Wellesca's second child) | Dec. 24, 1911 |
| Death of John Bernhardt Smith | Mar. 8, 1912 |
| Receives proofs of vol. 1 of "The Mosquitoes of North and Central America and the West Indies" | Apr. 2–4, 1912 |
| SINKING OF THE TITANIC | Apr. 15, 1912 |
| Barnes writes Sen. Wm McKinley about removing Dyar as Custodian at USNM | Oct. 1912 |
| 'Abdu'l Bahá visits 804 B Street (Bahá'í Hospice: Wellesca's home) | 1912 |
| Dyar attends his last meeting of Entomological Society of Washington as a member | Dec. 5, 1912 |
| Starts journal *Insecutor Inscitiae Menstruus* | Jan. 1913 |
| USDA position switched to "Expert" paid from "Preventing Spread of Moths" appropriation | Jun. 1, 1913 |
| Wallace Parsons "Joshan" born (Wellesca's last child) | Jul. 22, 1913 |
| Fieldwork on mosquitoes and moths: Washington, D.C., and Paramore's Island, VA | Jul. 15–Aug. 17, 1914 |
| Dyar story "The Man Who Did Return" has main character named John Smith | Aug. 5, 1914 |
| Under care of "nurse" Wellesca Allen, begins government approved trip to Nassau, Bahamas; Mrs. Leech watches Allen children in Washington, D.C. | Jan.–Feb. 1915 |
| U-BOAT SINKS LUSITANIA | May 7, 1915 |
| Separation agreement between Dyar and Zella | May 13, 1915 |

| Event Title (in Washington, D.C., unless mentioned otherwise) | Event Date |
|---|---|
| Leaves for Reno via Panama Canal with Wellesca, Mrs. Leech and children: claims move for health reasons | Jun. 16, 1915 |
| Dismisses Mrs. Leech, sends her and family back to Washington, D.C. | Jul. 1915 |
| 804 B Street house searched for evidence by Zella's attorneys | Sept.–Nov. 1915 |
| Zella files for divorce in Washington, D.C., Supreme Court, and for protection of real estate assets | Oct. 8, 1915 |
| Dyar attempts to negotiate uncontested divorce in Reno and to close case in Washington, D.C., with Zella's mother | Oct. 19, 1915 |
| Requests to protect his papers at museum from "Mrs. Dyar's agents" | Jan. 24, 1916 |
| Mosquito collecting: San Diego, CA | Mar.–May 1916 |
| Mosquito collecting: Yosemite, CA | May 14–18, 1916 |
| Dyars' divorce trial in Reno dismissed: fails Nevada residency requirements | Sept. 23, 1916 |
| Wellesca's divorce from Wilfred Allen denied in Reno, NV | Oct. 16, 1916 |
| Returns to Washington, D.C., via San Diego, CA | Jan. 1917 |
| UNITED STATES ENTERS THE GREAT WAR (WORLD WAR I) | Apr. 6, 1917 |
| U.S. Secretary of Agriculture Houston requests marriage license of Wilfred P. Allen and Wellesca Pollock; gives Dyar opportunity to refute charges of signing license as Mr. Allen | May 1917 |
| Begins collecting trip to Pacific Northwest: Steamboat, NV | May 13, 1917 |
| Report of tunnel collapse in Dupont Circle identifies Dyar, away on western travel, as digger | May 19, 1917 |
| Dyar dismissed from government service for posing as Wilfred Allen | Jun. 2, 1917 |
| Motion for new Allen divorce trial denied by second judicial district in Reno, NV | Jul. 27, 1917 |
| Offers to help Barnes with his Catalogue: "afterwards hostilities may be resumed according to circumstances" | Aug. 9, 1917 |
| Donates 44,000 specimens of Lepidoptera, mosquitoes, sawflies, and miscellaneous insects to USNM | Dec. 18, 1917 |
| Recuperating at Hotel Mahackemo, South Norwalk, CT, requests reinstatement from U.S. Secretary of Agriculture | Dec. 29–Jan. 21, 1918 |
| Canadian mosquito trip: White River, Ontario to Alberta | Jun. 13–Aug. 18, 1918 |
| Schaus recommended by Dyar for Assistant Custodian of Lepidoptera | Sept. 4, 1918 |
| Frederick Knab dies | Nov. 2, 1918 |
| Extensive mosquito fieldwork: British Columbia, Alaska, and Yukon | Apr. 27–Sept. 16, 1919 |
| Schaus appointed Honorary Assistant Curator, Division of Insects by Smithsonian Secretary Wolcott | Jul. 1, 1919 |
| PROHIBITION TAKES EFFECT IN THE UNITED STATES | Jan. 1920 |
| Mosquito collecting: Garfield, UT | Apr. 1920 |
| Brief stop in Watkins Glen, NY | May 1920 |
| Mosquito collecting: Tahoe City, Oregon, and Washington | Jun.–Jul. 1920 |
| U.S. CONGRESS RATIFIES WOMEN'S SUFFRAGE | Aug. 1920 |

| Event Title (in Washington, D.C., unless mentioned otherwise) | Event Date |
|---|---|
| Finds his limacodid specimens missing after Busck gives to Barnes | Oct. 11, 1920 |
| Dyar divorce final: Alameda Co., CA | Dec. 6, 1920 |
| Petition for Allen divorce rehearing denied by Nevada Supreme Court | Apr. 15, 1921 |
| "Marries" Wellesca Pollock Allen: Reno, NV (no public record) | Apr. 26, 1921 |
| Mosquito collecting: Lake Tahoe and Pacific NW, Montana, North Dakota, and Minnesota | May–Jul. 1921 |
| Makes $44,000 net per annum, but huge loans due | 1922 |
| Writes Barnes supporting sale of Barnes Collection to the USNM | Mar. 27, 1922 |
| Adopts Allen children: changes their last names to Dyar | Apr. 6, 1922 |
| Mosquito collecting: Winnipeg, Rochester, Yellowstone; goiter treatment in Long Beach, CA | May–Jul. 1922 |
| Begins editorship of independent Bahá'í magazine *Reality* | Oct. 1922 |
| Death of Harriet M. Peabody (1841–1922): Oxford, ME | Dec. 9, 1922 |
| Places date on arch of B St tunnels (Dyar's fifty-seventh birthday) | Feb. 14, 1923 |
| Mosquito collecting: around Grand Lake, CO | May 31–Jul., 1923 |
| Mosquito collecting with R. Shannon: Panama Canal Zone | Jul. 17–Aug. 21, 1923 |
| Writes approximately 70 short stories | 1923–1927 |
| Mosquito collecting: Pacific northwest and Yosemite | Apr. 29–Jun. 22, 1924 |
| J. EDGAR HOOVER HEADS FBI | May 1924 |
| Mystery tunnels discovered in Dupont Circle after street collapse: Dyar revealed as digger two days later | Sept. 25, 1924 |
| Abnormal heart tachycardia: Washington, D.C. | Oct. 9, 1924 |
| Purchases Sixteenth Street Mansion (The Chastleton) for $2,300,000 | Feb. 7, 1925 |
| Moth specimens missing after visit by W. Barnes | Oct. 3, 1925 |
| *Tentamen* ruled invalid by International Committee of Zoological Nomenclature | 1926 |
| Last trip to the West: Glacier Park, MT | Apr. 12–Jul. 31, 1926 |
| Row over Schaus bawling out Dyar's artists | Jan. 31, 1927 |
| Completes text of "Mosquitoes of the Americas" | May 20, 1927 |
| Submits manuscript naming limacodids for ex-wife and female relatives; completes limacodid manuscript for Seitz "Macrolepidoptera of the World" | Nov.–Dec. 1927 |
| Howard helps reinstatement effort, writes to new Bureau Chief C.L. Marlatt | Mar. 19, 1928 |
| Marlatt writes Dyar about his desire to reinstate him | Oct. 8, 1928 |
| Heads group of investors to purchase Parkside apartment building ($400,000) | Nov. 12, 1928 |
| Writes Stockberger of USDA about desperate situation for employment | Nov. 19, 1928 |
| Draft of job description for return to government service | Dec.? 1928 |

| Event Title (in Washington, D.C., unless mentioned otherwise) | Event Date |
| --- | --- |
| Group photo taken of Dyar with USNM lepidopterists and L.O. Howard | Jan. 4, 1929 |
| Gives Saturday morning lecture on mosquitoes to Howard University medical students at USNM | Jan. 12, 1929 |
| Suffers stroke while at U.S. National Museum on a Saturday | Jan. 19, 1929 |
| H.G. Dyar, Jr., dies | Jan. 21, 1929 |
| Wellesca admits that Dyar was Wilfred Allen by using Sept. 1906 marriage date on Columbia University alumni form for late husband | Feb. 14, 1929 |
| WALL STREET CRASH OF 1929 | Oct. 29, 1929 |
| Perle Nora Knopf dies: New York, NY | Dec. 24, 1931 |
| Zella injured by trailer at bench near Long Beach Municipal Pier | Aug. 1938 |
| Zella P. Dyar dies: Long Beach, CA | Dec. 8, 1938 |
| Wellesca Pollock Dyar dies: Washington, D.C. | Jun. 21, 1940 |
| Tunnel collapse in Dupont Circle: Washington, D.C. | May 24, 1958 |
| Dorothy Dyar Hill dies: Washington, D.C. | Jun. 1965 |
| Otis Peabody Dyar dies: Lancaster, CA | Sept. 1968 |
| Roshan Dyar dies: Oxnard, CA | Apr. 25, 1980 |
| Harrison Golshan Dyar dies: Pinellas, FL | Apr. 26, 1988 |
| Wallace Joshan Dyar dies, last survivor of Dyar, Jr.: Pinellas, FL | Mar. 6, 1999 |
| Lowell P. Dyar dies, Wellesca's oldest grandchild: Washington, D.C. | Dec. 30, 2003 |
| Roberta Hill dies of cancer, Zella's only grandchild: New York, NY | Jul. 18, 2009 |

# NOTES

### ABBREVIATIONS

| | |
|---|---|
| ACM | Archives of the Carnegie Museum of Natural History |
| AES | American Entomological Society Manuscript Collection at ANSP |
| AMNH, NYESA | American Museum of Natural History, New York Entomological Society Archives |
| ANSP | Library, Academy of Natural Sciences Philadelphia |
| AVA | Wellesca P. Allen v. Wilfred P. Allen, State of Nevada Supreme Court |
| CUA, WTMFP | Cornell University Archives, W.T.M. Forbes Papers, |
| CUBLA, TDAC | T.D.A. Cockerell Collection, Colorado University Boulder Library and Archives |
| DVD | Harrison G. Dyar v. Zella P. Dyar transcripts, defenders exhibit number, Washoe County Court, Reno, Nevada |
| FMNHIA, HSP | Herman Strecker Papers, Field Museum of Natural History Institution Archives |
| HFA | Hudson Family Archives, Plattsburgh State University |
| HGDP | Harrison Gray Dyar Papers, Smithsonian Institution Archives (copy of Acc. 95-006 to be deposited by author in Department of Entomology Archives, NMNH: folder numbers as cited herein) |
| LOC, MMC 2567 | the Papers of Harrison G. Dyar, Library of Congress |
| MOE | Memorandum of Errors |
| NBA | National Bahá'í Archives, Wilmette, Illinois |
| NARA | National Archives & Records Administration |
| NLB, SAKP | Sigard Adolphus Knopf Papers, History of Medicine Division, National Library of Medicine |
| NYSA, JALP | New York State Archives, Albany, Joseph A. Lintner Papers |
| RUL-SC&UA | Rutgers University Library, Special Collections and University Archives |
| SIA | Smithsonian Institution Archives |
| UMA, CHFP | Univerisity of Massachusetts Archives, Charles H. Fernald Papers |

## INTRODUCTION

1. Written by members of the Lepidopterists' Society between 1973 and 1976 to poke fun at its fellow enthusiasts, both living and deceased, one "pellet" of *Frass*, a term for insect feces, came out each year at the location of annual meeting during its four years of its existence. See Anonymous, "A Fabulous Man was H. Dyar," *Frass: an Occasional Journal of Para-lepidopterology* (1974).

## CHAPTER 1

1. Jones, *Historical Sketch*, 36.
2. Harrower, *In Memoriam: Harrison Gray Dyar*, 14.
3. Munroe, *Concord Antiquarian Society Papers*, 4.
4. Anonymous, "A New Era in Chronometry." *Christian Advocate*, 1/1/1824.
5. Jones, ibid.
6. Ibid., 35.
7. The biographical reference by Harrower mentions Dyar spending time in Paris and London, stating that Dyar was abroad for nearly thirty years rather than the twenty suggested herein. See Harrower, *In Memoriam*, 8; see account of Charles Walker in Jones, ibid.
8. Perhaps a family trait for tunneling. See: Anonymous, "List of Patents [Improvement of Tunneling]." *The London and Edinburgh Philosophical Magazine and Journal of Science* 1 (1832).
9. Anonymous, "New Patents (Sodium Bicarbonate)." *The Metropolitan Magazine* 23 (1838); Kiefer, DM, "Soda ash, Solvay style. Today's Chemist at Work." *Chemistry Chronicles* 11 (2002); see also http://www.pburch.net/dyeing/FAQ/sodaash.shtml; Anonymous, "Improvement in the Manufacture of White Lead (Charles Button and Harrison Grey [sic] Dyar of London)." *United States Patent Office* 1 (1839).
10. Munroe, ibid., 13.
11. Dyar, H.G., "Editorial: The Old New England Stock." *Reality* 7 (1924); Anonymous, "H.G. Dyar, president of New York and New Haven Railroad." *New Haven, Connecticut? City Directory* (1849), 268; Anonymous, "General Railroad Convention." *New York Times*, 1/17/1855, 4; Anonymous, *New York Daily Tribune*, 5/11/1855.
12. See Anonymous, "Report of the Commissioner of Patents for the year 1857." *Arts and Manufactures, in Three Volumes.* Serial Set Vol. No. 925; Report: S.Exec.Doc. 30 vol.1 (1858).
13. See, Ripley, G., and C.A. Dana, eds., "Steam Carriage." *American Cyclopaedia*. V. 15, 335-336 (D. Appleton & Co., New York, London, 1883).
14. United States Internal Rev. Serv. Assessment Lists 1862-1918.
15. For information on divorce date for Parnie Hannum see: Dyar and Allied Families, LOC, Rare Books, CS71, D996, 1903Q; for information about the Hannum family is in: 1860 United States Federal Census (Aaron C. Hannum), Berkshire, Adams, MA.
16. Listening to the oral history interviews with Roberta Hill, I noted her response to our question about there being "a medium in the family." She mentioned she had heard of a relative who was a medium at the Lincoln séances. Around the same time I found in Dyar's Lepidoptera Catalogue that "Aunt Parnie" had collected a tiger moth (*Arctia* [*Grammia*] *virgo*) for her nephew in 1889. Surmising that Parnie was a Hannum, a quick Internet search of "Parnie Hannum séance" led directly to accounts her participation in the Lincoln séances. See Henson, P.M. and M.E. Epstein, "Dyar Family Oral History Interviews (R. Hill–L. Dyar)," 1999, SIA, RU009570; Dyar Catalogue, "Aunt Parnie collects Arctia virgo," 6/29/1889.
17. Maynard, N.C., *Was Abraham Lincoln a Spiritualist?* 37–38.
18. Ibid., 79–80.

19. Ibid., 82–83.
20. Ibid., 100.
21. Today the riots have become known from "Gangs of New York," Martin Scorsese's 2002 movie loosely based on Herbert Asbury's book. See Phisterer, F., *New York in the War of the Rebellion*, 28th ed. (Albany: J.B. Lyon Company, 1912); see also Anonymous, *Twenty-eighth Battery Light Artillery. Annual Report of the Adjutant-General*, vol. 2. New York.
22. Harrower, C.S., ibid.; Dyar, H.G., "Editorial," ibid.
23. It is possible that prior to Perle Dyar's birth, the Dyars lived in Milton, Massachusetts. "The house of H.G. Dyer [sic?]" was robbed of a "small trunk containing notes, mortgages, certificates of Blue Hill and Continental National Bank stock, and also between $9,000 and $10,000 in United State bonds.... The trunk was found after a search, but the U.S. bonds had been abstracted by the thieves." According to Dyar's 1903 genealogy, there was no one named "H.G. Dyer" among the "ER" Dyers. The wealth of the victim suggests that this could have been H.G. Dyar, Sr. See Anonymous, *Albany Evening Journal*, 2/15/1867, 3; see also Dyar and Allied Families, ibid., and Dyar, H.G., ibid.
24. An undated map confirms the transfer of the 178 acres of land at Linwood Hill between Federal Vanderburgh to H. G. Dyar. See Rhinebeck Historical Society, item 1995-026.02, Callahan Collection: Maps, 1791-1954.
25. 1870 United States Federal Census (Harrison G. Dyar), New York, Dutchess, Rhinebeck, 56.
26. Harrower, ibid., 12–13.

## CHAPTER 2

1. Dyar, "Blue Book," 7/27/1882.
2. 1875 New York State Census, Dutchess Co., Rhinebeck.
3. The articles reported a heist of Eleonora Dyar's "diamonds and pearls" in December 1878, a theft that was perhaps the basis for later short stories by Dyar on the subject, including "Diamonds Going and Coming" and "An Involuntary Wedding." See: Anonymous, Kingston, NY, *Daily FreeMan*, 1/6/1879, [np]; Anonymous, Poughkeepsie, NY, *Daily Eagle*, 1/6/1879.
4. Anonymous, "A Good Entertainment," *The Rhinebeck Gazette*, 1/15/1880, 1.
5. The original "Blue Book" and other notebooks used by Dyar are in the Department of Entomology Archives, National Museum of Natural History, Smithsonian Institution, Washington, D.C.
6. Dyar Story, "The Rash Act of the Reverend Mott," 9/7/1915, SIA, HGDP, Acc. 95-006.
7. Dyar to P.N. Knopf, 7/26/1901, ibid.
8. Dyar's later comfort with spiritualism is suggested by his children staying at Watkins Glen, under the care of Mary Robbins Meade, who performed séances. See: Carr, V.S., *Paul Bowles: A Life* (New York: Scribner, 2004). Dyar also wrote a number of short stories about reincarnation in his last fifteen years of life. These findings suggest Dyar's religious views were complex, not simply materialistic as suggested by Epstein, M.E., and P.M. Henson, *American Entomologist* 38 (1992).
9. Dyar to P.N. Knopf, 7/26/1901, ibid.
10. Ibid.
11. Dyar shared the stories with a mutual acquaintance Annie. See ibid., 12/5/1913, SIA, HGDP, Acc. 95-006.
12. Ibid; "Cruikshank" likely was from Dickens illustrator George Cruikshank, who wrote about debunking apparitions. Having him fear the "Banshee" may have

entertained spiritualists; for Dyar story see ibid., and Cruikshank see Ruickbie, L, "A Brief Guide to Ghost Hunting," (Constable & Robinson, London, 2013).
13. Ibid.
14. In earlier decades Lucy Hudson's patients around Hartford, CT, regarded her "excellent judgment and remarkable success" at curing ailments in spite of no "doctor's diploma or medical degree." Her small "magical 'pellets' [were] unlike the old 'allopathic' pills, drugs and doses." See: Timlow, H.R., "Ecclesiastical and other sketches of Southington, Conn," http://www.ebooksread.com. "Lucy A. Hudson" was thirty-six years old and living in Hartford Co. (Southington), CT in 1850 and sixty-six when with the Dyars in 1880. See 1850 United States Federal Census (Lucy A. and Henry J. Hudson), CT, Hartford, Burlington, p. 6 and 1880, ibid. (Eleonora Dyar), New York, Dutchess Co., Rhinebeck, District 66, p. 25; King, W.H., *History of Homeopathy and Its Institutions in America*. Vol. 1 (New York: Lewis Publishing Co., 1905), 207; Anonymous, Poughkeepsie, NY, *Daily Eagle*, 1/6/1879.
15. For biographical information about the Hudson family, see Anonymous, Poughkeepsie, NY, *Daily Eagle*, 1/6/1879 [np]; Simpson, A., "Guide to the Hudson Family Papers 95.3 (revised 2009 by D. Kimok)," Special Collections, B.F. Feinberg Library, SUNY Plattsburgh, Plattsburgh, NY; Anonymous, "Henry James Hudson Collection," SC1998.39 M2A 6,4. Sibley Library, Eastman School of Music. Rochester, NY; Seeking to contact a departed sister, the Hudsons sought a medium, perhaps finding Mrs. Dyar. See H. Hudson Correspondence, Ms 90476, Connecticut Historical Society.
16. For records of hundreds of moths collected at electric lights by George H. Hudson for Dyar, see Dyar Catalogue, 1885–1891.
17. Charles Hudson's daughter Lucy, the grandniece of "Aunt Lucy," was a renowned violinist.
18. DeGarmo's other interests included ornithology, geology (possibly fossils), and early human inhabitants in North America. He also wrote on the topic of religion, including a book about the Hicksite Quakers. See: Anonymous, "DeGarmo Institute: Catalogue," *Consortium of Rhinebeck History, Rhinebeck Historical Society Library* data 1995-007.01 (1881–1882): 87; Cassino, S.E., *The Scientists International Directory* (Boston: S.E. Cassino, 1882), 31; Marquis, AN, *Who's Who in America, Volume 5.* (1908), 482; Anonymous, "Thirty-Fifth Regular Meeting 1/4/1887," *Transactions of the Vassar Brothers Institute and the Science Section* 4 (1885–1887), 87.
19. The permanence of Eleonora's move to Boston is suggested by the absence of receipts to workers at Linwood Hill from this point forward. In addition to Blue Books see: Anonymous, "Dyar, Mrs. E. R., house 123 Mt. Pleasant Avenue." *Boston City Directory*, 1883.
20. Anonymous, "Roxbury Latin School, Reports for the Years 1882–1883, 1883–1884, and 1884–1885, and Roxbury Latin School Graduates of 1885," Roxbury Latin School.
21. Beccaloni, G.W., M.J. Scoble, G.S. Robinson, and B. Pitkin (eds). (2003). "The Global Lepidoptera Names Index (LepIndex)"; Pape, T. and F.C. Thompson (2010). "Systema Dipterorum: The Biosystematic Database of World Diptera [version 1.0]"; Smith, D.R., *Transactions of the American Entomological Society*.
22. Anonymous, "Roxbury Latin School, Reports," ibid.
23. Anonymous, "Harvard College Class of 1889 secretary's report, issue 6, by Harvard College (1780–). Class of 1889 (Henry Newell Herman p. 102 and James Montgomery Newell p. 160)," 6/1909.
24. Listed as a cadet in: Anonymous, "Institute of Technology." *Boston Herald*, 12/2/1885, 3; Dyar referred to as "student, boards above" with his mother at the 170 W. Chester Park address in the Boston City Directory in 1885 and 1886; whereas he

is listed with Jacob Clough in 1887 and 1888, not as a boarder or a student, after his mother had married; Harrison Gray Dyar papers, LOC, Manuscript Div, MMC 2567, Dyar Genealogy, Book 4, Female Lines 2.
25. Dyar, H.G., "Investigation of a proposed synthesis of tartaric acid from butyric acid" (B.S., Massachusetts Institute of Technology, 1889), Institute Archives—Noncirculating Collection 3; Crosby, W.O., "Relations of the Pinite of the Boston Basin to the Felsite and Conglomerate." *Technology Quarterly and Proceedings of the Society of Arts, Massachusetts Institute of Technology* 2 (1889).
26. Anonymous, "Marcellus Ayer, Eugene Ayer Boarder," *Boston City Directory*, 1873, 67; Anonymous, "Marcellus Seth Ayer Founder, First Spiritual Temple October 8, 1839 to January 30, 1921," http://www.fst.org/msayer.htm; Anonymous, "Dedication of a Temple. Opening of the Magnificent Edifice, the Gift of Mr. M. S. Ayers." *Boston Daily Globe*, 9/28/1885, 2.
27. The trance discourse was to commemorate the thirty-seventh anniversary of the new spiritualist movement in the United States, which began with the Fox Sisters in Hydesville, New York. See Dyar, E.R., "First Spiritual Temple. "The New Dispensation" (2011), http://www.fst.org.
28. Anonymous, "Jacob N.M. Clough birth in Monmouth, ME," (1828), http://trees.ancestry.com; Anonymous, *Knights Templar (Masonic order), Knights Templar (Masonic order). Grand Encampment (MA and RI)* (Printed by John Wilson and Son, 1864); 1860 United States Federal Census (Jacob N.M. Clough), Massachusetts, Suffolk Co., Boston Ward 4, p. 156; Anonymous, "Clough, Jacob N.M., dry goods, 36 Hanover, boards 17 Hancock [Somerville]," *Boston City Directory*, 1873; 1880 United States Federal Census (Jacob N.M. Clough), Massachusetts, Suffolk Co., Boston District 663, 23.
29. Death of Eleonora R. Dyar Clough, "Massachusetts Death Records, 1841–1915," 5/18/1888, no. 33, index and images from Massachusetts State Archives.
30. Anonymous, "First Spiritual Temple. Our Church Pioneers. Mrs. E.R. Dyar," (2011), http://www.fst.org; Anonymous, "Deaths (Clough in this city)," *Boston Evening Transcript*, 5/18/1888, 4.
31. That Eleonora Dyar was among the "materialized spirits" at Nettie Colburn Maynard's séance is easy to overlook because her name appeared as "Mrs. Cora Dyer Clough" rather than "ER Dyar Clough," as she had been known in Boston. "Cora" was likely a misprint of "Nora," a nickname she was known to use, and it is plausible that her two sisters in attendance would have used it. See Anonymous, "Spirits in Glee. A Materialization That Was Surprising to the Natives." *Wheeling Register*, 4/6/1891; Anonymous, "Shadows of Departed Ones: Did Allen Putnam Appear in the Boston Spiritual Temple Wednesday?" *Boston Herald*, 10/31/1887.
32. Dyar Catalogue, 5/31–7/25/1888.
33. Dyar's physical appearance is from a 1923 passport application. See: Anonymous, 7/9/1923, NARA, Washington, D.C., Special Passport Applications, 1914–1925, ARC Identifier 1150696, MLR Number A1 536, b4211, v24; see also: Anonymous, "Institute of Technology," *Boston Herald*, 12/2/1885, 3.
34. 1870 United States Federal Census (Philo Peabody), Maine, Oxford, Canton, p. 18; Peabody, SH, *Peabody (Paybody, Pabody, Pabodie) genealogy* (Boston, Massachusetts: Charles H. Pope, 1909), 123, 245; 1850 United States Federal Census (Susanna Peabody), ibid., p. 4; 1860 United States Federal Census (Philo Peabody), ibid., p. 16.
35. Zella M. Peabody's 5' 10" is deduced from a photograph of her and Harrison Dyar taken in 1925 or 1926, in which they are similar in height (see Figure 25.2a).
36. Dyar, H.G., "The Old New England Stock." *Reality* 7 (1924).

37. Dyar Catalogue, 2/24–4/26/1888; ———, "Arthur Hudson sent eastern butterflies," 1888; ———, "Arthur Hudson brought from California various butterflies," 5/1889; ———, "Arthur and Susie collect at electric lights in Pasadena," 2/1889.
38. ———, 2/24–4/26/1888; Anonymous, *Pasadena Daily News*, 10/15/1889; Anonymous, "S. Adolphus Knopf." *New York Times*, 7/16/1940, 17; Anonymous, "Perle N. Dyar," ibid., 12/25/1931, 23.
39. See Dyar Catalogue, 2/24–4/26/1888.
40. James S. Miller, personal communication.
41. Anonymous, "Perle N. Dyar," *New York Times*, 12/25/1931, 23; Anonymous, "S. Adolphus Knopf," ibid., 7/16/1940, 17; Anonymous, "Weddings," *Los Angeles Daily Herald*, 10/20/1889, 3.
42. Dyar Catalogue, 10/26–30/1889.

## CHAPTER 3

1. The other book Dyar mentions appears to be lost. Dyar Blue Book, 6/28/1882.
2. Ibid., 6/17–10/6/1882.
3. Dyar refers to the woodbine-feeding caterpillar, an eight-spotted forester moth (*Alypia octomaculata*) (see Plate 5e), along with Syracuse. However, he mentions "Miss Lizzi's 2nd Woodbine cat." and "Miss Bowne's 1st Woodbine cat. Spun," which suggests it was from Rhinebeck because Eliza Sands Bowne was his neighbor there. Dyar apparently brought his caterpillars along with him to Syracuse.
4. Ibid., 6/28/1882; 1880 United States Federal Census (Eleonora Dyar), New York, Dutchess, Rhinebeck, District 66, 25.
5. Dyar's neighbors were second or more generation New Yorkers. Mrs. Asher in his notes was either Catherine (wife of George), or Louisa (wife of Philip). Young Dyar's new net proved useful the following day when collecting butterflies—"Argynnis Aphrodite, Eudemus Bathyllus, and Hipparchia Alope" (now *Speyeria aphrodite, Thorybes bathyllus,* and *Cercyonis pegala*)—with it. See Dyar Blue Book, 6/28/1882 and 1880 United States Federal Census (Eleonora Dyar), ibid.
6. Dyar Blue Book, 8/3/1882.
7. The swallowtail collected by Josiah Hannum was incorrectly identified in the Blue Book as the "thoaz" [sic] swallowtail. Giant swallowtails (*Papilio cresphontes*), while localized, are found in the northeast; whereas the thoas swallowtail (*Papilio thoas*) is a similar southern species.
8. Ibid., 8/13–14/1882.
9. In 1920, when Dyar figured out that William Barnes had removed some limacodids from the National Collection, he referred to them as his "most cherished treasures." See Dyar to L.O. Howard, 10/11/1920, NARA, RG7, E-34, b125, f "Dy5."
10. Records in Dyar's Catalogue of Lepidoptera began in 1885, a 6 × 8½-inch lined ledger book that lists the specimens in Dyar's collection. Entries have unique numbers on the left (corresponding with those placed on specimens), followed by locations, dates of collection or emergence (= OUT), and names of relatives and friends that donated specimens. Also included are records of specimens purchased, including the cost, from private collectors and auctions at NY entomology clubs and exchanges with other collectors. The catalogue now serves as a gazetteer for Dyar's travels, especially from 1885 to 1902. The final total of specimens reached in 1913 was 39,950. The first twelve years of this catalogue were transcribed from other notebooks, as they were hand written over "DR DEAN 1897," which had been rubber-stamped on the pages (see top of Plate 4b). Dr. Bashford Dean, an expert on the embryology and paleontology of fishes, was a professor

in the Dept. of Zoology at Columbia University in 1897 when Dyar was a faculty member in Pathology at Columbia College of Physicians and Surgeons. Apparently Dean either gave Dyar the spare notebook or Dyar grabbed a discarded one. See Anonymous, "Faculty (Bashford Dean)," *Columbia College in the City of New York. University Bulletin* (1896).

11. Louisa Hoff, whose finds are denoted in the catalogue as "Lou in Rhinebeck," would later live with Perle and Dyar's brother-in-law Dr. Knopf in Paris and New York. Dyar Catalogue, 8/1885.
12. Ibid.
13. Ibid., 5/1886.
14. ———, 6–8/1886; Dyar Blue Book, 8/23/1886, Entry 31
15. Dyar collected at sites near Linwood Hill and Rhinebeck in 1887 including Esopus Island, Pells Dock, Post Road, Montgomerys, Kaaterskill Falls (Greene Co.) and Shaupeneek Ridge (Ulster Co.). Dyar also mentioned *Catocala relicta* and *Desmia maculates* from Perle in Rhinebeck and in late 1887 reported purchases or exchanges of specimens with Dr. C.S. McKnight, of Saranac Lake, NY from throughout the United States. See Dyar Catalogue, 6–12/1887; for Hudson's collection see: Anonymous, "Vice-Principal and Professor of Natural Sciences. . ." *Elmira Daily Gazette*, 9/22/1893, 6.
16. Dyar's records of moth emergences in Boston occurred on the same dates in March and April as his collecting records in Florida. This suggests one of the siblings was in Boston, while the other was in Florida, or someone else was keeping track of the emergences in Boston. Because Dyar was still enrolled at M.I.T. at this time, it seems reasonable that he went back to Boston after a few weeks in Florida to attend classes. However, Perle's name is only mentioned by catalogue record 830 on February 25 (Altamont) and 883-884 for two specimens on April 5 (Rockledge).
17. The reference to "Perle got" is difficult to interpret because it is at the beginning of a string of records that have quotation marks following for the location. However, it is most likely that Perle only collected one specimen because there is not an extra quote mark for her name, only one each for Altamont Spring, and Florida. See Dyar Catalogue, 2/24–4/26/1888 Entry Nos. 898-1132.
18. The aquatic snout moths included *Nymphula* (now *Paraponynx*) *badiusalis* and *N*. (now *Elophila*) *obliteralis*. See Dyar, H.G., "The North American Nymphulinae and Scopariinae," *Journal of the New York Entomological Society* 14 (1906); ———, ibid. 10, 202–208; Dyar Catalogue, 5/31–7/25/1888.
19. ———, 6/25–6/29/1889.
20. ———, 6/8–7/11/1889.
21. Strecker, *Lepidoptera, Rhopaloceres, and Heteroceres, Indigenous and Exotic; with Descriptions and Colored Illustrations* (1872).
22. Dyar to J.A. Lintner, 6/6/1890, NYSA, Albany, JALP.
23. C.H. Fernald to Dyar, 3/12/1887, SIA, HGDP; this is the earliest known scientific correspondence of H.G. Dyar, Jr.
24. See Dyar to F.H. Strecker, 10/26/1888, FMNHIA, HSP, s1, b21, "Dyar, H.G."
25. Ibid.
26. Dyar, H.G., "Partial preparatory stages of *Dryopteryx rosea*, Wlk," *Entomologica Americana* 4 (1888); ———, "Preparatory stages of *Janassa lignicolor*, Walker," 5 ibid; ———, "Description of the larva of *Datana major*, G. & R.," *The Canadian Entomologist* 21 (1889); ———, "The larva of *Limacodes inornata*, G. & R.," ibid; ———, "Preparatory stages of *Euplexia lucipara*, Linn," ibid; ———, "Note on the larva of *Thyatira pudens*, Guen," ibid.

27. C.V. Riley, in his published reply to Dyar's correspondence in *Insect Life*, made the common misspelling of *Harpalus pensylvanicus* as *H. pennsylvanicus*. See ———, "Insects at electric lamps," *Insect Life* 1 (1889), 285.
28. J.B. Smith to Dyar, 1/24/1889, SIA, HGDP, b2, f6.
29. Dyar to F.H. Strecker, 4/6/1890.
30. Dyar to J.A. Lintner, 6/6/1890, NYSA, Albany, JALP.
31. C.V. Riley to Dyar, 5/29/1890, SIA, HGDP, b2, f3.
32. Dyar, H.G., *Entomologica Americana* 6.
33. The manuscript of these articles is found in NARA, RG7, b19, along with letters to Riley and Howard pertaining to *Insect Life*. See C.V. Riley to Dyar, 5/29/1890; Dyar, H.G., "Preparatory stages of *Syntomeida epilais* Walker and *Scepsis edwardsii* Grote," *Insect Life* 2; ———, "Description of Certain Lepidopterous Larvae," *Insect Life* 3 (1890).
34. Dyar to C.V. Riley, 11/29/1890; Dyar to L.O. Howard, 12/19/1890, ibid.
35. Dyar, H.G., "Description of Certain Lepidopterous Larvae," *Insect Life* 3 (1891), ibid.
36. ———, "A list of Sphingidae and Bombycidae Taken at Electric Lamp at Poughkeepsie, N.Y.," ibid.
37. See *Nola* manuscript and others in NARA, RG7, b19.
38. Dyar to J.A. Lintner, 6/6/1890, NYSA, JALP; ———, 8/30/1890.
39. ———, 6/6/1890; Dyar Catalogue, 6-7/1890.
40. Dyar, H.G., "Descriptions of the Preparatory Stages of Two Forms of *Cerura cinerea* Walk." *Psyche* 6 (1891).
41. Mallis, A., *American Entomologists* (New Brunswick, NJ: Rutgers, 1971), 292–294.
42. Ibid.
43. Henry Edwards to Dyar, 12/2/1890, SIA, HGDP, b1, f6.
44. Dyar to J.A. Lintner, 6/13/1891, NYSA, JALP.
45. Dyar, H.G. "The Number of Molts of Lepidopterous Larvae." *Psyche* 5 (1890), 175–176.
46. It can't be determined exactly when Dyar began measuring the head widths for the project. He could have measured preserved larvae in alcohol or shedded heads from Florida after returning to Rhinebeck in late March 1890. See Dyar Blue Book, 12/1889–9/1890.
47. Dyar, H.G. "The Number of Molts of Lepidopterous Larvae" (1890).
48. Dyar wrote that the head width of the sixth instar of *Hyphantria cunea* was not used because "its head is seen to have been dwarfed" and it later died. He used another larva for the measurement of the final instar (2.4 mm). See Dyar, H.G., (1890), 422; Dyar Blue Book, 6/27/1890, Entry No. 193.
49. Dyar, "The Number of Molts of Lepidopterous Larvae" (1890).
50. Ibid. Brooks independently developed a similar model, referred to as Brooks' Rule, based on the length of stomatopod Crustaceans, to differentiate species from the *HMS Challenger* in the 1870s rather than their number of molts. See Benson, K.R., "Brooks, Stomatopods, and Decapods: Crustaceans In Research And Teaching," in *History of Carcinology*, ed. Frank Truesdale (Rotterdam, Brookfield,: CRC Balkema, 1993). For unpublished calculations of limacodid body length from head-width ratios see Dyar Blue Book Entries 532–574.
51. The growth ratio for caterpillars is often given as between 1.25 and 1.4. However, Dyar's ratios in the 1890 paper shows a range of 1.3 to 1.7.
52. There are many instances in Dyar's notes in the Blue Books of egg masses found on particular plants in nature and rare notes of measurements from shed skins.
53. Ironically, it would prove difficult for Dyar to make head-width measurements of limacodid caterpillars because their heads are pulled inside the thorax, much like a

turtle pulls its head inside the shell. Furthermore, the old head skin is unavailable to measure because it's eaten after being shed. However, Dyar made head width measurements for several limacodid species in 1895.
54. Dyar, H.G., "The Number of Larval Stages in the Genus *Nadata.*" *Psyche* 6 (1892).
55. Resh, V.H., and R.T. Cardé, *Encyclopedia of Insects.*

## CHAPTER 4

1. Dyar to J.A. Lintner, 5/18/1891, NYSA, JALP.
2. Anonymous, "Meeting of NY Ent. Soc. 4/2/1895," *Journal of the New York Entomological Society* 3 (1895); Minutes of NY Ent. Soc., 4/16/1895, AMNH, NY Ent. Soc. Archives.
3. Dyar Catalogue, 7/23–24/1889; Dyar to C.H. Fernald, 4/23/1903, UMA, CHFP.
4. Dyar Catalogue, 9–10/1889; Anonymous, "Marriage Licenses." *Los Angeles Daily Herald,* 10/10/1889, 8.
5. Dyar, H.G., "Description of Certain Lepidopterous Larvae." *Insect Life* 3 (1891): 390.
6. In his Catalogue Dyar referred to collecting sites as Palm Beach in 1889–1890, and occasionally more specific localities such as the west shore of Lake Worth and Jupiter. He surveyed "Lake Worth" in 1900, but it is clear that this was the same as "Palm Beach," described as "[t]he strip of land between Lake Worth and the Atlantic Ocean." See Dyar Catalogue, 11/1889–3/1890 and Dyar, H.G., "Notes on the Winter Lepidoptera of Lake Worth, Florida." *Proceedings of the Entomological Society of Washington* 4 (1901).
7. *Arctia docta* is now *Notarctia proxima.* See Dyar, H.G., "Preparatory Stages of *Arctia docta* Walk," *Entomologica Americana* 6 (1890)
8. Dyar to F.H. Strecker, 3/26/1890, FMNHIA, HSP, s1, b21, "Dyar, H.G."
9. Dyar Catalogue, 3/1890.
10. Soon after their arrival in late March, Zella purchased groceries in the village of Rhinebeck. See Dyar Blue Book 14, 7/3/1890.
11. Dyar, H.G., "A list of Sphingidae and Bombycidae Taken at Electric Lamp at Poughkeepsie, N.Y," *Insect Life* 3 (1891).
12. Zella opened a grocery account at C. Steffens on Fifty-ninth Street in New York City. She would spend around $17 on groceries each month until the following spring. See Dyar Blue Book 10, 10/1890.
13. Dyar wrote Lintner from the European St. James Hotel in Washington, D.C. about identifying the moth (*Edema albicosta*; now *Symmarista albifrons*) at the U.S. Department of Agriculture: "I have examined Herrich-Schäffer's figure and it represents the form with the sharp narrow tooth projecting from the white costa band—so you are right in your separation of the forms and the species common at Rhinebeck . . . which I have taken in *Edema albiscosta.*" See Dyar to J.A. Lintner, 5/18/1891, NYSA, Albany, JALP.
14. Dyar Catalogue, 4/29–5/9/1891.
15. Dyar Catalogue, 5/22–9/24/1891.
16. Dyar, H.G., "Collecting butterflies in the Yosemite Valley," *Entomological News* 3 (1892): 30–31; Smith, D.R., "The Sawfly Work of H.G. Dyar (Hymenoptera: Symphyta)," *Transactions of the American Entomological Society* 112 (1986).
17. Dyar, H.G., "Choice of Food [Letter to editor]," *Psyche* 6 (1891).
18. Dyar to H. Skinner, 11/11/1891, ANSPL, AES, ms coll E76, no. 8.
19. Trexler, K.A., "The Tioga Road; a History 1883–1961," http://www.yosemite.ca.us/library/tioga_road/business_ventures.html.
20. Dyar, "Collecting butterflies in the Yosemite Valley," *Entomological News* 3, 32–33.

21. Ibid.
22. Dyar, H.G., "A New *Hepialus* from California," Ibid. 5 (1894); Dyar to F.H. Strecker, 11/4/1894, FMNHIA, HSP, s1, b21, "Dyar, H.G."
23. Bethune, C.J.S., "John B. Lembert," *The Canadian Entomologist* 28 (8).
24. Ibid., 217–218.
25. The Dyars left Yosemite by stagecoach. The "stage ride out of the [Yosemite] valley was more than they [moth larvae] could endure ..." See Dyar, H.G., "Notes on Bombycid Larvae—III." *Psyche* 6 (1891), 178.
26. Dyar Catalogue, 10–12/1891; Dyar Blue Book, cover, 1891–1892.
27. Dyar to H. Skinner, 11/11/1891, ANSPL, AES, ms coll E76, no. 8; Dyar Catalogue, ibid.
28. ———, ANSPL, AES, ms collection E76, no. 8; Dyar, H.G., "*Thecla californica*," *Entomological News* 5.
29. Dyar to J.A. Lintner, 5/18/1891, NYSA, Albany, JALP.
30. This is the same address and contact information mentioned in ads placed by Dyar in *Can. Ent.* and in Cassino's naturalist directory at the time. See Dyar to H. Skinner, 11/11/1891, ANSPL, AES, ms coll E76, no. 8; Cassino, S.E., *The Scientists International Directory*. (Boston: S.E. Cassino, 1894).
31. Anonymous, "Passengers." *Daily Bulletin* (Honolulu, Hawaiian Islands), 1/26/1892, 3; Dyar Catalogue, 2–3/1892.
32. Dyar, H.G., "Larvae from Hawaii—a Correction." *The Canadian Entomologist* 32 (1900); Dyar Blue Book, 1893–1902 Books 10–23.
33. Among the species from Rhinebeck in Dyar's baggage was the limacodid *Packardia elegans*. See Dyar Catalogue, 2–3/1892.
34. J. Lintner to Dyar, 5/13/1892, SIA, HGDP, b2, f 1.
35. Dyar Catalogue, 4/6–6/17/1892.
36. Ibid., 6/27/1892.
37. Ibid., 7/11–15/1892.
38. Dyar Blue Book, 1893–1902, Books 10–23; Smith, D.R., "The Sawfly Work of H.G. Dyar (Hymenoptera: Symphyta)," *Transactions of the American Entomological Society* (1986).
39. Ibid. Dyar noted sawfly larvae on two previous occasions, one at Yosemite and the other in Portland, Oregon, in 1892; however, at this juncture he either kept a separate notebook or transcribed his notes from other sources, numbering the rearing records alphabetically. Once through the alphabet, he added a numerical prefix and goes through the alphabet again. Dyar used the spelling "Woods Holl" in these and his Lepidoptera notebooks.
40. Ibid.
41. The moth was *Acronicta morula*. See Dyar Blue Book, 9/06/1894, Entry No. 412A.
42. Anonymous, "Adirondack Passtimes [sic]: Sojourners in the Summer Resorts of the North Woods Country Having a Gay Time," *New York Evening Telegram*, 7/29/1894, 5; Anonymous, "Savants in Brooklyn ...," *New York Daily Tribune*, 8/16/1894, 5; Anonymous, "Scientists' Day of Pleasure. Some Went to Pleasure Bay, Others Dredged the Cholera Banks," *New York Times*, 8/19/1894, 8.
43. Zella's description of the paddle caterpillar (*Acronicta funeralis*) began as follows: "Head reddish; lobed; the labium black. Body black (like carbon). On each segment dorsally, is a white raised patch divided in the centre." Dyar changed "labium" to "labrum"—an error he made himself several years earlier. See Dyar Blue Book, 7/10/1894, Entry No. 503.
44. The record of the sawfly caterpillars on W. Ninety-ninth Street is found in: Dyar, H.G., "On the Larvae of Some Nematoid and Other Sawflies from the Northern Atlantic States." *Transactions of the American Entomological Society* 22 (1895), 312;

While in Plattsburgh Dyar collected a rosy striped oak worm moth (*Anisota virginiensis*). See Dyar Catalogue, 6/10/1895.
45. R. Ottolengui to M. Perkins, 6/10/1895, SIA, HGDP, b2, f 3.
46. Smith, "The Sawfly Work of H.G. Dyar (Hymenoptera: Symphyta)" (1986); Dyar Catalogue, 6–7/1895.
47. Dyar Catalogue, 7/2/1895.
48. Dyar, H.G., *Journal of the New York Entomological Society* 5. No date given in Dyar (1897), but Aug. 22 corresponds with notebook for the species and host in the Dyar sawfly book cited by D. Smith (1986).
49. Smith, "The sawfly work of H.G. Dyar (Hymenoptera: Symphyta)" (1986); Minutes of NY Ent. Soc., 5/18/1897, AMNH, NY Ent. Soc. Archives.
50. Dyar to C.L. Marlatt, 4/13/1895, NARA, RG7, E-3, b11, f "Dra-Dy."
51. Marlatt, C.L., "Some New Nematids." *The Canadian Entomologist* 28 (10).
52. Ibid.

## CHAPTER 5

1. Dyar to J.A. Lintner, 8/23/1892, NYSA, Albany, JALP.
2. Gossel, P.P., "The Emergence of American Bacteriology, 1875-1900" (Ph.D. dissertation, Johns Hopkins University, 1988), 38.
3. Dyar Catalogue, 9–12/1892.
4. Anonymous, "Meeting of Cambridge Entomological Club," *Psyche* 6 (1893): 410, 462, 494. See also Anonymous, "Meeting of Cambridge Entomological Club," *Psyche* 8 (1897)
5. Previous biographical sketches have stated Dyar received postgraduate credit at M.I.T. for the summer course, but Woods Hole records show he was a graduate student at Columbia College.
6. There are no records available to determine which courses Dyar took at Columbia; however, Dyar was invited to attend a dinner for Osborn that was hosted by his former students. See Anonymous, "Osborn Dinner," 2/11/1911, SIA, HGDP, Acc. 95-006, b1, f4; Black, E, *War against the Weak: Eugenics and America's Campaign to Create a Master Race* (New York/London: Four Walls Eight Windows, 2003).
7. Dyar to F.H. Strecker, 12/28/1893, FMNHIA, HSP, s1, b21, f"Dyar, H.G."
8. Dyar to J.A. Lintner, 9/23/1893, NYSA, Albany, JALP.
9. *Antispila nyssaefoliella* is the species of fairy moth (Heliozelidae) reared by Dyar. See ibid.
10. C.V. Riley to Dyar, 9/30/1893, SIA, HGDP, b2, f3.
11. Dyar to F.H. Strecker, 12/28/1893, ibid.
12. Anonymous, "Death of Berthold Neumoegen," *New York Times*, 1/22/1895, 9.
13. B. Neumoegen to Dyar, 4/23 and 5/5/1891, SIA, HGDP, b2, f4.
14. Neumoegen, B., and H.G. Dyar. "Preliminary Revision of the Bombyces of America North of Mexico." *Journal of the New York Entomological Society* 1, 2 (1893).
15. B. Neumoegen to Dyar, 3/24/1893, ibid.
16. Neumoegen, B., "Description of a Peculiar New Liparid Genus from Maine," *The Canadian Entomologist* 25 (1893).
17. The species is now a synonym of *Coenodomus hockingii* Walsingham, obscuring the name, but not the myth.
18. Weismann, A., *Studies in the Theory of Descent*, vols. 1 and 2. Edited and translated and edited, by Raphael Meldola, F.C.S. and Prefatory notice by Charles Darwin, (London: Sampson, Low, Marston, Searle, & Rivington, 1882).
19. Ibid.

20. Dyar, H.G., "Book Notices: Evolution and Taxonomy: An Essay on the Application of the Theory of Natural Selection in the Classification of Animals and Plants, Illustrated by a Study of the Evolution of the Wings of Insects and by a Contribution to the Classification of the Lepidoptera, by John Henry Comstock, B.S. The Wilder Quarter-Century Book, 37-113," *The Canadian Entomologist* 26 (1894).
21. Comstock, J.H., *Evolution and Taxonomy* (Ithaca, NY: The Wilder Quarter-Century Book, 1893).
22. ———, "A Classification of Lepidopterous Larvae," *Annals of the New York Academy of Sciences* 8 (1894).
23. Forbes, W.T.M., "Obituary. Harrison Gray Dyar," *Entomological News* 40 (1929).
24. Dyar to C.V. Riley, 3/3/1894, NARA, RG7, E-3, b11, f "Dra-Dy."
25. L.O. Howard to Dyar, 3/5/1894, ibid.
26. Dyar, H.G., "Additional notes on the classification of lepidopterous larvae," *Transactions of the New York Academy of Sciences* 14 (1894).
27. Perle was living in Paris while her husband, S.A. Knopf, was attending medical school at the Sorbonne. See MacCracken, H.M., E.G. Sihler, and W.F. Johnson, *New York University: Its History, Influence, Equipment and Characteristics, with Biographical Sketches and Portraits of Founders, Benefactors, Officers and Alumni*, vol. 2 (R. Herndon Company, 1903). See Dyar, "A classification of Lepidopterous Larvae" (1894), for Dyar's acknowledgement of Perle's help.
28. C.H. Fernald to Dyar, 4/5/1895, SIA, HGDP, b1, f7.
29. Dyar to C.H. Fernald, 4/6/1895, UMA, CHFP.
30. A.R. Grote to Dyar, 12/2/1895, SIA, HGDP, b1, f8.
31. Packard, A.S., "Studies on the Life-History of Some Bombycine Moths, with Notes on the Setae and Spines of Certain Species," *Annals of the New York Academy of Sciences* 8 (1893).
32. Chapman, T.A., "On the Phylogeny and Evolution of the Lepidoptera from a Pupal and Oval Standpoint," *Transactions of the Royal Entomological Society of London* 44 (1896).
33. Dyar, H.G., "The Larva of *Butalis basilaris* Zell.: The Relations of Its Setae," *Psyche* 7 (1895); the same group of figures appeared again in the December 1895 *JNYES*. The transformation series started with *Plutella porrectella* L. (formerly Tineidae, now Yponomeutoidea), *Simaethis pariana* Dyar (Choreutidae), *B. basilaris* (now *Scythrus*: Gelechioidea), *Oxyptilus periscelidactylus* (now *Geina*: Pterophoridae), *Ino pruni* (now *Rhagades*: Zygaenidae), *Megalopyge crispata* (Megalopygidae), *Sibine stimulea* (now *Acharia*: Limacodidae).
34. A.C. Weeks to Dyar, 11/16/1895, SIA, HGDP, b2, f8; G. Hulst to Dyar, 11/26/1895, ibid., f1.
35. Anonymous, "Meeting of NY Ent. Soc. 2/18/1896," *Journal of the New York Entomological Society* 4 (1896).
36. Of those studied by Dyar—the arctiids, pericopids, lymantriids, and a number of noctuids form the new family Erebidae based on the latest molecular evidence. Dyar found large numbers of the pericopid caterpillars at Palm Beach over the Christmas holidays in 1895 on vines of the poisonous devil's potato (*Echites umbellate*). Mr. F. Kinzel both identified the vines and helped Dyar raise the pericopine larvae. See Dyar, H.G., "On the Probably Origin of the Pericopidae: *Composia fidelissima* H.-S.," ibid.
37. ———, "On the Larvae of the Higher Bombyces (Agrotides Grote)," *Proceedings of the Boston Society of Natural History* 27 (1896).
38. ———, "Notes on the Phylogeny of Saturnians," *The Canadian Entomologist* 28 (1896).

39. Winslow, C.E., "William Thompson Sedgwick 1855-1921," *Journal of Bacteriology* 6 (1922); Gossel, "The Emergence of American Bacteriology, 1875-1900."
40. Dyar, H.G., "On Certain Bacteria from the Air of New York City," *Annals of the New York Academy of Sciences* 8 (1895).
41. Ibid.
42. ———, "Notes on Certain Variations in the Biological Characters of Two Species of Bacteria," *Transactions of the New York Academy of Sciences* 14 (1895).
43. Dyar, H.G., "On Certain Bacteria from the Air of New York City."
44. Ibid., 367, 369, 375; ———, "On the Larvae of Some Nematoid and Other Saw-flies from the Northern Atlantic States," *Transactions of the American Entomological Society* 22 (1895).
45. Dyar, H.G., "Recent Notes on Bacteria," *Transactions of the New York Academy of Sciences* 15 (1896); Adami, J.G., "How Is Variability in Bacteria to Be Regarded?," *Public Health Papers and Reports* 20 (1894).
46. Dyar to L.O. Howard, 3/19/1927, NARA, RG 7, E-35, b83, f "Dyar."
47. Hadley, P., "Microbic Dissociation: The Instability of Bacterial Species with Special Reference to Active Dissociation and Transmissible Autolysis. Six plates." *Journal of Infectious Diseases* 40 (1927); L.O. Howard to Dyar, 3/19/1927, NARA, RG 7, E-35, b83, f "Dyar"; Howard, L.O., *Fighting the Insects: The Story of an Entomologist, Telling of the Life and Experiences.* (New York: The Macmillan Company, 1933), 122.
48. Dyar, "Notes on Certain Variations in the Biological Characters of Two Species of Bacteria" (1895); Epstein, M.E., and P.M. Henson, "Digging for Dyar: The Man behind the Myth," *American Entomologist* 38 (1992); John Cairns, J. Overbaugh, and S. Miller, "The Origin of Mutants" *Nature* 335 (1988); Wainwright, M, "Extreme Pleomorphism and the Bacterial Life Cycle: A Forgotten Controversy," *Perspectives in Biology and Medicine* (1997).
49. Anonymous, *Approved Lists of Bacterial Names*, ed. V.B.D. Skerman, V. McGowan, and P.H.A. Sneath, Amended ed. (Washington, D.C.: American Society for Microbiology, 1989). *Serratia plymuthica* was mistakenly attributed to Dyar in the 1980 version of the "Approved lists"; however, this was corrected in the 1989 version (Lehmann and Neumann 1996).

## CHAPTER 6

1. Dyar, H.G., "The Life-Histories of the New York Slug Caterpillars. (Conclusion)," *Journal of the New York Entomological Society* 7 (1899).
2. The instability of taxonomic names for Lepidoptera during the late nineteenth and early twentieth centuries resulted in alternative names for Limacodidae including: Eucleidae, Cochlidiidae, Cochliopodidae, Apodidae, etc. See Epstein, M.E., "Revision and Phylogeny of the Limacodid Group Families, with Evolutionary Studies on Slug Caterpillars (Lepidoptera: Zygaenoidea)," *Smithsonian Contributions to Zoology* (1996).
3. The debris that sticks to these caterpillars is often frass or fecula, the pellets of caterpillar feces. In nature the caterpillars are freer from this problem because they eject the pellets off of the plant, more difficult in confined and moist containers. See Dyar, (1899), ibid. and Epstein, (1996), ibid.
4. ———, "Evolution of Locomotion in Slug Caterpillars (Lepidoptera: Zygaenoidea: Limacodid group)," *Journal of Research on the Lepidoptera* 34 (1995 [1997]); Murphy, S., J. Lill, and M.E. Epstein, "Natural History of Limacodidae of the Washington, D.C. Region," *Journal of the Lepidopterists' Society* 65 (2011).

5. Dyar, H.G., and E.L. Morton, "The Life-Histories of the New York Slug Caterpillars.-I," *Journal of the New York Entomological Society* 3 (1895).
6. Newcomb, H.H., " Emily L. Morton," *Entomological News* 28 (1917); Anonymous, "Ellison-Morton Family Tree," ancestry.com.
7. E.L. Morton to Dyar, 4/20/1893, SIA, HGDP, b2, f2; Morton, E.L., "Notes from New Windsor. *Isa textula* H.-S," *Entomological News* 3 (1892); Dyar, H.G., "Notes and News: A Bit of History," *Entomological News* 3 (1892).
8. Included among the cocoons, as far as Morton knew, were saddlebacks (*Acharia stimulea*), skiff moths (*Prolimacodes badia*) or rose slugs (*Parasa indetermina*), and possibly the hag moth (*Phobetron pithecium*). See E.L. Morton to Dyar, 4/24/1893, SIA, HGDP, SIA, b2, f2.
9. Dyar, H.G. and E.L. Morton, ibid.: 145.
10. E.L. Morton to Dyar, 5/1/1899, SIA, HGDP, ibid.
11. Dyar, H.G., "The Life-Histories of the New York Slug Caterpillars. -XV," ibid., 6 (1898): 95.
12. Dyar, "The Life-Histories of the New York Slug Caterpillars. (Conclusion)," ibid., (1899): 242.
13. Roberta Hill was told by her mother, Dorothy Dyar Hill, that quince jelly in tumblers was a Dyar family favorite. See Henson, P.M., and M.E. Epstein, "Dyar Family Oral History Interviews (R. Hill–L. Dyar)", 1999, SIA, RU009570.
14. Ibid.
15. Ibid.
16. Ibid., plates vii and viii; Minutes of NY Ent. Soc., 4/20/1897, AMNH, NYESA
17. Dyar, H.G., "On the White Eucleidae and the Larva of *Calybia slossoniae* (Packard)," *Journal of the New York Entomological Society* 5 (1897); ———, "Life-History of a European Slug Caterpillar, *Cochlidion avellana*," ibid., 7 (1899).
18. Anonymous, "Meeting of NY Ent. Soc. October 20, 1896," ibid., 5 (1897).
19. Minutes of NY Ent. Soc., 2/2/1897, ibid.; Dyar Catalogue, 12/1896.
20. Perhaps his 1893 embryology course taken at Woods Hole tweaked Dyar's awareness of the developing embryos. In Lepidoptera it is only this type of thin, flat egg that enables us to see detailed embryonic development from outside the egg. These types of eggs are found in Limacodidae and Tortricidae. See Minutes of NY Ent. Soc., 2/2/1897, ibid.; Dyar, "On the White Eucleidae and the Larva of *Calybia slossoniae* (Packard)" (1897).
21. Dyar, ibid., (1897).
22. Minutes of NY Ent. Soc., 4/201897, ibid.; an article on this subject appeared two years later in Dyar, H.G., "Note on the Secondary Abdominal Legs in the Megalopygidae," *Journal of the New York Entomological Society* 7 (1899).
23. Minutes of NY Ent. Soc., 10/5/1897, ibid.
24. For genealogical trees of Dyar family, see: Harrison Gray Dyar papers, LOC, Manuscript Div., MMC 2567, Dyar Genealogy, Book 1; also see: Dyar, H.G., *A Preliminary Genealogy of the Dyar family* (Washington, D.C.: Gibson Bros., 1903).
25. Those sharing information with Dyar for his genealogy was cousin Charles Warren Dyar, the editor of the *Boston Globe* and Daniel Everett Dyar of Winthrop Center, Massachusetts. C.S. Hannum of Westfield, Massachusetts, Josiah (Jo) C. Hannum, Ruth Marble provided the most information about the Hannums. Letters in the Library of Congress along with Dyar's genealogy notebooks are the only existing record of his communication his aunts and uncle other than reports of Lepidoptera he received from them in the catalogue and in the Blue Book (see Chapter 2). Harrison Gray Dyar papers, LOC, Manuscript Div, MMC 2567, Dyar Genealogy, Book 1.

26. The other Hannum siblings to survive to past 1900 were Angeline (Angie) Nancy Barden, Mary Robinson Bushnell, and Helen Sophia Eastabrook. See LOC, MMC 2567, the Papers of Harrison G. Dyar, b1, f "1899."
27. See Parnie Colburn to Dyar and Z.P. Dyar, 8/19/1899, LOC, MMC 2567, the Papers of Harrison G. Dyar, ibid.

## CHAPTER 7

1. There are a number of correspondence letters from the address at 243 West Ninety-ninth St., beginning with Fernald. See Dyar to C.H. Fernald, 4/6/1895, UMA, CHFP.
2. Dyar, H.G., "Notes on Some Moths from the Collection of Mr. A. Bolter," *The Canadian Entomologist* 28 (1896).
3. The specimens in Dyar's Catalogue would grow to approximately 14,800 by October 1897. He obtained several thousand specimens from William Schaus, intended for the USNM collection.
4. A $12 purchase of "about 100 moths from Mexico and Central America as far south as the Isthmus" made from the Rev. H. Th. Heyde in 1895 included a limacodid Dyar would name *Parasa prasina* (11746–7 in his catalogue)—a homonym—and later rename it *Parasa wellesca* in 1900. Dyar received other specimens in 1895 from William Barnes, T.D.A. Cockerell (Las Cruces, New Mexico), J.B. Smith, J.J. Rivers (U. California), Annie T. Slosson, and purchases from auctions or foreign dealers including Mr. G. Ruscherveyh (Buenos Aires). See Dyar, H.G., "A New *Parasa*, with a Preliminary Table of the Species of the Genus," *Psyche* 8 (1898); Dyar Catalogue, 1895; Dyar, H.G., "Change of Preoccupied Names," *The Canadian Entomologist* 32 (1900).

   In 1896 Dyar's largest purchases were 100 specimens from Staudinger and Hass for 200 marks and around 140 specimens offered at auction by George Franck to benefit the Brooklyn Entomological Society, paying $27.61, roughly 20 cents a specimen. See Anonymous, "Meeting of NY Ent. Soc. 4/22/1896 [Auction]," *Journal of the New York Entomological Society* 4 (1896).
5. See Anonymous, "Real Estate," *New York Herald Tribune*, 4/15/1899, 19; ibid., 10/10/1899.
6. Anonymous, *New York Daily Tribune*, 12/6/1896, 2.
7. Augustus N. Allen designed for the lessee Knickerbocker syndicate "at a cost of $18,000." See Anonymous, "In The Real Estate Field," *New York Times*, 5/25/1909, 14.
8. Anonymous, "In the Real Estate Field," *New York Times*, 1/1/1896, 15; Anonymous, "Real Estate Transfers," *New York Herald Tribune*, 5/4/1899, 10.
9. Dyar to L.O. Howard, 3/5/1897, NARA, RG7, E-3. b11, f "Dra-Dy."
10. ———, 10/31/1895, ibid.
11. L.O. Howard to Dyar, 12/27/1895, ibid.
12. Dyar to/from L.O. Howard, 1/16/1896, 1/18/1896, ibid.
13. L.O. Howard to/from Dyar, 1/22/1896, 1/26/1896, ibid.
14. L.O. Howard to Dyar, 2/7/1896, ibid.
15. ———, 1/31/1896, ibid.
16. Dyar to L.O. Howard, 2/4/1896, ibid.
17. L.O. Howard to Dyar, 2/5/1896, ibid.
18. The Onteora Club was mentioned by Howard in his book *Fighting the Insects: The Story of an Entomologist* (New York: The Macmillan Company, 1933) as a place for "writers, artists, and other persons of note." Howard's aunt and uncle were siblings Candace Wheeler, notable in textile design, and Francis B. Thurber, who founded the club. See L.O. Howard to Dyar, 12/10/1896, ibid.

19. Dyar to L.O. Howard, 2/9/1897, NARA, RG7, E-3, b11.
20. Dyar wrote Howard about their failure to meet at the Department of Pathology at Columbia in early May: "I was sorry not to see you last week as I wanted to have our talk before deciding whether to continue here or not next year, but it will be too late for that now. I should be glad to see you when you come or again, however. After the 6th my classes stop and I have no regular hours at the laboratory, so will meet you by appointment." See L.O. Howard to Dyar and reverse, 3/8 & 5/3/1897, ibid.
21. Dyar to J.A. Lintner, 4/19/1897, NYSA, Albany, JALP.
22. Schwarz, E.A., "Martin Larsson Linell," *Proceedings of the Entomological Society of Washington* 4 (1899).
23. Dyar to L.O. Howard, 6/5/1897, NARA, RG7, E-2, B21.; Schwarz, "Martin Larsson Linell" (1899).
24. The letter from Boston that Howard mentioned, outlining his proposal to Dyar, has not been found. See L.O. Howard to Dyar, 6/7/1897, NARA, RG7, E-2, b 21.
25. Dyar to L.O. Howard, 6/9/1897, ibid.
26. ———, 3/5/1897, NARA, RG7, E-3, B11.
27. L.O. Howard to Dyar, and reverse, 3/8 & 3/10/1897, NARA, RG7, E-3, B11.
28. Anonymous, *The United States National Entomological Collections* (SI Press, 1976); Dyar's 1897 summer fieldwork was mostly in New York City area, possible that he remained in the bacteriology laboratory at Columbia at least part-time. Ashmead and Currie began their positions on July 1, 1897, so it is plausible that Dyar committed to remain at Columbia until fall, when he went to the USNM.
29. Dyar decided to visit Washington in the fall and asked Howard accept a package because, he was concerned about specimens of Australian limacodid moths "being lost in forwarding." See Dyar to L.O. Howard, 6/9/1897.

## CHAPTER 8

1. Henson, P.M., and M.E. Epstein, "Dyar Family Oral History Interviews (R. Hill–L. Dyar)," 1999, SIA, RU009570.
2. Although the exact date of purchase of the home at 1512 Twenty-first Street NW is not known, that Dyar had a February 1899 Entomological Society of Washington meeting at his home suggests he moved in before the end of 1898. In early 1898, soon after his arrival in the District of Columbia, Dyar rented a flat at 110 C Street, Southeast, the present location of the Madison Building of the Library of Congress. See Howard, L.O., "Curators' Annual Reports," 1899, SIA, RU 158.
3. Anonymous, "Embassy of Indonesia" (2011), http://www.embassyofindonesia.org/aboutembassy/building.htm.
4. Peabody was not counted with the Dyar family in the 1900 U.S. Census, but her home base was with them in D.C. See Anonymous, "An Indian Fair," *The Land of Sunshine: The Magazine of California and the West* (edited by Chas. F. Lummis) 1900.
5. 1900 United States Federal Census (Harrison G. Dyar), District of Columbia, Washington, District 42: 21; Henson and Epstein, "Dyar Family Oral History Interviews (R. Hill–L. Dyar)" (1999).
6. Ibid.
7. The adjoining house was purchased in 1904 from Harriet G. Ogden for $15,000. Information on the children roaming freely came from personal communication with Eleanor and Lewis Curd, owners of 1510 and 1512 Twenty-first Street in 1990. See 1900 United States Federal Census (Harrison G. Dyar), ibid., 45; Epstein, M.E. and P.M. Henson, "Digging for Dyar: the Man Behind the Myth," *American Entomologist* 38 (1992).

8. Anonymous, "Pays $17,910 for Garden. Dr. H.G. Dyar Purchases Lot Adjoining His Twenty-first Street Home," *Washington Post*, 11/22/1906; C. Early to Dyar, 12/8/1906, SIA, HGDP, Acc. 95-006, b2, f4.
9. Anonymous, "Meeting of Ent. Soc. Wash. 2/2/1899, 1/4/1900," *Proceedings of the Entomological Society of Washington* 4 (1901); Anonymous, "Meeting of Ent. Soc. Wash. 2/7/1901," *Proceedings of the Entomological Society of Washington* 5 (1902).
10. J.C. Hannum to Dyar, 5/23/1900, LOC, MMC 2567, The Papers of Harrison G. Dyar, b1, f"1900," 2 of 2.
11. Dyar to F.H. Strecker, 9/8/1900, FMNHIA, HSP, s1, b21, "Dyar, H.G."
12. Anonymous, *Woman's Who's Who of America. A Biographical Dictionary of Contemporary Women of the United States and Canada, 1914-1915*, ed. John William Leonard (New York: American Commonwealth Co, 1914); Engle, R.L., *In the Light of the Mountain Moon: An Illustrated History of Skyland* (Shenandoah National Park Association, 2003); Passport Applications (Wellesca Allen), 3/18/1907, NARA, 1/2/1906–3/31/1925; ARC Identifier 583830/MLR Number A1 534; NARA Series: M1490; Roll #32.
13. Engle, ibid., 28; Pollock, G.F. *Skyland: The Heart of the Shenandoah National Park* (Virginia Book Co., 1960).
14. Logan, M., *The Part Taken by Women in American History* (Wilmington, DL: Perry-Nalle Publ. Co., 1912); Henson, P.M., and M.E. Epstein, "Dyar Family Oral History Interviews (W.J.Dyar)," 5/17/1992, SIA, RU009570: 33–34; 1870 United States Federal Census (George H. Pollock), Massachusetts, Middlesex, Weston, p. 27; 1880 United States Federal Census (George H. Pollock), Washington, District of Columbia, District 40, p. 44.
15. Dyar, H.G., "The Life History of a Second Epiplemid (*Callizzia amorata* Pack.),", *Proceedings of the Entomological Society of Washington* 5 (1903); Pollock, G.F., *Skyland (the Eaton Ranch of the East) Situated on High Plateau in the Blue Ridge Near Grand Old Stony Man Peak, Overlooking Famous Shenandoah Valley* (Roanoke, VA: Stone Printing and Mfg., 1920).
16. Dyar, H.G., "A List of the North American Lepidoptera and Key to the Literature of This Order of Insects," *Bulletin of the United States National Museum* 52 (1903): vii.
17. The reprint with Dyar's note "call it Wellesca" is at the National Agricultural Library in Dyar's manuscript notes (vol. 13, pt. 1). See Dyar, H.G., "A New *Parasa*, with a Preliminary Table of the Species of the Genus," *Psyche* 8 (1898); Dyar, H.G., "Change of Preoccupied Names," *The Canadian Entomologist* 32 (1900); ———, "A List of American Cochlidian Moths, with Descriptions of New Genera and Species," *Proceedings of the United States National Museum* 29 (1905).
18. Anonymous, *Washington Times*, 4/9/1909, 8.
19. Anonymous, "Society," *Washington Herald*, 5/8/1910, 3.
20. Zella's deafness may have been caused by otosclerosis, a hereditary condition twice as common in women, in which bone grows over the middle ear bones. It has been shown that otosclerosis can become worse during pregnancy.
21. Henson and Epstein, "Dyar Family Oral History Interviews (R. Hill-L. Dyar)," 1999, SIA, RU009570.
22. Although this account was written in Dyar's short story "A Fool and His Friend," it is likely true that Zella read to Dyar in this fashion given that most of the story is well supported by other information, including Zella's being stone deaf. See Dyar Story, "A Fool and His Friend," 11/25/1915, SIA, Acc. 95-006, HGDP, b2, f6.
23. When the Peabody home closed in 1961, the money from its assets was given to Harvard Medical School, which to this day has the "Harriet M. Peabody Professor

of Orthopedic Surgery." Presented herein is the first known link of Peabody to H.G. Dyar. See Anonymous, "Newton History Museum at the Jackson Homestead (Harriet M. Peabody)," www.ci.newton.ma.us/Jackson/exhibitions/peabody-home.asp.
24. Peabody, S.H., *Peabody (Paybody, Pabody, Pabodie) genealogy* (Boston, MA: Charles H. Pope, 1909); Anonymous, "Peabody, Mrs. Hettie M., Matron, Children's Mission, 277 Tremont," *Boston City Directory*, 1882–1885; Anonymous, "Peabody, Mrs. Harriet M., House 546 Tremont Street," *Boston City Directory*, 1887; Anonymous, "City News in Brief (Life Among the Navajo Indians)," *Washington Post*, 2/10/1914, 5.
25. H.M. Peabody to Dyar, 11/16/1902, SIA, HGDP, Acc. 95-006, b3, f4; Dyar included the new moth from Utah along with others collected by Schwarz and Barber in Arizona. See Dyar, H.G. "List of Lepidoptera taken at Williams, Arizona. By Messrs. Schwarz and Barber. I. Papilionoidea, Sphingoidea, Bombycoidea, Tineioidea (in part)," *Proceedings of the Entomological Society of Washington* 5 (1903).
26. Dyar, H.G., "The Old New England Stock," *Reality* 7 (1924): 2–7.

## CHAPTER 9

1. S.P. Langley to Dyar, 11/12/1897, SIA, HGDP, b2, f1.
2. Dyar's wrote to Lintner to say he was mailing him adult specimens of an injurious sawfly (*Acordulecera dorsalis*) for the New York State Collection in Albany. See Dyar to J.A. Lintner, 10/19/1897, NYSA, JALP; Mallis, A, *American entomologists* (New Brunswick, NJ: Rutgers, 1971).
3. Specimens Dyar identified for agriculture included inchworm moths (Geometridae) from Frank Hurlbut Chittenden of the Bureau of Entomology (USDA). Chittenden, a transplanted New Yorker, as Dyar was to be, settled in Washington in 1891. See F.H. Chittenden to Dyar, 11/22/1897, NARA, RG7, E-2, b24, f "Dy."
4. Limacodid species, *Kronea minuta*, described by Tyrone Reakirt in 1864, was supposed to have been in Strecker's collection. The specimen was never to be found. See Dyar to F.H. Strecker, 12/24/1897 and 10/10/1893, FMNHIA, HSP, s1, b21, "Dyar, H.G."
5. L.O. Howard to Dyar, 12/30/1897, SIA, HGDP, b2, f1.
6. Dyar to L.O. Howard, 1/6/1898, ibid., E-2, b2.
7. Curators' Annual Reports, Division of Insects, 1898, SIA, RU 158, USNM, 1881-1964.
8. Dyar to F.H. Strecker, 10/23/1899, FMNHIA, ibid.
9. W.H. Ashmead to Dyar, 6/28/1898, SIA, HGDP, b1.
10. Howard, L.O., "Curators' Annual Reports," 1899, SIA, RU 158.
11. Ibid.
12. Cockerell, T.D.A., "Review of: A List of North American Lepidoptera, and Key to the Literature of this Order of Insects," *Science* (1903).
13. Anonymous, "Meeting of NY Ent. Soc. October 6, 1896," *Journal of the New York Entomological Society* 5 (1897).
14. Dyar to C.H. Fernald, 10/20/1896, UMA, CHFP.
15. ———, 9/18/1898, ibid.
16. Dyar, H.G., "A List of the North American Lepidoptera and Key to the Literature of This Order of Insects," *Bulletin of the United States National Museum* 52 (1903).
17. ———, 11/18/1899, ibid.
18. Dyar to L.O. Howard, 1/18/1900, NARA, RG7, E-2, b26.
19. Dyar, "A List of the North American Lepidoptera and Key to the Literature of this Order of Insects" (1903): v.

20. Dyar to C.H. Fernald, 2/24/1902, UMA, CHFP. The names of families and superfamilies were very unstable at this time because at least some authors believed that a family group name had to have a valid genus name (this was later ruled by the International Committee on Zoological Nomenclature to be unnecessary).
21. Dyar, "A List of the North American Lepidoptera and Key to the Literature of This Order of Insects" (1903): v-vi.
22. Ibid., 96.
23. ———, "New North America Lepidoptera with Notes on Larvae," *Proceedings of the Entomological Society of Washington* 5 (1903): 291.
24. Anonymous, "Over Six Thousand Kinds of Butterflies: Valuable Work on Lepidoptera of United States," *Washington Times*, 3/4/1903, 5.
25. Cockerell, ibid. (1903).
26. Anonymous, "Book Notices: A List of North American Lepidoptera, and Key to the Literature of this Order of Insects.—By Harrison G. Dyar, Ph. D. Bulletin of the United States National Museum, No. 52. Washington, D.C., Government Printing Office, 1902. 1 vol 8 vo.; pp., xix., 723," *The Canadian Entomologist* 35 (1903)
27. Smith, J.B., *Checklist of the Lepidoptera of Boreal America* (Philadelphia: American Entomological Society, 1903), v.
28. Dyar, H.G., "Life Histories of North American Geometridae. —I," *Psyche* 8 (1899); ———, "Life Histories of North American Geometridae. LVIII," *Psyche* 14 (1907).
29. Ibid., *Psyche* 8 (1899), 310.
30. See examples of these species from Mexico and Panama in Dyar, H.G., "Descriptions of New Species and Genera of Lepidoptera from Mexico," *Proceedings of the United States National Museum* 47 (2054) (1914); ———. "Report on the Lepidoptera of the Smithsonian Biological Survey of the Panama Canal Zone," ibid., 47 (2050).
31. Perhaps the report of Schaus's tenure at the USNM beginning 1895 rather than the actual onset in 1918 resulted from his expressed intention to donate large portions of his private collection in the 1890s. The 1895 date was reported in: Anonymous, *The United States National Entomological Collections* (SI Press, 1976), 23.
32. Beccaloni, G.W., M.J. Scoble, G.S. Robinson, and B. Pitkin, eds. (2003). "The Global Lepidoptera Names Index (LepIndex)."
33. Dyar, H.G., "The North American Nymphulinae and Scopariinae," *Journal of the New York Entomological Society* 14 (1906); Nymphulinae are now referred to as Acentropinae.
34. Beccaloni et al., ibid.
35. Dyar to L.O. Howard, 10/18/1920, NARA, RG7, E-34, b125, f "Dy 3 of 5."
36. At the April 1898 meeting Dyar postulated that a sawfly from Rhinebeck (*Emphytus canadensis*) was parthenogenic and concurrently, Dyar—noting that sawflies are typically found in a cluster—had an odd *Pleuroneura* sawfly raised from a single, free feeding larva. See Anonymous, "Meeting of Ent. Soc. Wash. 1/6/1898 (p. 262), 4/7/1898 (p. 302), 3/3/1898 (p. 296)," *Proceedings of the Entomological Society of Washington* 4 (1899); Dyar, H.G., "Identification of the Euclid [sic] Larvae Figured in Glover's "Illustrations of North American Entomology," ibid.
37. Anonymous, "Meeting of Ent. Soc. Wash.," ibid., 10/20/1898 (p. 337), 11/03/1898 (pp. 339–340), ibid.
38. ———, 1/12/1899 (pp. 347–348), ibid.
39. Howard, L.O., "The Entomological Society of Washington [History]," ibid., 11 (1909): 14
40. Anonymous, "Meeting of Ent. Soc. Wash. 2/2/1899," ibid., 4 (1900).

## CHAPTER 10

1. Grote, A.R., "Our Quarter Centenary," *The Canadian Entomologist* 26 (1894).
2. For the debate on the Tentamen in the nomenclature committee of the American Association for the Advancement of Science (AAAS), see Sorensen, WC, *Brethren of the Net: American Entomology, 1840–1880* (Tuscaloosa, AL: University of Alabama, 1995).
3. Leach, W., *Butterfly People: An American Encounter with the Beauty of the World* (New York: Pantheon Books, 2012).
4. Hulst, G.D., "Some Remarks Upon the Catocalae, in Reply to Mr. A.R. Grote," *Papilio* 1 (1881): 218.
5. Dyar, H.G., "Dedication," *Insecutor Inscitiae Menstruus* 1 (1913).
6. A.R. Grote to Dyar, 6/15/1894, SIA, HGDP, b1, f8.
7. Ibid.
8. Ibid.
9. Sorensen, *Brethren of the Net: American Entomology, 1840–1880*.
10. Dyar, H.G., "The Lepidoptera of the Kootenai District of British Columbia," *Proceedings of the United States National Museum* 27 (1904): 868; Smith, J.B., "New Noctuids for 1903, No. 4, with Notes on Certain Described Species," *Transactions of the American Entomological Society* 29 (1903): 214.
11. Dyar, H.G., "Book Notice. List of Lepidoptera of Boreal America. by John B. Smith, Sc. D., etc. Philadelphia, American Entomological Society, 1891," *The Canadian Entomologist* 24 (1892).
12. Smith, J.B., "Correspondence. Prof. J.B. Smith's List of Lepidoptera," ibid.
13. Dyar, H.G., "Notes on *Cerura*, with Descriptions of New Species," *Psyche* 6 (1892).
14. Smith, J.B., *List of the Lepidoptera of Boreal America* (Philadelphia: American Entomological Society, 1891).
15. J.B. Smith to L.O. Howard, 11/5/1897, RG7, E-2, b96, f "Smith (A-Z) to So."
16. Full of evolutionary zeal from Comstock, Smith wrote Dyar about going to the Washington Society to speak on the origins of insect mouthparts. Dyar was in Florida during Smith's talk, but those present disagreed with Smith's hypothesis that the beak of true bugs evolved from maxilla rather than labium. His critics proved correct. See J.B. Smith to Dyar, 1/1/1898, SIA, HGDP, b2, f6; Anonymous, "Meeting of Ent. Soc. Wash. 2/10/1898," *Proceedings of the Entomological Society of Washington* 4 (1899): 294.
17. Smith, J.B., and H.G. Dyar, "Contributions Toward a Monograph of the Lepidopterous Family Noctuidae of Boreal North America. A Revision of the Species of *Acronycta* (Ochsenheimer) and of Certain Allied Genera," *Proceedings of the United States National Museum* 21 (1898).
18. See J.B. Smith to L.O. Howard, 11/5/1897, RG7, E-2, b96, f "Smith (A-Z) to So."
19. Smith and Dyar, "Contributions toward a monograph of the lepidopterous family Noctuidae of boreal North America. A Revision of the Species of *Acronycta* (Ochsenheimer) and of Certain Allied Genera" (1898): 6.
20. Anonymous, "Meeting of Ent. Soc. Wash. 10/20/1898," *Proceedings of the Entomological Society of Washington* 4 (1899): 337.
21. Dyar, HG, "On Certain Identifications in the Genus *Acronycta*," *The Canadian Entomologist* 33 (1901).
22. J.B. Smith to L.O. Howard, 4/29/1900, RG7, E-2, b97, "f SH-SM Oct. 1899–July 1900"; Smith mentioned to Dyar that Howard wrote him that "it was looked upon in Washington as a good joke."
23. Smith, J.B., "Types and Synonymy," *The Canadian Entomologist* 33 (1901): 147.
24. Dyar to L.O. Howard, 5/21/1901, NARA, RG7, E-2, b 24, f "Dy."

25. Dyar, H.G., "Further about the Types of *Acronycta*," *The Canadian Entomologist* 33 (1901).
26. Smith, J.B., "*Acronycta* and Types," ibid: 232–234.
27. J.B. Smith to L.O. Howard, 6/4/1901, NARA, RG7, E-2, b98, "f Smith (A-Z) to SY April 1901- Dec. 1901."
28. Dyar to L.O. Howard, 5/21/1901, ibid., b24, f"Dy."
29. Dyar, H.G., "Papers from the Harriman Alaska Expedition. XII. Entomological Results (6): Lepidoptera," *Proceedings of the Washington Academy of Sciences* 2 (1900).
30. Dyar to L.O. Howard, 5/21/1901, NARA, RG7, E-2, b 24, f"Dy."
31. J.B. Smith to L.O. Howard, 4/29/1900, ibid., E-2, b 97, "f SH-SM Oct. 1899–July 1900"; ———, 6/4/1901, ibid., E-2, b98, "f Smith (A-Z) to SY April 1901–Dec. 1901."
32. Dyar to L.O. Howard, 5/21/1901, ibid., E-2, b24, f"Dy."
33. G. Hulst to Dyar, 4/23/1900, SIA, HGDP, b2, f1.
34. Dyar would become known as being extremely reluctant to lend type specimens, in spite of the loan policy he mentioned to Smith in 1900; J.B. Smith to L.O. Howard, 4/29/1900, ibid., E-2, b97, "f SH-SM Oct. 1899-July 1900."
35. Anonymous, "Meeting of Ent. Soc. Wash. 1/8/1903," *Proceedings of the Entomological Society of Washington* 5 (1903); Dyar, H.G., "Annual Address of the President: Some Recent Work in North American Lepidoptera," *Proceedings of the Entomological Society of Washington* 5 (1903).
36. ———, "Annual Address of the President: Some Recent Work in North American Lepidoptera" (1903): 170–171.
37. J.B. Smith to Dyar, 10/24/1901, SIA, HGDP, b2, f6.
38. Morris, F.A., "History of the Laboratory Club," *The Scarlet Letter: Annual Publication of the Greek Letter Fraternities* 33 (1903): 168; Patterson, G., *The Mosquito Crusades. A History of the American Anti-Mosquito Movement from the Reed Commission to the First Earth Day* (New Brunswick, NJ: Rutgers University Press, 2009), 26–27.
39. Dyar to J.B. Smith, 4/03/1903, RUL-SC&UA.
40. J.B. Smith to Dyar, 4/1/1903, ibid; J.B. Smith to L.O. Howard, 5/16/1903, ibid.
41. ———, 4/1/1903, ibid.
42. Dyar to J.B. Smith, 4/03/1903, ibid.
43. J.B. Smith to L.O. Howard, 6/4/1901, NARA, RG7, E-2, b 98, "f Smith (A-Z) to SY April 1901- Dec. 1901."
44. Dyar, H.G., "Notes on Mosquitoes on Long Island, New York," *Proceedings of the Entomological Society of Washington* 5 (1903): 45.
45. Howard, L.O., *Fighting the Insects: The Story of an Entomologist* (New York: Macmillan, 1933). Howard mistakenly gives the year as 1902 and the location as Woods Hole, Massachusetts.
46. Although this publication was presented at the Nov. 14, 1901 meeting of the Washington Entomological Society, it postdates two others by Dyar in the volume 9 of the *Journal of the New York Entomological Society* (Dec. 1901). Dyar, H.G., "Notes on mosquitoes on Long Island, New York," ibid. (1902).
47. Epstein, M.E., and P.M. Henson, "Digging for Dyar: The Man behind the Myth," *American Entomologist* 38 (1992); Howard, L.O., *Mosquitoes; How They Live; How They Carry Disease; How They Are Classified; How They May Be Destroyed* (New York: McClure, Phillips & Co., 1901).
48. Dyar, H.G., and F. Knab, "Diverse Mosquito Larvae That Produce Similar Adults," *Proceedings of the Entomological Society of Washington* 6 (1904).
49. These totals include new species (2,821) and genera (347). See Anonymous (2005), "LepIndex: The Global Lepidoptera Names Index," The Natural History

Museum, London; Pape, T., and F.C. Thompson. (2010). "Systema Dipterorum: The Biosystematic Database of World Diptera [Version 1.0]."
50. The eventual publication in the *Proceedings* mentioned by Smith in his letter was Dyar, H.G., ibid (1902). See J.B. Smith to Dyar, 10/24/1901, SIA, HGDP, b2, f6.
51. Dyar to J.B. Smith, 3/28/1903, RUL-SC&UA.
52. J.B. Smith to Dyar, 4/1/1903, ibid; Dyar, H.G., "Illustrations of the Larvae of North American Culicidae—III," *Journal of the New York Entomological Society* 11 (1903): 26–27.
53. Dyar to J.B. Smith, 4/03/1903, RUL-SC&UA.
54. ———, 4/22/1905, ibid.
55. Patterson, *The Mosquito Crusades. A History of the American Anti-Mosquito Movement from the Reed Commission to the First Earth Day*, 28; J.B. Smith to Dyar, 4/1/1903, RUL-SC&UA.
56. J.B. Smith to L.O. Howard, 6/4/1901, RG7, E-2, b98, "f Smith (A-Z) to SY April 1901- Dec. 1901."

## CHAPTER 11

1. H.G. Dyar, AVA, MOEV2, p. 506 (Oct. 1916).
2. Weekapaug, Rhode Island. See Dyar Catalogue 7-8/1904 and Dyar Blue Book Entries 1364–1372.
3. For Lepidoptera records from Chesapeake Beach see Dyar Catalogue, 6/1904.
4. In 1917 Dyar donated 15,000 Lepidoptera, approximately the number of specimens in his catalogue prior to arriving in Washington, to the USNM.
5. Dyar Blue Book, 6/10–8/3/1898, Entries 700–745.
6. ———, 6/9–13/1899, 7/10–8/27/1899, Entries 742–747, 754–767.
7. ———, 7/28/1899, Entries 758, 760.
8. ———, 8/1–24/1900, Entries 915–922, Entries 915–922.
9. ———, 5/11–7/30/1901, Entries 930–1101.
10. ———, 6/21–9/5/1902, Entries 1117–1134.
11. ———, 5/29–8/19/1903, Entries 1141–1367.
12. ———, 9/24–26/1904, Entries 1373–1374.
13. ———, 4/19–25/1905, Entries F62–F72; 8/18–19/1905, Entry 1379.
14. ———, 5/10–6/25/1906, Entries G7–G10.
15. ———, 8/8–11/1907, Entries 1386–1388.
16. ———, 8/15–17/1908, Entries 1395–1397.
17. ———, 7/20–25/1909 and Dyar, H.G., "A Report on Mosquitoes at Dublin, New Hampshire, Particularly on the Occurrence of *Mansonia perturbans* Walker [Diptera, Culicidae]," *Proceedings of the Entomological Society of Washington* 11 (1909).
18. Dyar Blue Book, 3/1–21/1905, Entries F1–F38.
19. ———, 1/18–2/11/1898, Entries 691–704.
20. Smith, D.R., "The Sawfly Work of H.G. Dyar (Hymenoptera: Symphyta)," *Transactions of the American Entomological Society* 112 (1986): 384.
21. Records of Lepidoptera on the winter trip to Nassau are found in: Dyar Catalogue, 1/31–2/5/1898; Dyar wrote his rearing notes Royal Victoria Hotel stationary. See Dyar Blue Book 3, 2/8–22/1898, Entry 178.
22. ———, 1/11–3/6/1900, Entries 775–895.
23. Moths reared on the 1900 Florida trip included *Blastobasis guilandinae* on gray nicker bean, at the time *Guilandina bonducella* (hence the moth name) and now *Caesalpina bonduc*. See: Dyar Catalogue, 1/11/–3/5/1900 and Busck, A., "New Species of Moths of the Superfamily Tineina from Florida," *Proceedings of the United States National Museum*

23 (1901); Several Snout Moths (*Nymphula* species) Collected as Adults on This Trip Have Aquatic Caterpillars. See Dyar, H.G., "The North American Nymphulinae and Scopariinae," *Journal of the New York Entomological Society* 14 (1906).
24. Around Bellport Dyar collected caterpillars, including *Nola*, in "all the swamp lands of Long Island in Brookhaven, Southhaven, Quogue, etc." See ———, "On the larvae of North American Nolidae, with descriptions of new species," *The Canadian Entomologist* 31 (1899): 63; Dyar Blue Book, 6/14/1898, Entries 714–715.
25. Caudell's unpublished catalog is in Entomology Archives, NMNH. Howard, L.O. and A. Busck, "In Memorium: Andrew Nelson Caudell," *Proceedings of the Entomological Society of Washington* 38 (1936): 34.
26. Ibid., 35.
27. The "green fool" and extensive Colorado Orthoptera records are found in: Caudell, A.N., "Notes on Orthoptera from Colorado, New Mexico, Arizona, and Texas, with Descriptions of three new Species," *Proceedings of the United States National Museum* 26 (1903).
28. Dyar Catalogue, 5/22-7/23/1901.
29. Dyar, H.G., "Descriptions of the Larvae of Some Moths from Colorado," *Proceedings of the United States National Museum* 25 (1903).
30. E. Oslar to Dyar, 5/29/1901, 6/14/1901, SIA, HGDP, b2, f3; Dyar to L.O. Howard, 5/14/1901, NARA, RG7, E-2, b24, f"Dy"; Dyar Blue Book, 5/12/1901, Entry 932.
31. A.N. Caudell to Dyar, 7/7/1901, SIA, HGDP, b1, f4.
32. ———, 8/9/1901, ibid., b1, f3.
33. Dyar Catalogue, 7/18/1901.
34. Dyar, H.G., "New Lepidoptera from the United States," *Journal of the New York Entomological Society* 12 (1904).
35. Dyar to P.N. Knopf, 7/26/1901, SIA, HGDP, Acc. 95-006, b1, f3.
36. A.N. Caudell to Dyar, 7/7/1901, SIA, HGDP, b1, f4.
37. Along with her Bahá'í name, Aseyeh, Wellesca was referred to as Lesca and Sadie by her brother George Freeman Pollock, whereas Dyar used the names Lesca, Kittie, Kit, and mama.
38. See also Anonymous, "Leader in Kindergarten Work. Death of Mrs. Louise Pollock. Who Was Long a Resident of Washington.," *Washington Post*, 7/26/1901, 7.
39. Coquillett, D.W., "Three New Species of Nemoatcerous [sic] Diptera," *Entomological News* 13 (1902): 85.
40. Dyar to F.H. Strecker, 10/26/1901, FMNHIA, HSP, s1, b21, "Dyar, H.G."; Beutenmueller, W., "Herman Strecker," *Journal of the New York Entomological Society* 9 (1901).
41. Dyar Catalogue, 6/19/–9/25/1902.
42. Marlatt, C.L., "Notes on the Periodical Cicada in the District of Columbia in 1902," *Proceedings of the Entomological Society of Washington* 5 (1903).
43. Dyar Catalogue, 6/19/–9/25/1902.
44. Dyar, H.G., "Notes on Mosquitoes in New Hampshire," *Proceedings of the Entomological Society of Washington* 5 (1903); ———, "Mosquitoes of the United States," *Proceedings of the United States National Museum* 62 (1922); Dyar Catalogue, 6/19/–9/25/1902.
45. Dyar Blue Book, 10/3/1902 1123–1374; A record for *Culex territans* Walker was reportedly collected by Dyar on November 2 at Plummers Island in Dyar, H.G., *Proc U.S. National Mus* 62 (1922): 19. However, no year was given; Dyar Catalogue, 6/19/–9/25/1902.
46. A.N. Caudell to Dyar, 7/17/1902, SIA, HGDP, b1, f4.

47. The USNM accessioned 20,000 specimens of Lepidoptera as a private donation on Dyar's return. See also Currie, R, "An Insect Collecting Trip to British Columbia," *Proceedings of the Entomological Society of Washington* 6 (1904): 24.
48. Among the first moths collected by this method were *Hadena binotata* Walker, *Mamestra laudabilis* (Guenée), and *Leucana roseola* Smith (Noctuidae).
49. Currie, R., "An Insect Collecting Trip to British Columbia," ibid.: 27–28.
50. Dyar to L.O. Howard, 6/20/1903, NARA, RG7, E-2, b25.
51. Dyar, H.G., "Notes on the Mosquitoes of British Columbia," *Proceedings of the Entomological Society of Washington* 6 (1904): 37; Ibid., 39.
52. Ibid.
53. Ibid., 40.
54. W. Schaus to Dyar, 8/9/1905, SIA, HGDP, b2, f3.
55. Henson, P.M. and M.E. Epstein, "Dyar Family Oral History Interviews (R. Hill-L. Dyar)", 1999, SIA, RU009570.
56. Currie, "An Insect Collecting Trip to British Columbia" (1904); Clarke, J.F.G., "Revision of the North American Moths of the Family Oecophoridae, with Descriptions of New Genera and Species," *Proceedings of the United States National Museum* 90 (1941); Dyar Catalogue, 9/5–9/15/1903.
57. Dyar, H.G., "The Lepidoptera of the Kootenai District of British Columbia," *Proceedings of the United States National Museum* 27 (1904): 898; The caterpillar of *Thera otisi* feeds on juniper, information unbeknownst to Dyar at the time.
58. Dyar Catalogue, 5/23–28/1904; Dyar Blue Book, 5/29/1904, Entries 1123–1374; Dyar, "Mosquitoes of the United States" (1922): 14–15; Chesapeake Beach in Dyar Catalogue, 5/23–28/1904; Dyar, "Mosquitoes of the United States" (1922) 6/17/1904; ———, "Brief notes on Mosquito Larvae," *Journal of the New York Entomological Society* 12 (1904).
59. Dyar Catalogue, 5/23–28/1904; Clarke, "Revision of the North American Moths of the Family Oecophoridae, with Descriptions of New Genera and Species" (1941): 202; Dyar, H.G., "A New Tortricid from the Sea Shore," *Proceedings of the Entomological Society of Washington* 6 (1904); ———, "Mosquitoes of the United States" (1922); ———, "The North American Nymphulinae and Scopariinae," *Journal of the New York Entomological Society* 14 (1906); Dyar Blue Book, 5/29/1904, Entries 1123–1374; Dyar, H.G., "A New Mosquito," *Journal of the New York Entomological Society* 13 (1905).
60. ———, "Mosquitoes of the United States" (1922): 19, 21, 30, 35–36, 50–51, 71, 88–90, 92, 98; Howard, L.O., H.G. Dyar, and F. Knab, *The Mosquitoes of North and Central America and the West Indies. Systematic Description*, vol. 4 (part II) (Washington, D.C.: Carnegie Institution of Washington, 1917), 588, 699, 1026.
61. Dyar Blue Book, 3/1–3/6/1905, Entries 1375–1376; Dyar, "A New Mosquito" (1905); See write-up about Evelyn Mitchell in Anonymous, "Here's a Noted Washington Scientist." 1909. *Washington Post*, 7/11/1909, M8.
62. ———, "Mosquitoes of the United States" (1922): 54, 79.
63. Dyar, H.G., and F. Knab, "Diverse Mosquito Larvae That Produce Similar Adults," *Proceedings of the Entomological Society of Washington* 6 (1904); ———, "The Larvae of Culicidae as Independent Organisms," *Journal of the New York Entomological Society* 14 (1906); Dyar, "Mosquitoes of the United States" (1922).
64. At Big Tuppers Lake in late August Dyar found aquatic snout moth *Elophila gyralis*. See Dyar Blue Book, 9/28/1905 1381; Dyar, H.G., "Description of the Larva of *Tortricidia fiskeana* Dyar," *Journal of the New York Entomological Society* 15 (1907),
65. H.G. Dyar, AVA, MOEV2, p. 506 (Oct. 1916); Henson and Epstein, "Dyar Family Oral History Interviews (R. Hill- L. Dyar)", 1999, SIA, RU009570.

66. H.G. Dyar, AVA, MOEV2, ibid.
67. Ibid.
68. Dyar, "Mosquitoes of the United States" (1922): 20, 86; Dyar Catalogue, "Arroyo Seco," 5/23/1906; ———, "Los Angeles," 5/30/1906.
69. Dyar, H.G., "Report on the Mosquitoes of the Coast Region of California, with Descriptions of New Species," *Proceedings of the United States National Museum* 32 (1907).
70. Aitken, T.H.T., "Contributions Toward a Knowledge of the Insect Fauna of Lower California. No. 6 Diptera: Culicidae," *Proceedings of the California Academy of Sciences* 24 (1942): 165, 169.
71. In addition to southern California Dyar reported a trip to Salinas, some distance to the north on 6/25/1906 (p. 20), perhaps an error in the literature. See Dyar, "Mosquitoes of the United States" (1922): 20, 28, 82, 88, 106: 6/14–7/2/1906.
72. Anonymous, "Meeting of Ent. Soc. Wash. 10/4/1906," *Proceedings of the Entomological Society of Washington* 8 (1906).
73. Dyar, "Mosquitoes of the United States" (1922): 14–15, 59, 67, 70, 75, 86, 94, 104: 7/27–8/11/1906.
74. Dyar Blue Book, 8/18–23/1906, Entries C137–C139; Howard, Dyar, and Knab, *The Mosquitoes of North and Central America and the West Indies. Systematic Description*, 698 (Moose Jaw, SK); Dyar Blue Book, "North Portal, Saskatoon, SK," 8/26/1906 p. 12, Entry C140.
75. Caudell, A.N., "An Insect Ventriloquist," *Entomological News* 18 (1907).
76. Cancellation of phone service for the remainder of the summer indicates that the Dyars remained in Maine or the northeast for that amount of time. See ibid. 7/11/1907.
77. Dyar Blue Book, 8/9–14/1907, Entries 1387–1389.
78. C.D. Walcott to Dyar, 4/5/1907, SIA, RU 45, b17, f14; Dyar, H.G., "Distribution of Mosquitoes in North America," *Proceedings of The Seventh International Zoological Congress, Boston, 19–24 August, 1907* (1912).
79. Collins, C.W., "The Oriental Moth (*Cnidocampa flavescens* Walker) and Its Control," *United States Department of Agriculture* (1933).
80. Dyar, H.G., "The Life History of an Oriental Species of Cochlidiidae Introduced into Massachusetts (*Cnidocampa flavescens* Walk.)," *Proceedings of the Entomological Society of Washington* 11 (1909): 165; Fernald, H.T., and J.N. Summers, "The Early Stages of the Oriental Moth," *Entomological News* 18 (1907).
81. L.O. Howard to Dyar (and reverse), 1/18/1910, NARA, RG7, E-34, b 125, f "Dy 5 of 5"; Dyar had received live cocoons of this species from Japan several years before his attempt at establishing the oriental limacodid. The adults emerged, presumably confined, in Denver, Colorado (Dyar Catalogue, 6/20/1901).
82. Dyar collected three species of mosquitoes in Lincoln, Maine during August 1908: *Culex testaceus*, *Mansonia perturbans*, and *Aedes sollicitans*, See Dyar v. Dyar, 1916; Dyar, "Mosquitoes of the United States" (1922), 14–15, 31, 90; ———, "A report on Mosquitoes at Dublin, New Hampshire, Particularly on the Occurrence of *Mansonia perturbans* Walker [Diptera, Culicidae]" (1909).
83. Dyar Catalogue, "Plattsburgh and Valcor Island, New York, 7/20–25/1909."

### CHAPTER 12

1. Anonymous, "Meeting of NY Ent. Soc. 2/2/1904," *Journal of the New York Entomological Society* 12 (1904).
2. Ibid.; Dyar's tenure as editor of the *Journal of the New York Entomological Society* began with volume 12, the first number dated March 1904.

3. Dyar, H.G., "Review of ' Synonymic Catalogue of the North American Rhopalocera. Supplement No. 1, by Henry Skinner,'" ibid. 13 (1905).
4. Skinner, H., "A Review of a Review," *Entomological News* 16 (1905); Dyar, H.G., "A Review of the Hesperiidae of the United States," *Journal of the New York Entomological Society* 13 (1905).
5. Skinner, ibid.
6. Dyar to H. Skinner, 12/8/1905, ANSPL, AES, ms. coll. E76, no. 60.
7. H. Skinner to Dyar, 12/11/1905, SIA, HGDP, b2.
8. Dyar, H.G. "New Facts That Are Not New," *Entomological News* 16 (1905): 310.
9. Soule, C.G., "Letter to Editor," *Entomological News* 16 (1905): 333; Sherman, F., "Letter to Editor," *Entomological News* 17 (1906): 32.
10. H. Skinner to Dyar, 12/11/1905, SIA, HGDP, b2.
11. Dyar, H.G., "Letter to Editor," *Entomological News* 17 (1906): 32.
12. Skinner, H., "Letter to Editor Polyphemus Cocoons," ibid. 17 (1906): 32–33; Soule, ibid.
13. Dyar to H. Skinner, 6/19/1906, ANSPL, AES, ms coll E76, no. 6.
14. Ibid.
15. Skinner, H., "On Dr. Dyar's Review of the Hesperidae," *Entomological News* 17 (1906).
16. Skinner, H., "The infallible errs," ibid.: 115.
17. Dyar to H. Skinner, 6/19/1906, ANSPL, AES, ms coll E76, no. 6.
18. Skinner, H., "Impressions Received from a Study of our North American Rhopalocera," *Journal of the New York Entomological Society* 4 (1896): 113.
19. Dyar, H.G., "Notes on the Phylogeny of Saturnians," *The Canadian Entomologist* 28 (1896): 305.
20. H. Skinner to Dyar, 1/29/1897, SIA, HGDP, b2, f6.
21. Dyar to H. Skinner, 12/7/1900, ANSPL, AES, ms coll E76, no. 45; Dyar, H.G., "A New Cochlidian of the Palearatic [sic] Group [author as Dyer]," *Entomological News* 11 (1900).
22. Dyar to H. Skinner, 12/7/1900, ANSPL, AES, ms coll E76, no. 45.
23. Howard, L.O., *Yearbook No. 6 of Carnegie Institute of Washington* (1908); Anonymous, *The United States National Entomological Collections* (SI Press, 1976), 24; Rathbun, R., *Report on the Progress and Condition of the U.S. National Museum for the Year Ending June 30, 1911* (1912).
24. Dyar, H.G. and F. Knab, "The Larvae of Culicidae as Independent Organisms," *Journal of the New York Entomological Society* 14 (1906).
25. Ibid.
26. Skinner, H., "Up to Date Methods in Entomological Publications," *Entomological News* 17 (1906).
27. Of 56 species named from larvae in their 1906 "Independent Organisms" paper, only 23 are considered synonyms. See Gaffigan, T.V. et al. (2013). "Systematic Catalog of Culicidae." Walter Reed Biosystematics Unit.
28. Coquillett, D.W., "Letters: Dr. Dyar's Square Dealing," *Entomological News* 17 (1906).
29. Dyar to H. Skinner, 6/19/1906, ANSPL, AES, ms coll E76, no. 6_?
30. Dyar, H.G., and F. Knab, "Book Notice: A Monograph of the Culicidae of the World by F. V. Theobald: London, 1907. Volume IV," *Journal of the New York Entomological Society* 15 (1907); Dyar, H.G., "Book Notice: Mosquito Life. By Evelyn Groesbeeck Mitchell, A.B., M.S.; G. P. Putnam's Sons, New York and London. The Knickerbocker Press, 1907," *The Canadian Entomologist* 40 (1908); Mitchell, E.G., *Mosquito Life: The Habits and Life Cycles of the Known Mosquitoes of the United States; Methods for*

Their Control; and Keys for Easy Identification of the Species in Their Various Stages (Putnam, 1907).
31. Mitchell, E.G., "Validity of the Culicid Subfamily Deinoceritinae," *Psyche* 14 (1907): 11
32. ———, "A Reply to Dr. Dyar," *The Canadian Entomologist* 40 (1908)
33. Dyar, H.G., "Book Notice: Mosquito Life. By Evelyn Groesbeeck Mitchell, A.B., M.S.; G. P. Putnam's Sons, New York and London. The Knickerbocker Press, 1907," ibid.
34. Coquillett, DW, "Dr. Dyar's criticism of 'Mosquito Life,'" ibid.
35. Anonymous, "Author Defends Book Attacked by Critic," *Washington Times*, 4/29/1908, 8.
36. L.O. Howard wrote a letter of introduction for Dyar to Worthington. See A.S. Worthington to Dyar, 4/30/1908, SIA, HGDP, Acc. 95-006, b3, f2; Anonymous, "No. 128 Mitchell vs. Dyar Attorneys, Ambrose-Worthington," *Washington Post*, 10/18/1910, 12; Anonymous, "Mitchell vs. Dyar: dismissed for failure to prosecute. Attorney W.E. Ambrose; defendants attorney A.S. Worthington," *Washington Post*, 11/16/1911, 9; Anonymous, "Libel Suit Abandoned," *Washington Herald*, 11/17/1911, 6.
37. Mitchell, "A Reply to Dr. Dyar" (1908): 94; Leonard, J.W., "'Evelyn G. Mitchell' in Woman's Who's Who of America: A Biographical Dictionary of Contemporary Women of the United States and Canada," 567; Ogilvie, M.B., and J.D. Harvey, "Evelyn Groesbeeck Mitchell," in *The Biographical Dictionary of Women in Science: L-Z* (2000); Dyar's story "Taming of a Suffragette" is found written over the following letter: A.S. Worthington to Dyar, SIA, RU 7101, Acc. 95-006, HGDP, b3, f2.
38. Torre-Bueno in Mallis, A, *American Entomologists* (New Brunswick, NJ: Rutgers, 1971), 315.
39. Dyar, H.G., "A Report on Mosquitoes at Dublin, New Hampshire, Particularly on the Occurrence of *Mansonia perturbans* Walker [Diptera, Culicidae]," *Proceedings of the Entomological Society of Washington* 11 (1909).
40. Ibid.
41. Smith, J.B., "Concerning *Culex perturbans* at Dublin, New Hampshire," *Entomological News* 20 (1909).
42. Dyar, H.G., "Concerning Dr. John B. Smith at Dublin, New Hampshire," ibid. 21 (1910).
43. Smith, J.B., "Letter: Once More *Culex perterbans* in Notes and News," ibid.
44. Skinner, H., "On an Episode: The Charge," ibid. 21 (1910): 85.
45. Besides indicating other noctuid species he considered synonyms created by Smith, Dyar named a pyralid moth from Stony Man, collected ten years earlier by himself on July 31, 1900, and named it *Polloccia alticolalis*, the genus after Wellesca Pollock and family. See Dyar, H.G., "Some Moths from Claremont, California with Notes on Certain Allied Species," *Pomona College Journal of Entomology* 2 (1910): 375–376; Smith, J.B., "Revision of the Species of *Pleonectyptera* Grt.," *Transactions of the American Entomological Society* 33 (1907): 377.
46. Smith, J.B., "New Species of Noctuidæ for 1911. No. 1," *Journal of the New York Entomological Society* 19 (1911); for a noctuid specialist's view that history has not been kind towards the quality of Smith's work see: Todd, E.L., "The Noctuid Type Material of John B. Smith (Lepidoptera)," *U.S. Department of Agriculture Technical Bulletin* (1982): 1
47. Anonymous, "Meeting of Ent. Soc. Wash. 3/14/1912," *Proceedings of the Entomological Society of Washington* 14 (1912); Mallis, *American Entomologists*, 315–320.

48. For Dyar's lapse in attendance see: Anonymous, "Meeting of Ent. Soc. Wash. 4/4/1912, 5/2/1912, 6/6/1912," *Proceedings of the Entomological Society of Washington* 14 (1912)
49. Dyar Story completed in Washington, D.C., "The Man Who Did Return," 8/5/1914, SIA, HGDP, Acc. 95-006, b2, f1.
50. Skinner, ibid.

## CHAPTER 13

1. Anonymous, "Meeting of Ent. Soc. Wash. 10/3/1912," *Proceedings of the Entomological Society of Washington* 14 (1912): 330–332; Dyar's personal copy of Stratton's "Moth's of Limberlost" is in the National Agricultural Library in Beltsville, MD; Stratton-Porter, G., *Moths of the Limberlost* (Garden City, NY: Doubleday, Page & company, 1912).
2. Dyar attended one meeting of the Entomological Society of Washington as a visitor on a mosquito-related topic in April 1913. See Anonymous, "Meeting of Ent. Soc. Wash. 4/3/1913," *Proceedings of the Entomological Society of Washington* 15 (1913):107–108.
3. Dyar to W.C. Cook, 9/14/1923, SIA, HGDP, b1, f4.
4. See Terry Carpenter's translation of *Insecutor Inscitiae Menstruus* in Carpenter, T.L. and T.A. Klein, "2011 AMCA Memorial Lecture Honoree: Dr. Harrison Gray Dyar Jr.," *Journal of the American Mosquito Control Association* 27 (2011).
5. Grote, A.R., ed., *The North American Entomologist* (Buffalo, New York: Reinecke & Zesch, 1879–1980).
6. Dyar, H.G., "Notes on Cotton Moths," *Insecutor Inscitiae Menstruus* 1 (1913); Grote wrote that the name *argillacea* of Hübner had priority over the later name *Anoma xylina* that he had given it in 1864 in Grote, A.R., "List of the Noctuidae of North America," *Bulletin of the Buffalo Society of Natural Sciences* (1874): 24.
7. L.O. Howard to Dyar, 10/10/1913, NARA, RG7, E-34, b125 f "Dy 5 of 5"; Dyar, H.G., "Report on the Lepidoptera of the Smithsonian Biological Survey of the Panama Canal Zone," *Proceedings of the United States National Museum* 47 (1914).
8. Dyar to W.C. Cook, ibid.
9. Dyar, "Report on the Lepidoptera of the Smithsonian Biological Survey of the Panama Canal Zone," *Proceedings of the United States National Museum* (1914).
10. ———, "Descriptions of New Species and Genera of Lepidoptera from Mexico," ibid.
11. Meyrick's numbers of species and subspecies are from Lindroth, C, *Systematics specializes between Fabricius and Darwin: 1800-1859*, ed. Smith, Mittler, and Smith, The History of Entomology (Annual Review Inc. and The Entomological Society of America, 1973). Estimates of the numbers of species and genera named by Dyar and Schaus were tabulated from Beccaloni, G.W., M.J. Scoble, G.S. Robinson, and B. Pitkin (eds.) (2003). "The Global Lepidoptera Names Index (LepIndex)."
12. ———, "Descriptions of New Species and Genera of Lepidoptera from Mexico," *Proceedings of the United States National Museum* 47 (1914): 389.
13. Dyar to J.D. Gunder, 12/3/1928, SIA, HGDP, b1, f9.
14. While Dyar continued to preach conciseness, he also proposed a return to Latin for species descriptions as in the time of Linnaeus. This was to solve the problem of taxonomic works now being in many tongues. See Dyar, H.G., "A Necessity for Taxonomic Workers," *Proceedings of the Entomological Society of Washington* 30 (1928).
15. The rules of nomenclature allow the "first reviser" to chose a single lectotype, to stand as the sole representative specimen that defines a species. See Dyar to T.D.A. Cockerell, 3/20/1912, CUBLA, T.D.A. Cockerell Collection, B5, F74.

16. Packard, A.S., "Monograph of the Bombycine Moths of North America," ed. T.D.A. Cockerell, vol. 12, *Memoirs of the National Academy of Sciences* (1914).
17. Melissa, B., "Study Nature, Not Books: Cockerell and the Amateur Naturalist," *The Journal of Young Investigators* 22 (2011).
18. Anonymous, "Doings of Societies," *Entomological News* 18, no. 10 (1907).
19. Dyar to T.D.A. Cockerell, 11/24/1911, CUBLA, b. 5, f. 74.
20. Dyar was never able to publish Joutel's drawings of limacodid caterpillars, which are in the Illustration Archives, Department of Entomology, NMNH. See Dyar to T.D.A. Cockerell, ibid.
21. ———, 3/20/1912, ibid; W. Barnum to Dyar, 4/4/1912, SIA, HGDP, Acc. 95-006, b4, f2; Dyar to T.D.A. Cockerell, 4/11/1912, CUBLA, b5, f74.
22. Dyar to T.D.A. Cockerell, 7/10/1912, ibid.
23. ———, 7/26/1912, ibid.
24. Ashmead did not step down form his USNM position until around March 1908, seven months before his death.
25. C.D. Walcott to Dyar, 4/5/1907, SIA, RU 45, b17, f14, "7th Int. Cong. Zool."
26. See Anonymous, "A New Type of Patriot," *Paducah Evening Sun* (Paducah, KY) 12/26/1906, 2; Bache, R., "The Nation's New Volunteers," *Technical World Magazine*, 1907.
27. Dyar to L.O. Howard, 4/24/1908, NARA, RG7, E-2, Ltrs. Rec., b3.
28. Anonymous, *The United States National Entomological Collections* (SI Press, 1976).
29. Dyar to T.D.A. Cockerell, 1/9/1918, NARA, RG7, E-34, b125, f "Dy 4 of 5."
30. Contrary to the widely held view that Dyar worked gratis in his thirty-one years in Washington, his salaried positions, including two years substituting for Ashmead, ran from 1906 through 1917. His 1913 USDA funding was under "General Expenses" and his title was "Expert" at $1800 per year. The appropriation for the salary was "Preventing Spread of Moths [and other] Miscellaneous Insects." The following year, his title changed to "Entomological Assistant," which as Howard explained, was to comply with the government's Civil Service Commission and the title of "Expert" was eliminated to "bring as many [government] employees as possible into the classified service." H. Morrison to W.P.A. Dyar, 2/28/1929, NARA, RG7, E-35, b83, f "Dyar"; L.O. Howard to Dyar, 2/17/1914, NARA, RG7, E-34, b125, f "Dy 5 of 5"; Acting Bureau Chief to Dyar, 6/1/1913, ibid.
31. Mallis, A., *American Entomologists* (New Brunswick, NJ: Rutgers, 1971), 330–332.
32. Charlotte Barnes had a sister Amaryllis T. Gillett, part of Washington social circles and in real estate. Howard wrote about crossing paths with Barnes's sister-in-law. See L.O. Howard to W. Barnes, 12/18/1916, NARA, RG7, E-66, b11; Barnes, W., and J. McDunnough, *Checklist of Lepidoptera of Boreal North America* (Decatur, IL: Herald Press, 1917). For more information about the Gillett sisters see Logan, M., *The Part Taken by Women in American History* (Wilmington, DL: Perry-Nalle Publ. Co., 1912), 895–896.
33. W. Barnes to W.B. McKinley, 12/7/1912, NARA, RG7, E-66, b3, f "Ba 1 of 3." An earlier letter to McKinley, unrelated to Dyar, sought help on a customs-related matter to get the Rev. Taylor's collection out of Nanaimo, B.C.
34. L.O. Howard to Dyar, 8/18/1913, NARA, RG7, E-34, b125, f "Dy 5 of 5."
35. Dyar, H.G., "Notice on Volume II, no. 4, of Barnes and McDunnough's "Contributions to the Natural History of the Lepidoptera of North America," *Insecutor Inscitiae Menstruus* 1 (1913).
36. W. Barnes to L.O. Howard, 9/8/1913, NARA, RG7, E-66, b11.
37. ———, 12/28/1914, ibid.

## CHAPTER 14

1. W.P.A. Dyar to A.B. McDaniel, 1936, NBA.
2. Ibid; Dyar, H.G., "Editorial: The Old New England stock," *Reality* 7 (1924).
3. W.P.A. Dyar to A.B. McDaniel, ibid.
4. Ibid.
5. Dyar to Z.P. Dyar, 4/1909?, DVD, DENo26.
6. Ibid.
7. Ibid.
8. W.P. Allen to Chicago Bahá'í Community, 5/28/1907, NBA.
9. Ibid.
10. W.P.A. Dyar to A.B. McDaniel, 1936, ibid.
11. W.P. Allen, AVA, MOEV1, p. 74.
12. Anonymous, "Business Property at 910 Fourteenth Street Bought by Mrs. W.P. Allen for $22,500—Block Rapidly Being converted for Business Purposes—New Dwellings for Monroe Street," *Washington Post*, 2/27/1908, 16.
13. Anonymous, "Real Estate Transfers," *Washington Post*, 2/16/1906, A4; Anonymous, "Real Estate," *Washington Times*, 2/10/1906, 2. The Dyars' real estate activities were once again in evidence during late December 1907 when they received a property at 1924 N Street NW from Charles Early. The building, in turn, was exchanged for the Winchester apartment house on Eighteenth Street NW. The three-story building of brick with stone trimmings between T and U Streets was acquired in March 1908 for a sum of $25,000. The Winchester was on a lot 44 by 80 feet. See Anonymous, "Real Estate," *Washington Herald*, 3/1/1908.
14. C. Early to Dyar, 7/6/1907, SIA, HGDP, Acc. 95-006, b2, f1.
15. ———, 12/7/1908, ibid., b3, f4.
16. Ibid.; Anonymous, "Real Estate Transfers in the District," *Washington Times*, 11/15/1908, 8.
17. Anonymous, "Business Property at 910 Fourteenth Street Bought by Mrs. W.P. Allen for $22,500—Block Rapidly Being Converted for Business Purposes," *Washington Post*, 2/27/1908, 16; Anonymous, "Boom in Real Estate," *Washington Post*, 3/1/1908, ARF6.
18. Gough was referring to eight units at 615–617 Fourteenth Street NW. See E. Gough to W.P. Allen, 6/11/1909, SIA, HGDP, Acc. 95-006, b2, f3.
19. ———, ibid., b2, f1.
20. Anonymous, "Hyattsville Lots Sold," *Washington Post*, 6/4/1911, C2.
21. Alfred Higbie was listed as the manager of the Munsey office building in 1907. It was not until May 1909 that we have Higbie in the real-estate business. See Dyar to Z.P. Dyar, 3/8/1915, DVD, DENo19; Anonymous, "Munsey Building (Alfred Higbie Manager)," *Washington Times*, 6/7/1907, 11; Anonymous, "Toronto Apartments," *Washington Times*, 5/10/1909, 13.
22. Engle, R.L. 2003. *In the Light of the Mountain Moon: An Illustrated History of Skyland*: Shenandoah National Park Association.
23. Wellesca was listed as an owner of a cottage and patron of Skyland in Pollock, G.F., "Skyland: Situated on High Plateau in the Blue Ridge . . ."
24. Anonymous, "Real Estate Transfers in the District," *Washington Times*, 11/15/1908, 8; Anonymous, "Realty Transfers," ibid., 11/22/1908, 8.
25. Anonymous, "Invitation: Feast of the Declaration of the Bab & party for Baby Roshan," 5/23/1909, NBA; May 23, 1844 is the date that the Bab, the progenitor of the Bahá'í faith announced "the coming of the Day of God" and on that same day his son 'Abdu'l Bahá was born.

Notes (267)

26. Correspondence from 'Abdu'l Bahá's daughter Monever praised Wellesca for her willingness to give these stones away to those in the American Bahá'í community Monever to W.P. (Aseyeh) Allen, undated, NBA.
27. W.P. Allen to Z.P. Dyar, 5/25/1909, AVA, MOEV4, p. 955.
28. Dyar to Z.P. Dyar, 4/1909?, DVD, DENo26.
29. Wellesca's five surviving sisters in 1909 were Susan, Amia, Loue, Uila, and Mazel. See Engle (2003).
30. Z.P. Dyar to Dyar, 1909?, DVD, DENo32.
31. Ibid.
32. A specimen of a male limacodid *Isa textula*, collected by Dyar at Skyland on July 11, 1911, is in the Lepidoptera collection of the National Museum of Natural History, Smithsonian Institution. Only the label (see Figure 14.4) and wings remain.
33. Dyar sought *Monoleuca semifascia* in Morris Plains, NJ in 1898 and 1899 because Neumoegen in 1883—and later Fiske in Tryon, NC—found adults in each place. See Dyar, H.G., "The Life Histories of the New York Slug-Caterpillars. XX" *Journal of the New York Entomological Society* 22 (1914); Barber, H.S., "A Simple Trap Light Device," *Proceedings of the Entomological Society of Washington* 13 (1911); Anonymous, "Meeting of Ent. Soc. Wash. 10/6/1910," ibid. Other 1911 finds included the "red-cross" limacodid caterpillar (*Tortricidia*) at Skyland and the black salt marsh mosquito (*Ochlerotatus taeniorhynchus*) and *Anopheles crucians*—a malaria vector at Virginia Beach found in late summer or early fall. The *Anopheles*, an indoor species according to both Dyar and J.B. Smith, did not require fieldwork. See Dyar Catalogue, "Skyland," 6/12/1912; Dyar, H.G., "Mosquitoes of the United States," *Proceedings of the United States National Museum* 62 (1922): 88; Howard, L.O., H.G. Dyar, and F. Knab, *The Mosquitoes of North and Central America and the West Indies. Systematic Description*, vol. 4 (part II) (Washington, D.C.: Carnegie Institution of Washington, 1917), 1026.
34. Katydid records from Stony Man are found in Rehn, J.A.G., and M. Hebard, " Studies in American Tettigoniidae (Orthoptera) VII," *Transactions of the American Entomological Society* 42 (1916); Dyar Catalogue, "Washington, D.C.," 5/29/1913
35. Finding Dyar specimens of moths and mosquitoes at the National Museum of Natural History—the only known repository—is difficult, particularly for those without published records or those in his catalogue, because they are not held in a separate collection. Furthermore, labels do not always have years or other information. The numbers from Dyar's Catalogue still remain on many Lepidoptera specimens, but there are few records at the end of his numbered catalogue between 1911 and 1914. The mosquito *Aedes solicitans* was reported from Paramore's Island in Dyar, "Mosquitoes of the United States" (1922): 90.
36. Dyar Catalogue, "Skyland/Stony Man," 7/18–28/1912; Paramore's Island mosquito record 7/15/1914 in Dyar, "Mosquitoes of the United States" (1922): 90; Dyar Story completed in Wachapreague Virginia, "How Burnham-Wood Came to Dunsinane," 7/19/1914, SIA, HGDP, Acc. 95-006, b1, f1; Chain Bridge mosquito record 8/17/1914 in Dyar, "Mosquitoes of the United States" (1922): 85.

## CHAPTER 15

1. Z.P. Dyar to Dyar, 2–4/1912?, DVD, DENo12.
2. 1910 United States Federal Census (Wilfred P. Allen), 4/16/1910, District of Columbia, Washington, Precinct 4, District 5, p. 7.
3. W.P. Allen to Z.P. Dyar, 5/12/1911, AVA, MOEV4, pp. 948–952.
4. Anonymous, "B Street Southwest," *Washington Post*, 4/16/1910, 11; Wellesca testified about having a miscarriage a year and a half after Roshan was born in W.P. Allen, AVA, MOEV1, pp. 75–76.

5. W.P. Allen to Z.P. Dyar, 5/12/1911, AVA, MOEV4, pp. 948–952.
6. Ibid.
7. Ibid.
8. In his 1912 visits to Washington, D.C., 'Abdu'l Bahá stayed the home of Agnes and Arthur Parsons in April, at the G.H. Ripley home in May, and at 1901 Eighteenth Street NW in November. See Allen, W.P., and H.G. Dyar, *Short Talks on the Practical Application of the Bahai Revelation* (Washington, D.C., 1922).
9. Anonymous, "DC Bahá'í Tour 2012 - Bahá'í Hospice of Aseyeh Allen," http://www.dcbahaitour.org/place/?id=43.
10. W.P. Allen to Z.P. Dyar, 5/12/1911.
11. Henson, P.M., and M.E. Epstein, "Dyar Family Oral History Interviews (R. Hill-L. Dyar)," 1999, SIA, RU009570.
12. Z.P. Dyar to Dyar, 9–11/1911?, DVD, DEN025.
13. Catherine Booth placed Dyar at the Allen home, tending the furnace, when Harrison Golshan Allen was born. See C. Booth, AVA, MOEV2, pp. 550–560; for 'Abdu'l Bahá's naming of the "Allen" Children, see W.P.A. Dyar to A.B. McDaniel, 1936, NBA; and Henson, P.M., and M.E. Epstein, "Dyar Family Oral History Interviews (W.J. Dyar)," 5/17/1992, SIA, RU009570, p. 3.
14. Dyar to Z.P. Dyar, 2–4/1912?, DVD (filed Sept. 21, 1916).
15. Ibid.
16. Ibid.
17. Z.P. Dyar to Dyar, ibid.
18. Dyar to Z.P. Dyar, ibid.
19. Z.P. Dyar to Dyar, ibid.
20. Ibid.
21. Ibid.
22. Although the photograph of Dyar and the children on muleback from the Handy Family photographs is undated, 1912 fits the discussion about travel at around the time of the International Congress of Entomology. See ibid.
23. Dyar to Z.P. Dyar, ibid.
24. Z.P. Dyar to Dyar, ibid.
25. Dyar, H.G., "Address for 'Subscriptions and Matter for Publication,'" *Insecutor Inscitiae Menstruus* 1 (1913); Anonymous, "Friends School Exercises," *Washington Post*, 6/6/1913, 2.
26. Testimony of household on 7/23/1913 was given by M.A. Stover, AVA, MOEV2, p. 525 (began Oct. 3, 1916); W.P. Allen, AVA, MOEV1, pp. 79–80. The cook when Wallace Parsons Allen was born was perhaps Josephine Campbell. See 1910 United States Federal Census (Wilfred P. Allen), 4/16/1910, District of Columbia, Washington, Precinct 4, District 5, p. 7.
27. Although it has been stated from a number of sources that 'Abdu'l Bahá selected the names Roushan, Golshan, and Joushan for the boys, it was suggested by Mrs. Leech that Dyar wrote a fictitious letter to Wellesca from 'Abdu'l Bahá on a Hammond typewriter, inventing the Persian names for the children. However, translated tablets from the National Bahá'í Archives show this to be false. See W.P.A. Dyar to A.B. McDaniel, 1936, NBA.
28. As shown in the 1890s (see chap. 6), Harrison G. Dyar grew up keenly aware of his genealogy as reflected in his son having the name Otis (e.g., Stephen Otis married into the Dyar family in Ohio). Dyar was inclined to invent names of insects by combining syllables from different sources (e.g., *Venadi*o, Mexico + genus *Limacodes* = *Venadi-codia*). For Wallace Dyar's interpretation of his name, see Henson

and Epstein, "Dyar Family Oral History Interviews (W.J. Dyar)," 5/17/1992, SIA, RU009570.
29. Dyar to P.N. Knopf, 12/28/1913, SIA, HGDP, Acc. 95-006, b3, f3.

## CHAPTER 16

1. Mary Parsons, 7/24/1916, AVA, MOEV3, pp. 683-685.
2. H.G. Dyar, 7/1915, AVA, MOEV4, p. 871.
3. Dyar to L.O. Howard, 2/16/1914, NARA, RG7, E-34, b 125, f"Dy 5 of 5"; L.O. Howard to Dyar, 12/14/1914, ibid.
4. See Z.P. Dyar to Dyar, 4/22/1914, DVD, DENo29.
5. Zella collected *Anopheles maculipennis* Meigen for her husband. See: Dyar, H.G., "Mosquitoes of the United States," *Proceedings of the United States National Museum* 62 (1922): 106.
6. Dyar story "How Burnham-Wood came to Dunsinane" (completed in Wachapreague, VA), 7/19/1914. See SIA, HGDP, Acc. 95-006, b1, f1.
7. Z.P. Dyar to Dyar, 4/22/1914, DVD, DENo29.
8. A. Busck to L.O. Howard, 9/21/1914, NARA, RG7, E-34, b68, f "Busck-Caesar."
9. F. Knab to Dyar, 11/27/1914, SIA, HGDP, Acc. 95-006, b2, f1.
10. L.O. Howard to Dyar, 12/14/1914, NARA, RG7, ibid; Mary Parsons, 7/24/1916, AVA, MOEV3, pp. 683–685.
11. L.O. Howard to Dyar, 1/2/1915, SIA, HGDP, Acc. 95-006, b1, f1.
12. Wellesca found Mrs. Maud A. Leech by placing a want ad. See W.P. Allen, AVA, MOEV4, pp. 965–974.
13. Henson, P.M., and M.E. Epstein, "Dyar Family Oral History Interviews (W.J. Dyar)," 5/17/1992, SIA, RU009570.
14. Dyar to L.O. Howard, 1/25/1915, NARA, RG7, E-34, Gen. Corr., b 125, f "Dy."
15. The species were found in the Bahamian rock holes were *Culex similis* Theobald, *Culex aseyehae*, new species, and *Culex sphinx* Howard, Dyar, and Knab. See Dyar, H.G., and F. Knab, "Notes on the Species of *Culex* of the Bahamas," *Insecutor Inscitiae Menstruus* 3 (1915): 112; Howard, L.O., H.G. Dyar, and F. Knab, *The Mosquitoes of North and Central America and the West Indies. Systematic Description*, vol. 3 (part 1) (Washington, D.C.: Carnegie Institution of Washington, 1915).
16. Dyar, H.G., and F. Knab, "Notes on the Species of *Culex* of the Bahamas" (1915),
17. H.R. Durant to W.P. Allen, 3/3/1915, SIA, HGDP, Acc. 95-006, b3, f4.
18. American Literary Bureau to Dyar, 2/10/1915, ibid.
19. Z.P. Dyar to Dyar, 2-4/1912?, DVD, DENo12.
20. Zella refers to a long letter from Dyar—now missing—that clearly dealt with preliminary negotiations for a separation agreement. According to Dyar's version of events that led to the separation in his short story "A Fool and His Friend," Zella accused him of desertion and made a demand for legal separation. Dyar also refers to this letter on March 8, 1915. See: Dyar to Z.P. Dyar, 3/8/1915, DVD, DENo19.
21. Z.P. Dyar to Dyar, 3/2/1915, DVD, DENo20.
22. Dyar to Z.P. Dyar, 3/8/1915, DVD, DENo19.
23. Ibid.
24. Ibid.
25. Ibid.
26. Dyar Story, "The Meeting," 6/9/1917, SIA, HGDP, Acc. 95-006, b1, f3.
27. Dyar to O.P. Dyar and D. Dyar, 3/23/1915, DVD, DENo21,22.
28. DVD, "Separation Agreement," 5/13/1915, Plaintiff's ex. "a"
29. Dyar to L.O. Howard, 1/25/1915, NARA, RG7, E-34, Gen. Corr., b125, f"Dy."

30. L.O. Howard to Dyar, 4/29/1915, ibid.

## CHAPTER 17

1. Dyar to H.M. Peabody, 10/21/1915, DVD, DENo16.
2. DVD transcript, "Separation Agreement," 5/13/1915, Plaintiff's ex. "a"; Wellesca knew Zella's Berkeley cousin Zinie Kidder as an art educator, as Wellesca wrote Zella suggesting she meet baby Roshan during her visit to Washington in 1909.
3. For information about the passengers on the *SS Finland* from New York City to San Francisco, see deposition by Karl Vaupel, 7/12/1916, AVA, MOEV3, 607–625.
4. W.P. Allen, 1916?, pretrial statements to R. Ford or A.G. Miller, HGDP, Acc. 95-006, b1, f5.
5. W.P. Allen, 1916, AVA, MOEV3, p. 756.
6. Ibid., pp. 757–759.
7. Dyar to Coray Leech, 7/7/1915, DVD, DENo5.
8. Dyar mentions Leech's blackmail in a letter to Howard, 11/26/1925, NARA, RG7, E-35, b83, f"Dyar"; W. McLean to Dyar, 8/4/1915, SIA, HGDP, Acc. 95-006, b2, f6.
9. M.A. Leech to A.G. Miller, 12/14/1915, AVA, MOEV3, pp. 584–585.
10. Anonymous, "District Court News." *Washington Post*, 10/8/1915, 9.
11. Dyar to Z.P. Dyar, 10/8/1915, DVD, Defendants ex. 12.
12. Dyar to Z.P. Dyar, 11/25/1915, AVA, MOEV4, p. 894.
13. R. Ford to W.P. Allen, 11/10/1915, SIA, HGDP, Acc. 95-006, b1, f5.
14. M.A. Leech, 7/10/1916, AVA, MOEV2, pp. 481–485.
15. Dyar to H.M. Peabody, 10/21/1915, DVD, DENo16.
16. Ibid.
17. Ibid.
18. Dyar to H.M. Peabody, 11/5/1915, DVD, DENo17, pp. 64–65.
19. Ibid.
20. Dyar to Z.P. Dyar, 11/25/1915, AVA, MOEV4, p. 894.
21. Dyar, H.G., "New Species of American Lepidoptera of the Families Limacodidae and Dalceridae," *Journal of the Washington Academy of Sciences* 17 (1927); Oppewall, J, "From the Gold Rush to the Hollywood Rushes: Some Notes on the History of Butterfly Farming in California," *Terra [members magazine of the Natural History Museum of Los Angeles County]* 1979.
22. W.P. Allen to Z.P. Dyar, 12/5/1915, AVA, MOEV1, pp. 215–222.
23. Ibid.
24. Dyar to Z.P. Dyar, 11/25/1915, AVA, MOEV4, p. 894.
25. W.P. Allen to Z.P. Dyar, 12/5/1915, ibid., pp. 215–222.
26. W.P. Allen to A.G. Miller, ibid., pp. 214–215.
27. Dyar to A. Busck, 3/5/1916, SIA, RU 140, DER, b16
28. A. Busck to Dyar, 1/28/1916, SIA, HGDP, Acc. 95-006, b4, f "3 of 3."
29. Robinson White to J.A. Lynham, "Invoice for steno reporting and type written transcript," 7/26/1916, DVD, Notice of Motion, (filed Sept. 13, 1916).
30. Karl Vaupel and Joseph Davis (depositions), 7/12/1916, AVA, MOEV3, pp. 607–625.
31. H.G. Dyar, AVA, MOEV4, pp. 902–904.
32. Dyar Story, "A Fool and His Friend," 1915, SIA, HGDP, Acc. 95-006, b2, f6.

## CHAPTER 18

1. Dyar Story, "An Old Manuscript," 10/23/1923, SIA, HGDP, Acc. 95-006, b2, f5.
2. For Zella's filing for divorce and a newspaper article see: Anonymous, "District Court News." *Washington Post*, 10/8/1915, 9; Anonymous, "Rich Man Who Works Sued for

Divorce: Harrison Dyar. Whose Income Is $18,000 and Salary $1,800. Accused by Wife of Loving Another," *Washington Herald*, 10/8/1915, 4.
3. Anonymous, "Entomologist of Renown asks Divorce," *Reno Evening Gazette*, 9/19/1916, 8.
4. Anonymous, "Mrs. Wellesca Allen Seeks Divorce," *Washington Post*, 1/24/1916, 7.
5. Dyar to A. Busck, 3/5/1916, SIA, RU 140, DER, b16; Anonymous, "Divorce Defendant an Imaginary One," *Nevada State Journal (Reno)*, 5/23/1916, 8.
6. Anonymous, "Mrs. Dyer [sic] Asks Funds-Wife of Smithsonian Expert Wants Share in Estate: Mrs. Wellesca P. Allen, of Washington, Named as 'the Other Woman,'" *Washington Post*, 3/4/1916, 9.
7. Dyar Story, "Uxores Defendam," 7/3/1916, SIA, HGDP, ibid., b3, f4.
8. Anonymous, "Is Terrible Affair, Declares Mrs. Dyar," *San Francisco Chronicle*, 3/4/1916, 1.
9. Dyar to A. Busck, 3/5/1916, ibid.
10. W.P. Allen, "The Universality of Religion and the Abolition of Sectarianism," 6/18/1916, ibid., b3, f4.
11. Z.P. Dyar to L.O. Howard, 7/8/1916, NARA, RG7, E-34, b125, f "Dy 4 of 5."
12. Dyar to L.O. Howard, 9/10/1916, ibid.
13. Anonymous, "Women Try Divorce Case. Six on Panel Which Hears Capital Man's Suit at Reno," *Washington Post*, 9/21/1916, 3.
14. Anonymous, "Halts Dyar Divorce Case. Court Dismisses Suit, Holding Plaintiff Nonresident," ibid., 9/24/1916, A13.
15. Corrections to Zella's letter by Dyar suggest it was a joint effort. The letter also explains, in an insert in Dyar's hand, that Dyar was in Reno for job-related reasons, and "he hoped would benefit his health. See Z.P. Dyar (draft letter), 1916?, SIA, HGDP, Acc. 95-006, b1, f5.
16. W. McLean to Dyar, 3/14/1918, ibid., b1, f4.
17. Anonymous, "Mrs. Z. P. Dyar Seeks Divorce from Scientist," *Oakland Tribune*, 2/5/1919, 1.
18. Z.P. Dyar v. H.G. Dyar, Defendant, Final Decree of Divorce, Superior Court, Alameda County, California, no. 57236 (filed Dec. 9, 1920).
19. W. McLean to Dyar, 3/14/1918, SIA, HGDP, Acc. 95-006, b1, f4
20. G. Brown to Hon. T.C. Hart, AVA, MOEV1: 30.
21. W.P. Allen, AVA, MOEV1, pp. 42–43.
22. Ibid., 44–52.
23. Ibid.
24. Ibid.
25. Ibid.
26. W.P. Allen to Roshan Allen and M. Leech, AVA, pp. 96–100, DENo3.
27. Brief of Amici Curiae (G. Brown), 1/27/1920, Allen v. Allen, Supreme Court of the State of Nevada, pp. 14–16, (filed Jan. 30, 1920).
28. Opinion on petition for rehearing by J. Coleman, 4/13/1921, AVA No2304 Supreme Court of the Sate of Nevada.
29. George F. Pollock had plenty of reason to rejoice about his sister's marriage to Dyar, who was the major benefactor of Pollock's Skyland resort—someone he most assuredly wanted in the family. One gets the sense of Dyar's importance when George wrote his sister several years earlier that it was "absolutely necessary to have his [Dyar's] cooperation in consummating this deal in order that nothing may transpire to thwart or prevent its being carried out." See G.F. Pollock to W.P.A. Dyar, 5/28/1921, SIA, HGDP, Acc. 95-006, b2; ———, 1918, ibid.

30. In various short stories Dyar has male protagonist getting or seeking a marriage license, even in cases where there has been no consent of the other party. Examples include "The Head of Madour," "Matching Wits with Jessamine," "Time for Three," "His One Sin," "The Adventure of Common Law," "Dame Fortune Smiles," and "Uxores Defendam."
31. NARA, Record Group 21, Records of the District Courts of the United States, District of Columbia, Adoption Case Files, File #449, Box 10. The file at NARA has five items in it: (1) HGD's petition, (2) Wellesca's consent, (3) affidavit of Richard A. Ford, (4) affidavit of Alfred Higbie, and (5) the adoption order.
32. Dyar Story, "A Saucepan for Breakfast," 10/17/1917, SIA, HGDP, Acc. 95-006, b1, f5; Dyar, H.G.; "An Anecdote of the Law-A Story," *Reality* 13 (1927).
33. Ruggles, S., "The Rise of Divorce and Separation in the United States, 1880-1990," *Demography* 34 (1997)
34. Wellesca claimed that Dr. Fedora Armstrong in Los Angeles found that she needed an operation to facilitate childbearing because of a "thick membrane" and recommended a specialist, Dr. Armstead, do the operation. However, no such doctors are listed in the Los Angeles city directory in 1906. W.P. Allen, AVA, MOEV1, pp. 42–43.
35. Dyar to Z.P. Dyar, 2–4/1912?, DVD (filed Sept. 21, 1916).
36. Henson, P.M., and M.E. Epstein, "Dyar Family Oral History Interviews (R. Hill-L. Dyar)," 1999, SIA, RU009570.
37. Dyar, H.G., *A Preliminary Genealogy of the Dyar family* (Washington, D.C.: Gibson Bros., 1903).
38. Wilfred Allen's fictitious parents' names Susan and George were also the names of Wellesca's oldest and closest siblings in age, Susan Plessner Pollock and George Freeman Pollock.
39. Logan, M., *The Part Taken by Women in American History* (Wilmington, DL: Perry-Nalle Publ. Co., 1912).
40. Dyar to Z.P. Dyar, 3/8/1915, DVD, DEN019.
41. Pamela M. Henson (personal communication, 1990s).

## CHAPTER 19

1. Dyar to L.O. Howard, 4/28/1917, NARA, RG7, E-34, b125, f"Dy4,"
2. D.F. Houston or Office of Sec. of Ag. to Clerk Hustings Court, 5/1/1917, RG16, Records of the Office of the Sec of Ag, 1917 CH-Gre, Box no. 159 5-1 Sol 106-221; Office of Inspection to Mr. Harrison, 12/27/1916, RG16, Records Sec Ag, Gen Corresp, Office of Sec Charges—Criticism 1 of 2, PI 191, E17, Box No 274.
3. D.F. Houston to Dyar, 5/8/1917, NARA, microfilm 400, reel #497, 406.
4. Dyar to L.O. Howard, 5/24/1917, ibid.
5. Dyar to A. Busck, 6/9/1917, SIA, RU 140, Dept. of Entomology Records (DER), b16, f1.
6. P.M. Henson suggested Dyar kept his position as Honorary Custodian of Lepidoptera by donating his collection. See Henson, P.M., "Oral History Interviews with John Frederick Gates Clarke", 1986, SIA, RU009555. (p. 88); The following report overestimated the number of Lepidoptera specimens Dyar donated to be 17,000 compared to the 15,000 in the accession record. See Accession #6 1931, 12/18/1917, SIA RU 305, Registrar, 1834-1958; Anonymous, *The United States National Entomological Collections* (SI Press, 1976).
7. B.P. Clark to L.O. Howard, 7/21/1917, NARA, RG7, E34, b81, f "CL 1".
8. L.O. Howard to B.P. Clark, 8/3/1917, ibid.

9. Busck continued to be a source that passed the Dyar legend to a younger generation of USNM scientists, including J.F. Gates Clarke. See Henson, P.M., ibid.
10. According to Schaus, Clark had written him on July 21, 1917 about the National Museum severing ties with Dyar. See W. Schaus to B.P. Clark, 8/20/1917, ACM, B. Preston Clark Corresp., f12.
11. W. Schaus to L.O. Howard, 3/31/1916, NARA, RG7, E-34, Gen. Corresp, b320, f "Schaus, William."
12. Dyar to L.O. Howard, 8/13/1917, ibid, b125, f "Dy4."
13. Ibid., 9/4/1918.
14. A. Busck to H.G. Dyar (draft), 6/26/1912, SIA, RU 140, DER, 1909-1963, b16, f1.
15. Ibid.; L.O. Howard to Dyar, 8/9/1917, ibid.
16. Dyar to L.O. Howard, 8/13/1917, ibid.
17. L.O. Howard to C. Heinrich, 8/25/1917, NARA, ibid; Dyar to O'Leary, 8/28/1917, NARA, RG7, E-34, Gen. Corr., b125, f "Dy."
18. Barnes, W., and J. McDunnough, *Checklist of Lepidoptera of Boreal North America* (Decatur, IL: Herald Press, 1917).
19. Dyar, H.G., "The Barnes & McDunnough "List" (Lepidoptera)," *Insecutor Inscitiae Menstruus* 5 (1917).
20. Dyar to W. Barnes, 8/9/1917, NARA, RG7, E-34, b125, f4.
21. D.F. Houston to Dyar, 1/3/1918, NARA, RG16, Records of the Office of Secr. Ag., Charoal-Charges Box No. 375.
22. Dyar to T.D.A. Cockerell, 5/30/1918, CUBLA, TDAC. Earlier Dyar had named a Costa Rican limacodid moth after Crawford's wife and Wellesca's niece, Emily Baker, who had worked at the National Museum.
23. H.G. Dyar to T.D.A. Cockerell, 1/9/1918, NARA, RG7, E-34, b125, f "Dy4."
24. As a friend of Dyar's Knab was described as the "one exception [as] his associates at the Museum had little to do with him.... They seemed to understand each other well; and, as you know, they collaborated in the development of some excellent work in the taxonomy of mosquitoes." See Mallis, A., *American Entomologists* (New Brunswick, NJ: Rutgers, 1971).
25. F. Knab to Dyar, 7/20/1914, SIA, HGDP, Acc. 95-006, b4, f3.
26. ———, 12/4/1914, ibid., b4, f3.
27. ———, 1/4/1915 and 3/2/1916, ibid., b4, f3.
28. Forbes was known as "pink whiskers," according to Cornell Professor Jack Franclemont, who knew him (Richard Brown, pers. comm., 2014). See F. Knab to Dyar, 9/9/1915, ibid., b4, f3.
29. F. Knab to Dyar, 10/4/1915, ibid.
30. ———, 8/20/1915, ibid.
31. ———, 9/13/1915, ibid.
32. Howard, L.O., H.G. Dyar, and F. Knab, *The Mosquitoes of North and Central America and the West Indies. Systematic Description*, vol. 4 (part II) (Washington, D.C.: Carnegie Institution of Washington, 1917).
33. Cockerell, T.D.A., "Reviewed work(s): The Mosquitoes of North and Central America and the West Indies by L. O. Howard; H. G. Dyar; F. Knab," *Science* 45 (1917)
34. Dyar to T.D.A. Cockerell, 9/20/1917, CUBLA, TDAC, b5, f74; See also letters from Dyar to Cockerell relating mostly to Colorado or western mosquitoes or their mail shipments in the same archival collection from 9/20–12/15,1917 and 4/1–9/12, 1918.
35. Dyar to T.D.A. Cockerell, 9/20/1917, ibid.

36. Dyar to L.O. Howard, 1/14/1918, NARA, RG7, E-34, b125 f "Dy4."
37. F. Knab to Howard, 2/9/1918, ibid., f "Knab." Harbach, R.E., "The Culicidae (Diptera)," Zootaxa 1668 (2007); see also Dyar, H.G. and F. Knab, "Book Notice: A Monograph of the Culicidae of the World by F. V. Theobald: London, 1907. Volume IV," *Journal of the New York Entomological Society* 15 (1907).
38. Caudell, A.N., A. Busck, and L.O. Howard. "Frederick Knab." *Proceedings of the Entomological Society of Washington* 21 (1919).

## CHAPTER 20

1. A. Busck to W. Barnes, 10/12/1920, SIA RU 140, DER, 1909-1963, b3, f6.
2. The Lepidoptera collection was further subdivided into main (or line) and duplicate collections, the latter comprising specimens of less than optimal condition.
3. W. Barnes to A. Busck, 2/1/1919, SIA RU 140, DER, 1909-1963, b3, f6.
4. A. Busck to Dyar, 7/17/1920, SIA.
5. Dyar to L.O. Howard, 10/11/1920, NARA, RG7, E-34, b125, f5.
6. Ibid.
7. L.O. Howard to Dyar, 10/15/1920, ibid.
8. Dyar to L.O. Howard, 10/15/1920, ibid.
9. Ibid.
10. ———, 10/16/1920, ibid.
11. A. Busck to W. Barnes, 10/12/1920, SIA RU 140, DER, 1909-1963, b3, f6.
12. L.O. Howard to A. Busck, 11/11/1920, NARA, RG7, E-34, b67, f "Busck."
13. Dyar to L.O. Howard, 12/4/1920, NARA, RG7, E-34, b124, f "Dy."
14. Ibid.
15. Dyar supported having Schaus to be his Assistant Custodian (not Curator) of Lepidoptera, even with a salary because Schaus's investments in Europe suffered during the war. See Dyar to L.O. Howard, 9/4/1918, ibid.; Ravenel, W.D., "Report on the Progress and Condition of the United States National Museum for the Year Ending June 30, 1919," (1920): 127; ———, "Report on the Progress and Condition of the United States National Museum for the year ending June 30, 1920 p. 7" (1921).
16. Dyar to L.O. Howard, 12/2/1920, NARA, RG7, E-34, b124, f "Dy."
17. L.O. Howard to Dyar, 12/3/1920, ibid.; Henson, "Oral History Interviews with John Frederick Gates Clarke," 1986, SIA, RU009555: 93.
18. L. Stejneger to Dyar, 1/15/1924, NARA, RG7, E-34, b124, f "Dy."
19. Dyar to L.O. Howard, 2/11/1924, ibid.; L.O. Howard to Dyar, 1/18/1924, ibid.
20. Dyar to L.O. Howard, 5/5/1925, NARA, RG7, E-35, b83, f "Dyar"; L.O. Howard to Dyar, 5/6/1925, ibid.
21. Dyar to L.O. Howard, 2/11/1924, ibid.
22. L.O. Howard to Dyar, 1/18/1924, ibid.
23. Dyar to L.O. Howard, 2/11/1924, ibid.
24. Dyar to J.M. Aldrich, 1/2/1919, SIA, RU 140, DER.
25. The association of the names Schaus and Barnes has often confused lepidopterists because after the Barnes collection finally was incorporated into the National Museum in the 1930s, specimens labeled "Coll. Wm. Barnes" were side by side with those collected by Schaus and John "Jack" Barnes and labeled "Schaus and Barnes Coll."
26. Henson, "Oral History Interviews with John Frederick Gates Clarke," 1986, SIA, RU009555: 93.
27. Dyar to L.O. Howard, 1/31/1927, NARA, RG 7, E-35, b83, f "Dyar."

28. The National Geographic visit related to an article six months later—acknowledging the "16 butterfly and moth plates in four colors" made possible by the USNM and "Dr. Harrison G. Dyar, whose work in the classification of Lepidoptera has been an outstanding achievement in the activities of this great institution." See Showalter, W.J., "Strange Habits of Familiar Moths and Butterflies," *National Geographic Magazine*, July 1927.
29. Dyar to J.M. Aldrich, 3/2/1927, SIA, RU 140, DER.
30. M. Carpenter to J.M. Aldrich, 2/28/1927, ibid.
31. Dyar to J.M. Aldrich, 3/2/1927, ibid.
32. Ibid.
33. Dyar to C.W. Johnson, 12/18/1926, SIA HGDP, b2, f1.
34. Brown referred to Dyar as being "an odd duck." F.M. Brown to M.E. Epstein, "Visit to USNM during the 1920's," 1991.
35. L.M. Russell to M.E. Epstein and P.M. Henson, "Comment at Henson and Epstein lecture on H.G. Dyar" (1990s); Miller, D.R., and G.L. Miller, "Louise M. Russell (1905–2009)," *Proceedings of the Entomological Society of Washington* 113 (2011).

## CHAPTER 21

1. Dyar to L.O. Howard, 1/14/1918, NARA, RG7, E-34, b125, f "Dy."
2. Dyar's notebook "Matters referred from Bur. Ent. Incoming mosquitoes and lepidopterans" is currently in the Department of Entomology Archives, NMNH, Smithsonian Institution.
3. Dyar to L.O. Howard, 4/14/1919, ibid.
4. Ibid.
5. L.O. Howard to Dyar, 4/16/1919, ibid.
6. Dyar to L.O. Howard, ibid.
7. L.O. Howard to Dyar, ibid.
8. Dyar to L.O. Howard, 11/1/1920, ibid.
9. Ibid.
10. L.O. Howard to Dyar, 11/1/1920, ibid.
11. Dyar to L.O. Howard, 11/4/1920, ibid.
12. ———, 5/17/1921, ibid.
13. Dyar, H.G., "Mosquitoes of Canada," *Transactions of the Royal Canadian Institute* 13 (1921).
14. L.O. Howard to Dyar, 5/23/1921, ibid.
15. Dyar, H.G., and F. Knab, "Notes on Mosquito Work," *The Canadian Entomologist* 40 (1908).
16. Ludlow, C.S., "Mosquito Comment," ibid., 41 (1909).
17. Dyar to L.O. Howard, 10/4/1918, NARA, RG7, E-34, b125, f "Dy."
18. ———, 3/11/1920, ibid.
19. Dyar to L.O. Howard, 2/13/1925, ibid.
20. ———, 3/8/1925, NARA, RG 7, E-35, b83, f "Dyar"; L.O. Howard to Dyar, 2/14/1925, ibid.
21. Dyar to L.O. Howard, 3/8/1925, ibid.
22. ———, 6/22/1924, NARA, RG7, E-34, ibid.
23. That Wellesca was on these trips is based on evidence from Baháʾí lectures she gave, a passport application, and correspondence.
24. ———, 3/24/1916, Dyar v. Dyar, Defendant's ex. 17 (filed Sept. 1916); For Dyar's records for mosquitoes and moths from San Diego (3/26–5/7/1916), Yosemite and the Lake Tahoe region (5/14–7/23/1916) and Roseville, CA (8/20/1916), see Dyar,

H.G., "Mosquitoes at San Diego, California," *Insecutor Inscitiae Menstruus* 4 (1916); ———, "Notes and New Species of American Moths of the Genus *Scoparia* Haworth," *Proceedings of the United States National Museum* 74 (1929); Howard, L.O., H.G. Dyar, and F. Knab, *The Mosquitoes of North and Central America and the West Indies. Systematic Description*, vol. 4 (part II) (Washington, D.C.: Carnegie Institution of Washington, 1917); Dyar, H.G., "Mosquitoes of the United States," *Proceedings of the United States National Museum* 62 (1922).

25. Dyar, H.G., "The Mosquitoes of the Mountains of California," *Insecutor Inscitiae Menstruus* 5 (1917): 17; For more information on Price, see Fisher, W.K. "William Wightman Price," *The Condor* 25(2) (1923), 50–57; Lekisch, B., *Tahoe Place Names: The Origin and History of Names in the Lake Tahoe Basin* (Lafayette, CA: Great West Books, 2007), 42.
26. Dyar, "Mosquitoes of the United States" (1922): 18.
27. From early- to mid-July 1917 Dyar collected in Montana at Missoula, Evaro, Drummond, Glen, Whitehall, Bozeman, Big Timber, Park City, Youngs Point, and Laurel. See ibid.
28. In July and August 1918 Dyar also collected at Lake Minnewanka, Red Deer, Lochearn, and Lamoral, Alberta. See ibid.
29. ———, "The mosquitoes of British Columbia and Yukon Territory, Canada (Diptera, Culicidae)," *Insecutor Inscitiae Menstruus* 8 (1920): 3.
30. Back and forth between the Yukon and Alaska in late June and July. See ———, "Mosquitoes of the United States" (1922): 5/13–09/18/1919.
31. Dyar, H.G., "The Mosquitoes of British Columbia and Yukon Territory, Canada (Diptera, Culicidae)" (1920).
32. Through July 1, 1920, Dyar collected in the Tahoe area at Lake Tahoe, Summit, Truckee, Gold Lake (Sierra County); in Plumas Co., including Bear Lake, Gold Lake Camp, Lakes Center Camp, and Camp Elwell, followed by Washington State at Lake Cushman, Hoodsport, Sumas, and Oroville; and in Oregon at Montavilla and Crater Lake. See Dyar, "Mosquitoes of the United States" (1922).
33. San Juan Capistrano and Texas records see Dyar, ibid.
34. The crambid moth was previously known only from Henry Edwards's specimens from California mountains. Dyar to L.O. Howard, 10/16/1920, NARA, RG7, E-34, ibid.
35. See Dyar, "Mosquitoes of the United States" (1922): 5/6–07/27/1921; Dyar, A.W.P.A., and H.G. Dyar, *Short Talks on the Practical Application of the Bahai Revelation* (Washington, D.C. 1922). *Aedes varipalpus* Coquillett recorded by Dyar at Orr's Hot Springs, California (Dyar 1922: 62) for July 1921 is almost certain to be a date error—it is likely to have been in May between going to San Francisco and Medford. See ibid., 4/10–08/21/1921.
36. Dyar, "Mosquitoes of the United States" (1922): 90; Dyar Story, "Conway House Waits", 8/25/1921, SIA, HGDP, Acc. 95-006, b1, f4.
37. Dyar to L.O. Howard, 8/16/1922, NARA, RG7, E-34, ibid.
38. Dyar, H.G., "The Mosquitoes of the Yellowstone National Park (Diptera, Culicidae)," *Insecutor Inscitiae Menstruus* 11 (1923); Dyar to L.O. Howard, 8/16/1922, ibid. See also records for black flies Dyar found in Yellowstone in: Dyar, H.G. and R.C. Shannon, "The North American Two-winged Flies of the Family Simuliidae," *Proceedings of the United States National Museum* 69 (1927).
39. A. Aiello (pers. comm., 2015)
40. Dyar to L.O. Howard, 3/15/1923, NARA, RG7, ibid.
41. L.O. Howard to Dyar, 3/16/1923, ibid.
42. In 1923 James Zetek (1886–1959) was early in his tenure as Resident Manager of the Canal Zone Biological Area on Barro Colorado Island. Primarily interested

in termites and termite control, he impressed Dyar with his caterpillar rearing leading Dyar to request he prepare specimens for him, much to his boss Altus Lacy Quaintance's chagrin. See Dyar to R.C. Shannon, 6/23/1923, SIA, HGDP, b2, f5 "Shannon"; and Zetek, J, SIA, RU 7462, James Zetek Papers, circa 1921–1951; Dyar to L.O. Howard, 6/22/1924, NARA, RG7, E-34, ibid.
43. See Dyar, H.G, "The Mosquitoes of Panama (Diptera, Culicidae)," *Insecutor Inscitiae Menstruus* 13 (1925)
44. Ibid.
45. Dyar, H.G., and R.C. Shannon, "The American Chaoborinae (Diptera: Culicidae)," ibid. 12 (1924); Dyar, H.G., *The Mosquitoes of the Americas*, Carnegie Institution of Washington, Publication no. 387 (Carnegie Institution of Washington, 1928), p. 206; Dyar to L.O. Howard, 6/22/1924, NARA, RG7, ibid.

## CHAPTER 22

1. Dyar Story, "The Ghost Makers," 12/19/1913, SIA, HGDP, Acc. 95-006, b1, f3.
2. Anonymous, "Old Tunnel Here Believed to Have Been Used by Teuton War Spies and Bootleggers," *Washington Post*, 9/26/1924, 1, 8.
3. Anonymous, "'Mystery' Tunnels Built by Scientist 'Merely as Pastime,'" *Washington Post*, 9/27/1924, 1–2.
4. Anonymous, "Great Tunnel Mystery Leads Back to Little Hollyhock Bed," *Washington Evening Star*, 9/27/1924, 1–2.
5. The cartoon was about Senator Elect Charles Bryan, brother of William Jennings Bryan, who was believed to have been part of a strategy with La Follette to become president if the election was deadlocked. The label on the rear of Bryan's trousers "Socialist Calif Patch" refers to a decision by the California Supreme Court that required La Follette to be listed as a member of the socialist party rather than an independent, as he desired, on the ballot. See Rogers, W.A., "That Mysterious Tunnel in Washington," political cartoon, *Washington Post*, 9/30/1924, 6; Fontaine, S.S., "Wall Street Lonesome; Many Friends Are Absent," *Washington Post*, 9/30/1924, 17.
6. Brown, G.R., "Post-Scripts," ibid., 9/27/1924, 1.
7. Anonymous, "Pays $17,910 for Garden. Dr. H.G. Dyar Purchases Lot Adjoining His Twenty-first Street Home." *Washington Post*, 11/22/1906; for getting neighbors' permission for alley, see C. Early to Dyar, 12/8/1906, SIA, HGDP, Acc. 95-006, b2, f4.
8. Anonymous, "'Mystery' Tunnels Digger Burrows New Double Maze. More Elaborate Subterranean Labyrinth Opens from Cellar of Scientist's Home, Ending in Square Well 24 Feet Deep," *Washington Post*, 9/28/1924, 1–2; Henson, P.M., and M.E. Epstein, "Dyar Family Oral History Interviews (R. Hill-L. Dyar)," 1999, SIA, RU009570.
9. Anonymous, "Great Tunnel Mystery Leads Back to Little Hollyhock Bed," *Washington Evening Star*, 9/27/1924 1–2; Henson, P.M., and M.E. Epstein, "Dyar Family Oral History Interviews (W.J. Dyar)," 5/17/1992, ibid.
10. ———, ibid., 1999, SIA, RU009570.
11. Anonymous, "'Mystery' Tunnels Digger Burrows New Double Maze." *Washington Post*, 9/28/1924, 1–2; Anonymous, "Tunnel-Digging as a Hobby," *Modern Mechanix* 1932.
12. The Gerrity article in the *Post* describes the tunnel ceilings as about "8 feet" on the upper level reached from the basement, another article on the same day in the *Washington Star* has a photograph that agrees with the description. See Gerrity, J.F., "Mysterious Diggins? No! Just a Scientist's Hobby," *Washington Post*, 3/4/1942, 12; Anonymous, "New Mystery Tunnels Found under House on Independence Ave:

Sinister Legends Survive, But It Seems Dr. Dyar Just Liked to Dig," *Washington Star*, 3/4/1942.
13. Although Dyar's explanation that he packed all of the dirt for this massive tunnel in his backyard would have made sense at Twenty-first Street, where he had the land, it doesn't at the much smaller lot at B Street. Perhaps he dumped it elsewhere. See Anonymous, "'Mystery' Tunnels Digger Burrows New Double Maze" Anonymous, "Tunnel-Digging as a Hobby," *Modern Mechanix*, 1932; Henson and Epstein, ibid., 5/17/1992. See also Gerrity, J.F., ibid.
14. Anonymous, "Tunnel-Digging as a Hobby."
15. Henson and Epstein, ibid., 5/17/1992.
16. Kelly, J., "A Final Look at D.C.'s Tunnel-Digging Bug Man," *Washington Post*, 11/7/2012.
17. Robinson, L., "Our Editor," *Reality* 17 (1929); Anonymous, "'Mystery' Tunnels Digger Burrows New Double Maze"; Anonymous, "Tunnel-Digging as a Hobby"; Henson and Epstein, ibid., 25.
18. Ibid., 19.
19. Ibid., 6.
20. Anonymous, "Mystery Tunnel Joins Two Homes: Underground Passageway Bared in Wealthy Section; Built by Owner's Hands," *Washington Times*, 5/19/1917.
21. Ibid.
22. J. Lintner to Dyar, 5/13/1892, SIA, HGDP, b2, f1.
23. Harrower, C.S., *In memoriam: An address by Rev. Charles S. Harrower, at the obsequies of Harrison Gray Dyar, in Rhinebeck, N.Y. February 3, 1875* (New York: Jones, 1875), 9.
24. Anonymous, "Great Tunnel Mystery Leads Back to Little Hollyhock Bed," *Washington Evening Star*, 9/27/1924 1–2.
25. Anonymous, "Inspector Approves Tunnels Dug by Dyar," *Washington Post*, 10/8/1924, 9; Dyar to L.O. Howard, 10/9/1924, NARA, RG 7, E-35, b83, f "Dyar."
26. Bringing the B Street tunnel to the attention of the public brought out District building inspector Maj. John W. Oehmann a few days later. However, Dyar was granted permits to continue his digging in early October 1924. See: Anonymous, "Dyar's Tunnel Hobby May Be Curtailed. Scientist Must Get Permits in Future; Officials Investigate Possible Danger," *Washington Post*, 9/29/1924; and ———, "Inspector Approves Tunnel Dug by Dyar," *Washington Post*, 10/8/1924.
27. Anonymous, "List of Patents [Improvement of Tunneling]," *The London and Edinburgh Philosophical Magazine and Journal of Science* 1 (1832).
28. Dyar, "The Ghost Makers."
29. Kelly, "A Final Look at D.C.'s Tunnel-Digging Bug Man," ibid.

## CHAPTER 23
1. H.M. Peabody to S.A. Knopf, 12/22/1921?, NLB, SAKP, MS C41.
2. Ibid.
3. Mrs. Peabody returned home for seven months prior to her death. Zella may have been there because she is no longer listed in Berkeley at that time. See Death of Henich [sic] M Peabody, "Maine Death Records, 1617-1922," 12/9/1922.
4. Anonymous, "Berkeley Cal.-First Unitarian Church Rev. Harold E.B. Speight," *The Christian Register*, 3/23/1918.
5. California Voter Registrations, "Alameda County, Roll 12, 395." 1918.
6. Dyar, D., "Youth's Challenge to the Church," *Reality* 6 (1923); Anonymous, "1927," *Alumni of the Union Theological Seminary in the City of New York* (1958).
7. W. McLean to Dyar, 5/5/1920, SIA, HGDP, Acc. 95-006, b2, f2; information about Otis Dyar and Catherine Ross was obtained by oral interview of Handy Family to M. Epstein, 2/24/2010.

8. W.P.A. Dyar to Woozie Watson, 7/3/1921, NBA.
9. Dyar to L.O. Howard, 3/13/1924, NARA, RG7, E-34, b125, f "Dy 1 of 5."
10. Henson, P.M., and M.E. Epstein, "Dyar Family Oral History Interviews (W.J. Dyar)," 5/17/1992, 7.
11. "Dak" is the spelling given to me by Wallace J. Dyar. In the taped interview it was transcribed as "Dack." See Henson and Epstein, "Dyar Family Oral History Interviews (W.J. Dyar)," ibid. For a description of Wellesca being called "Mrs. Allen" and Mrs. Leech as "mama," see Leech's testimony in AVA, MOEV2.
12. Anonymous, "Will Adopt Wife's Sons: H.C. [sic] Dyar Files Petition to Change Boy's Names," *Washington Post*, 4/7/1922, 2.
13. Henson and Epstein, ibid., 7–8.
14. Ibid.
15. Robinson, L, "Our Editor," *Reality* 17 (1929).
16. C.R. Leech, AVA, MOEV2; Henson and Epstein, ibid., 12, 13.
17. Ibid., 18.
18. Ibid., 17–18; Alan Stone to M. Epstein, 2/27/1993, M.E. Epstein files.
19. Ibid., 11.
20. Ibid., 9–10.
21. Meade's pen name left off the "e" at the end of her last name. See also Carr, V.S., *Paul Bowles: A life* (New York: Scribner, 2004); Henson and Epstein, "Dyar Family Oral History Interviews (W.J. Dyar)," 5/17/1992, SIA, RU009570.
22. M.R. Meade to W.P. Allen, 7/26/1916, SIA, HGDP, Acc. 95-006, b3, f3.
23. ———, 7/15/1919, ibid., b3, f4.
24. H.G. Dyar and W.P. Dyar, "Last Will and Testament," 4/12/1922.
25. H.D. Dyar to Edward Young, 3/20/1923, NBA, Agnes Parsons Papers, b18.
26. Epstein, M.E., and P.M. Henson, "Digging for Dyar: The Man Behind the Myth," *American Entomologist* 38 (1992).
27. Ibid., 146.
28. This is the version of the twelve Bahá'í principles found on the inside cover of *Reality Magazine* by the time H.G. Dyar was the editor in 1923.
29. There are a number of letters by Dyar and Wellesca (as Aseyeh) in the National Bahá'í Archives, which reflect on their views of keeping blacks and whites separate at Washington Bahá'í meetings.
30. Aseyeh at some point signed an agreement to follow 'Abdu'l-Bahá's policy against segregated meetings in America.
31. W.P. Allen to Agnes Parsons?, 1918–1919, NBA.
32. 'Abdu'l Bahá to W.P. Allen, Tablet transl. by Shoghí Effendí Rabbání, home of 'Abdu'l Bahá, Haifa, Palestine, 3/12/1919, NBA.
33. Aseyeh Allen, "Notice of lecture: The Great Peace," 9/4/1919, SIA, HGDP, Acc. 95-006, b3, f2.
34. 'Abdu'l Bahá to H.G. Dyar, Tablet transl. by Aziz'O'llah S. Bahadur, Bahju, 5/5/1921, NBA.
35. Anonymous, "The Passing of Abdu'l Baha," *Bahai Topics. An information resource of the Bahai International Community*, http://info.bahai.org/article-1-3-4-4.html.
36. Dyar, A.W.P.A., and H.G. Dyar, *Short Talks on the Practical Application of the Bahai Revelation* (Washington, D.C., 1922).
37. Ibid.; R.D. Heiner to W.P.A. Dyar, 3/25/1922, SIA, HGDP, Acc. 95-006, b4, f4.
38. Anonymous, "Letter to subscribers of Reality Magazine," 11/14/1919, ibid., b2, f1.
39. Smith, P. *Reality Magazine: Editorship and Ownership of an American Bahá'í Periodical* (1984).
40. Dyar, H.G., "Mr. Bryan and His Monkeys," *Reality* 6 (1923).
41. ———, "The Old New England Stock," ibid., 7 (1924).

42. Haiati, A., "The Fallacy of Gender Equality in The Bahai Faith: Bigamy Allowed for Men Only," January (2008), http://www.bahaiawareness.com/bahai12.html.
43. Dyar, H.G., "A Dream of the Pope," *Reality* 6 (1923); ———, "As a Man Thinketh," ibid.; ———, "The Sixth Race," *Reality* 9 (1925).
44. Smith, ibid.
45. *Reality* ceased publication in March 1929, volume 16.

## CHAPTER 24

1. W. Barnes to Dyar, 3/22/1922, NARA, RG7, E-66, b9.
2. Dyar to F.H. Benjamin, 8/7/1925, ibid.
3. F.H. Benjamin to Dyar, 11/23/1925, SIA, HGDP, b1, f2.
4. Dyar wrote of past conflicts with James H. McDunnough, probably dating back to when McDunnough worked for Barnes, writing: "It is true that in the course of amenities McDunnough thought his corns were stepped on; but that is where he attributes more importance to himself than I was doing. He was purely an incident in my estimation which naturally he did not appreciate." See Dyar to F.H. Benjamin, 12/14/1925 SIA, HGDP, b1, f2.
5. F.H. Benjamin to Dyar, ibid.; Dyar to F.H. Benjamin, ibid.
6. Dyar to F.H. Benjamin, 12/14/1925, ibid.
7. Dyar to L.O. Howard, 10/3/1925, NARA, RG7, E-35, b83, f "Dyar."
8. L.O. Howard to Dyar, 10/3/1925, ibid.
9. Dyar to W. Barnes, 10/6/1925, ibid.; Dyar to L.O. Howard, 10/6/1925.
10. W. Barnes to Dyar, 10/13/1925, ibid.
11. A. Busck to F.H. Benjamin, 2/3/1927, NARA, RG7, E-66, b6, f "Be 1 of 4."
12. Ibid.
13. F.H. Benjamin to R. Ottolengui, 8/15/1924, ibid., f "Be 5 of 6." The history of the *mihi itch* has been reviewed in: Evenhuis, N.L., "The 'Mihi Itch'—a Brief History," *Zootaxa* 1890 (2008).
14. Dyar to F.M. Jones, 11/20/1922, 11/19/1927, SIA, HGDP, b2, fl.
15. There are numerous pieces of correspondence related to the sale of the Barnes Collection in the William Barnes papers. See NARA, RG7, E-66.
16. Epstein, M.E., and P.M. Henson, "Digging for Dyar: The Man Behind the Myth," *American Entomologist* 38 (1992); J.M. Aldrich to C.D. Walcott, 12/14/1921, NARA, RG7, E-66, b3, f "Ba 3 of 3."
17. W.C. Barnes to W. Barnes, 1922?, ibid.
18. W. Barnes to Dyar, 3/22/1922, NARA, RG7, E-66, b9.
19. A.F. Moore to C.D. Walcott, 12/19/1921, ibid.
20. J.M. Aldrich to C.D. Walcott, ibid.
21. W. Barnes to A.F. Moore, 5/12/1922, NARA, RG7, ibid.
22. Howard's letter responded to Dyar's request that he and Foster Benjamin be salaried—with Benjamin as his assistant—when purchase of the Barnes Collection was approved, in which Dyar wrote: "[Y]ou will please take note and PIN THIS IN A CONSPICUOUS PLACE (in memory) it will be duly appreciated and remembered as another of your many acts of consideration toward the undersigned." See L.O. Howard to Dyar, 5/19/1924, NARA, RG7, E-34, b 124, f "Dy"; Dyar to L.O. Howard, 5/13/1924, ibid.
23. Dyar to W. Barnes, 10/6/1925, ibid.
24. J. O'Connor to W. Barnes, 12/20/1927, ibid., b5, f "Ba 3 of 3."
25. Epstein and Henson, "Digging for Dyar: The Man Behind the Myth" (1992); Anonymous, *The United States National Entomological Collections* (SI Press, 1976).

## CHAPTER 25

1. Dyar to L.O. Howard, 1/18/1925, NARA, RG7, E-35, b83, f "Dyar."
2. Dyar was also assisted by G. Allen Mail, an assistant to extension entomologist W.B. Mabee of the Bozeman Experiment Station and previously for Canadian mosquito expert Eric Hearle "during the campaign at Banff." See ———, 5/24/1926, ibid; Dyar, H.G., "A New Species of Mosquito from Montana, with Annotated List of the Species Known from the State," *Proceedings of the United States National Museum* 75 (1929); ———, "Three Psychodids From the Glacier National Park (Diptera, Psychodidae)," *Insecutor Inscitiae Menstruus* 14 (1926).
3. W. McLean to Dyar, 7/26/1926, SIA, HGDP, Acc. 95-006, b2, f5.
4. Dyar to L.O. Howard, 5/24/1926, NARA, RG7, E-35, b83, f "Dyar."
5. Dyar to J.H. Horton, 4/19/1926, ibid; Dyar to F.H. Benjamin, 10/14/1926, NARA, RG7, E-66, b 9.
6. Dyar to L.O. Howard, 8/16/1922, NARA, RG7, E-34, b125, f "Dy 1 of 5."
7. Dyar, H.G., "The Mosquitoes of the Yellowstone National Park (Diptera, Culicidae)," *Insecutor Inscitiae Menstruus* 11 (1923); Dyar, H.G., "New *Aedes* from the Mountains of California," *Insecutor Inscitiae Menstruus* 4 (1916).
8. Dyar to L.O. Howard, 3/26/1924, NARA, RG7, E-34, b125, f "Dy 1 of 5."
9. ———, 12/17/1927, NARA, RG 7, E-35, b83, f "Dyar."
10. Dyar to M.J. Bronn, 11/26/1927, SIA, HGDP, b1, f3.
11. L.O. Howard to Dyar and reverse, 12/17/1927, NARA, RG7, E-35, b83, f "Dyar."
12. Dyar identified specimens from Dr. D. P. Curry (Panama), Dr. M. Nunez Tovar, and Dr. F. M. Root (Venezuela). Dyar to L.O. Howard, 1/9/1928, ibid.
13. ———, 3/26/1924, NARA, RG7, E-34, b 125, f "Dy 1 of 5." For Shannon's concern about Knab see: R.C. Shannon to F. Knab, 3/8/1917, SIA, HGDP, b1.
14. ———, 12/17/1927, NARA, RG7, E-35, b 83, f "Dyar."
15. L.O. Howard to Dyar, 5/23/1928, ibid.
16. Dyar to L.O. Howard, 10/10-11/1927 and reverse, ibid.
17. Dyar's manuscript for the New World Limacodidae in Seitz's "Macrolepidoptera" has keys—cut from the published version—to all of the new world genera and most of the species, under each genus. See Department of Entomology Archives, National Museum of Natural History.
18. M. Draudt to Dyar, 4/28/1927, SIA, HGDP, b1, f6.
19. Carl Heinrich made male genitalia preparations to help Dyar identify and name new species of *Sibine* (now *Acharia*). The information gathered from these slides was used in the Seitz publication, though the illustrations, now in the Dept. of Entomology Archives, NMNH, were never published; See also E.M. Hering to Dyar, 12/12/1927-2/6/1928, SIA, HGDP, b1.
20. Dyar to W.T.M. Forbes, 11/4/1927, CUA, WTMFP, b1, 1927.
21. Dyar, H.G., "New Species of American Lepidoptera of the Families Limacodidae and Dalceridae," *Journal of the Washington Academy of Sciences* 17 (1927); In addition to his own relatives, in 1905 Dyar named another limacodid *Semyra zinie* after Zella's cousin in Berkeley, California, Zinie Kidder: an artist and educator. See ———, "A list of American Cochlidian Moths, with Descriptions of New Genera and Species," *Proceedings of the United States National Museum* 29 (1905).
22. ———, "New Species of American Lepidoptera of the Families Limacodidae and Dalceridae," *Journal of the New York Academy of Sciences* 17 (1927): 550; ———, "Descriptions of Four South American Moths," *Proceedings of the Entomological Society of Washington* 30 (1928): 10.

23. On a four-month trip Zella took with Dorothy in 1926, both gave Perle Knopf's address as "16 W. Nineteenth Street, NY, NY." By the 1930 Census, Zella had moved to Seattle with her daughter.
24. Dyar to W.J. Holland, 6/9/1928, SIA, HGDP, b1, f9; While Dyar was unable to find *Speyeria nokomis* in the USNM collection, William Barnes was said to have problems collecting it in southern Utah on three failed expeditions. See Anonymous, "Butterfly Collection Bought by Smithsonian: Looking back at Central Illinois/Celebrate 2000," *Herald & Review (Decatur, Il)*, B6.
25. Clarke never personally met Dyar and did not recall receiving any correspondence from him (M. Epstein, personal comm., 1989–1990). See Dyar to J.F.G. Clarke, 10/19/1928, SIA, RU 189, b111; Henson, P.M., "Oral History Interviews with John Frederick Gates Clarke," 1986, SIA, RU009555: 25.

## CHAPTER 26

1. Dyar to L.O. Howard, 12/15/1927, NARA, RG7, E-35, b83, f "Dyar."
2. Anonymous, "Harry Wardman," http://en.wikipedia.org/wiki/Harry_Wardman; Anonymous, "H.G. Dyar Buys Cavanaugh. Nevada Man Said to Have Paid $215,000 for 17th St. Apartment House," *Washington Post*, 10/7/1917, R1.
3. Many letters from Dyar's advisors McLean and lenders J.J. Lampton and others suggest that Dyar was not getting rental prices he sought or was in debt for notes that came due. See SIA, HGDP, Acc. 95-006.
4. Anonymous, "Sixteenth Street Mansions Bought by Scientist-Digger," *Washington Post*, 2/7/1925, 1, 8.
5. For other purchases of Wardman buildings, see "Anonymous, H.G. Dyar Buys Cavanaugh." *Washington Post*, 10/7/1917, R1; and Anonymous, "Rutland Courts Sold," *Washington Times*, 6/9/1917, 8.
6. Collateral appears to have come from a loan Dyar secured through a deed of trust on the property at square 2526, lot 195 through David C. Finney to Harry J. Robb and James J. Lampton. See Anonymous, "Daily Legal Record," *Washington Post*, 1/19/1925, 11.
7. Prior to the purchase the property it received attention from the U.S. Senate's Ball Committee "on housing in the District of Columbia" the previous spring as one of the "stormy centers of [a] tenant-landlord row." See Anonymous, "Sixteenth Street Mansions Bought by Scientist-Digger," ibid., 2/7/1925, 1, 8.
8. Anonymous, "Former 'Chastleton' Sold; Hotel Planned," *Washington Post*, 3/6/1925, 1.
9. Henson, P.M., and M.E. Epstein, "Dyar Family Oral History Interviews (W.J. Dyar)," 5/17/1992, SIA, RU009570.
10. See "Nevada Adventures," http://www.nevadadventures.com.
11. Ibid.
12. H. Robinson to W.P.A. Dyar, 2/11/1927, SIA, HGDP, Acc. 95-006, b3, f1-3.
13. C.H. Pope to Dyar, 1/19/1927, ibid., b1, f5; Higbie died of pneumonia in December 1926. See Anonymous, "Alfred Higbie Dead: Pneumonia Victim," *Washington Post*, 12/14/1926, 11.
14. Higbie's last known letter includes this financial information: "I am enclosing you herewith the deed from The Old Virginia Land Co. to Wm. B. Duvall and wife conveying the Maycroft Apartment House, subject to the three trusts, which as you will notice the parties of the second part assume and agree to pay." See A. Higbie to Dyar, 7/16/1926, SIA, HGDP, Acc. 95-006, b2, f2.
15. Dyar, H.G., "Valedictory," *Insecutor Inscitiae Menstruus* 14 (1926).

16. ———, "An Anecdote of the Law-a Story," *Reality* 13 (1927); Anonymous, "Diamond Theft Basis of Detective Story," *Richmond Times Dispatch*, 1/31/1926.
17. Dyar to W.W. Stockberger, 11/19/1928, NARA, RG7, E-35, b83, f "Dyar"; Henson and Epstein; W.P.A. Dyar to C.L. Marlatt, 1/30/1929, NARA, RG7, E-35, b83, f "Dyar."
18. Dyar to L.O. Howard, 12/15/1927, ibid.
19. L.O. Howard to Dyar, 12/16/1927, ibid.
20. ———, 3/19/1928, ibid.; Dyar to C.L. Marlatt, 11/24/1928, ibid.
21. W.W. Stockberger to C.L. Marlatt, 10/1928, ibid.
22. Dyar to L.O. Howard, 10/6/1928, ibid.
23. L.O. Howard to C.L. Marlatt, 10/5/1928, ibid.
24. Dyar to L.O. Howard, 10/6/1928, ibid.
25. C.L. Marlatt to Dyar, 11/13/1928, ibid.
26. Dyar to C.L. Marlatt, 11/24/1928, ibid.
27. C.L. Marlatt to Dyar, 11/13/1928, ibid.
28. Dyar to W.W. Stockberger, 11/19/1928, ibid.
29. W.W. Stockberger to C.L. Marlatt, 10/1928, ibid.; Dyar to C.L. Marlatt, 11/24/1928, NARA, RG7, E-35, b83, f "Dyar."
30. Anonymous, "District Investors Buy Parkside for $400,000," *Washington Post*, 11/13/1928, 20; Anonymous, "$184,000 Is Paid For Parkside Hotel," ibid., 1/16/1931, 4.
31. Anonymous, "A. Joseph Hower vs. Harrison G. Dyar," *Washington Post*, 11/24/1928, 12; Engle, R.L., *In the Light of the Mountain Moon: An Illustrated History of Skyland* (Shenandoah National Park Association, 2003).
32. Anonymous, "Draft of Job Description for Dyar," 12/1928-1/1929?, NARA, RG7, E-35, b83, f "Dyar."
33. Anonymous, "Report on the Progress and Condition of the United States National Museum for the Year Ended June 30, 1929, Serial Set Vol. No. 9312 (see also same report for 1927 and 1928).
34. Dyar to F. Haimbach, 1/16/1929, SIA, HGDP, b1, f9.
35. Henson and Epstein, 31.
36. W.P.A. Dyar to C.L. Marlatt, 3/7/1929, NARA, RG7, E-35, b83, f "Dyar."
37. Reverend Ulysses Grant Baker Pierce—who read Dyar's eulogy—was President William Howard Taft's pastor. See Anonymous, "Dr. Dyar's Funeral Set for Tomorrow: Services to be Conducted at Lee's Chapel by Rev. U.G.B. Pierce for Mosquito Expert," *Washington Star*, 1/23/1929, 9.
38. Anonymous, "Cave Found Under 22d Street," *The Washington Post*, 2/4/1950, 3.
39. "Tunnel-Digging as a Hobby," *Modern Mechanix* 1932.
40. Henson, Pamela M., and Marc E. Epstein. Dyar Family Oral History Interview (W.J. Dyar), 5/17/1992.
41. Dyar's executor, according to the article, was Dr. Edward J. Irvine. See Gerrity, J.F. (1942), "Mysterious Diggins? No! Just a Scientist's Hobby." *Washington Post*, 12.
42. "New Mystery Tunnels Found under House on Independence Ave. Sinister Legends Survive, But It Seems Dr. Dyar Just Liked to Dig." *Washington Star*, 3/4/1942.
43. Anonymous, "Mid-City Cavern Find Recalls Weird Hobby," *The Washington Post and Times Herald*, 5/25/1958, A3.
44. Ibid.; The photograph in Plate 16b appeared in "Digging for Dyar." When Associated Press did a story on the article, interviewing Epstein and Henson, this image appeared in a number of newspapers in 1992. In October of that year NBC's *Saturday*

*Night Live* used the photograph in its "Weekend Update" segment, but it had nothing to do with Dyar. Rather it was used to commemorate the fifth anniversary of the cable media sensation of when Baby Jessica McClure fell into a well and was rescued in Texas.

45. Henson and Epstein, "Dyar Family Oral History Interviews (R. Hill–L. Dyar)," 1999, SIA, RU009570.
46. ———, "Dyar Family Oral History Interviews (W.J. Dyar)," 5/17/1992, SIA, RU009570.

## EPILOGUE

1. W.J. Dyar to M.E. Epstein (pers. comm. 1990).
2. "Dr. H.G. Dyar Dies; A Noted Biologist: Recognized as the Foremost Authority on American Mosquitoes." *New York Times*, 1/22/1929, 29; Alumni Federation Card (1929). Information questionnaire for Harrison Dyar, sent 1/24/1929, completed by Wellesca P. Dyar on 2/14/1929, received by Columbia on 2/23/1929, Columbia University Archives.
3. Ibid.
4. Activities by Stone and Benjamin are reported in: Wetmore, A. (1932). "Report on the Progress and Condition of the United States National Museum for the Year Ended June 30, 1932, 53."
5. Anonymous, "Injured Woman Found in Hotel," *Los Angeles Times*, 8/15/1938; Anonymous, "Obituary 8," *Los Angeles Times*, 12/9/1938, 15; Henson, P.M. and M.E. Epstein, "Dyar Family Oral History Interviews (R. Hill-L. Dyar)," 1999, SIA, RU009570.
6. In the interview with Lowell Dyar he said that his grandmother Wellesca lived at the "edge of Rock Creek Park," but this is incorrect based on her address of 3441 Seventeenth Street NW; ———, "Dyar Family Oral History Interviews (W.J. Dyar)," 5/17/1992, SIA, RU009570; ———, "Dyar Family Oral History Interviews (R. Hill–L. Dyar)," 1999, SIA, RU009570.
7. Anonymous, "Sister of G. Freeman Pollock dies of heart ailment," *Page News and Courier*, 6/25/1940.
8. Terry L. Carpenter (pers. comm. 2012) found that George Freeman Pollock purchased a plot for his family at the Glenwood Cemetery in Section D, Lot 204. The plot book shows the people and dates of internment as follows: 1. Wellesca Dyar, 29 Jun 1940; 2. Louisa[e] Pollock, 28 Jul 1901; 3. Amia L. Baker, 24 Oct 1936; 4. Harrison G. Dyar, 24 Oct 1936; 5. George H. Pollock, 2 Jan 1894; 6. James C. Crawford, 31 May 1952; 7. Uila P. Bradway, 25 Apr 1927; 8. John Baker, 30 Sep 1918; For information about George F. Pollock's cremation see Engle, R.L., *In the Light of the Mountain Moon: An Illustrated History of Skyland* (Shenandoah National Park Association, 2003), 101.
9. Terry L. Carpenter (pers. comm., 2015).
10. Anonymous, "Obituary 8," *Los Angeles Times*, 12/9/1938 15.
11. Henson, P.M. and M.E. Epstein (1999). Dyar Family Oral History Interviews (R.Hill-L. Dyar).

# SELECTED BIBLIOGRAPHY

Allen, W.P., and H.G. Dyar. (1920). *Introductions to the Baha'i Revelation*. Washington, D.C.

Barnes, W., and J. McDunnough. (1917). *Checklist of Lepidoptera of Boreal North America*. Decatur, IL: Herald Press.

Beccaloni, G.W., Scoble, M.J., Robinson, G.S., and Pitkin, B., eds. (2003). "The Global Lepidoptera Names Index (LepIndex)." World Wide Web electronic publication. http://www.nhm.ac.uk/entomology/lepindex. [Accessed August 11, 2011.]

Benson, K.R. (1993). "Brooks, Stomatopods, and Decapods: Crustaceans in Research and Teaching." In *History of Carcinology*, edited by Frank Truesdale. Rotterdam: CRC Balkema.

Bethune, C.J.S. (1896). "John B. Lembert." *The Canadian Entomologist* 28 (8): 217–218.

Beutenmueller, W. (1901). "Herman Strecker." *Journal of the New York Entomological Society* 9: 200.

Busck, A. (1901). "New Species of Moths of the Superfamily Tineina from Florida." *Proceedings of the United States National Museum* 23 (1208): 225–254.

Busck, A. (1903). "A Revision of the American Moths of the Family Gelechiidae, with Descriptions of New Species." *Proceedings of the United States National Museum* 25 (1304): 767–938, Plates XXVIII–XXXII.

Carpenter, T.L. (2005). "Notes on the Life of Dr. Clara Southmayd Ludlow, Ph.D., Medical Entomologist." *Proceedings of the Entomological Society of Washington* 107 (3): 657–662.

Carpenter, T.L., and T.A. Klein. (2011). "2011 AMCA Memorial Lecture Honoree: Dr. Harrison Gray Dyar Jr." *Journal of the American Mosquito Control Association* 27 (3): 336–343.

Caudell, A.N., A. Busck, and L.O. Howard. (1919). "Frederick Knab." *Proceedings of the Entomological Society of Washington* 21 (3): 41–52.

Chapman, T.A. (1896). "On the Phylogeny and Evolution of the Lepidoptera from a Pupal and Oval Standpoint." *Transactions of the Royal Entomological Society of London* 44 (4): 567–587.

Clarke, J.F.G. (1974). "Presidential Address-1973: The National Collection of Lepidoptera." *Journal of the Lepidopterists' Society* 28: 181–204.

Cockerell, T.D.A. (1903). "Review of: A List of North American Lepidoptera, and Key to the Literature of this Order of Insects." *Science* (3/27/1903): 501–505.

Comstock, J.H. (1893). *Evolution and Taxonomy: An Essay on the Application of the Theory of Natural Selection in the Classification of Animals and Plants, Illustrated by a Study of the Evolution of the Wings of Insects and by a Contribution to the Classification of the Lepidoptera*. Ithaca, NY: Wilder Quarter-Century Book.

Currie, R. (1904). "An Insect Collecting Trip to British Columbia." *Proceedings of the Entomological Society of Washington* 6: 24–37.

Dyar, H.G. (1888). "Partial Preparatory Stages of *Dryopteryx rosea*, Wlk." *Entomologica Americana* 4: 179.
———. (1889). "Preparatory stages of *Janassa lignicolor*, Walker," *Entomologica Americana* 5: 91–92.
———. (1889). "Description of the Larva of *Datana major*, G&R." *The Canadian Entomologist* 21: 34–35.
———. (1889). "Investigation of a Proposed Synthesis of Tartaric Acid from Butyric Acid." BS, Massachusetts Institute of Technology.
———. (1889). "The Larva of *Limacodes inornata*, G&R." *The Canadian Entomologist* 21 (4): 77–78.
———. (1890). "The Genus *Datana* Walker." *Entomologica Americana* 6: 127–132.
———. (1890). "A New Form of *Cerura* from California." *The Canadian Entomologist* 22 (12): 253–255.
———. (1890). "Notes on Two Species of *Datana* with Descriptions of Their Larval Stages." *Psyche* 5: 414–420.
———. (1890). "The Number of Molts of Lepidopterous Larvae." *Psyche* 5: 420–422.
———. (1890). "Preparatory Stages of *Dilophonota edwardsii* Butl. and *D. ello* Linn." *Entomologica Americana* 6 (8): 141–143.
———. (1890). "Preparatory Stages of *Syntomeida epilais* Walker and *Scepsis edwardsii* Grote." *Insect Life* 2 (11–12): 360–362.
———. (1891). "Choice of Food." Letter to Editor. *Psyche* 6: 196.
———. (1891). "A List of Sphingidae and Bombycidae Taken at Electric Lamp at Poughkeepsie, N.Y." *Insect Life* 3: 322–325.
———. (1891). "Notes on Bombycid Larvae—I–III." *Psyche* 6 (183): 110–112; 145–147; 177–179.
———. (1891). "A Revision of the Species of *Euclea*, *Parasa* and *Packardia*, with Notes on *Adoneta*, *Monoleuca* and *Varina ornata* Neum." *Transactions of the American Entomological Society* 18 (2–3): 149–158.
———. (1891). "On the Specific Distinctness of *Halisidota harrisii*, with Notes on the Preparatory Stages of the Species of *Halisidota* Inhabiting New York." *Psyche* 6 (186): 162–165.
———. (1892). "Book Notice. List of Lepidoptera of Boreal America. by John B. Smith, Sc. D., etc. Philadelphia, American Entomological Society, 1891." *The Canadian Entomologist* 24 (2): 47–48.
———. (1892). "Collecting Butterflies in the Yosemite Valley." *Entomological News* 3: 30–33.
———. (1892). "Life History of *Orgyia cana* Hy. Edw." *Psyche* 6: 203–205
———. (1892). "The Number of Larval Stages in the Genus *Nadata*." *Psyche* 6: 337–340.
———. (1893). "Descriptions of Certain Lepidopterous Larvae." *The Canadian Entomologist* 25 (6): 158–160.
———. (1893). "The Larvae of the *Clisiocampa*." *The Canadian Entomologist* 25 (1): 37–44
———. (1893). "Notes on Two Species of Tenthredinidae, from Yosemite, Cal." *The Canadian Entomologist* 25 (8): 195–196
———. (1893). "Synonymic and Structural Notes." *Entomological News* 4 (1): 33–36.
———. (1894). "Additional Notes on the Classification of Lepidopterous Larvae." *Transactions of the New York Academy of Sciences* 14: 49–62.
———. (1894). "A Classification of Lepidopterous Larvae." *Annals of the New York Academy of Sciences* 8: 194–232.
———. (1894). "Description of Certain Lepidopterous Larvae." *Proceedings of the Boston Society of Natural History* 26: 394–403.
———. (1894). "*Thecla californica*." *Entomological News* 5: 329.

———. (1895). "On Certain Bacteria from the Air of New York City." *Annals of the New York Academy of Sciences* 8: 322–380.
———. (1895). "The Larva of *Butalis basilaris* Zell.: The Relations of its Setae." *Psyche* 7: 252–253.
———. (1895). "Notes on Certain Variations in the Biological Characters of Two Species of Bacteria." *Transactions of the New York Academy of Sciences* 14: 94–99.
———. (1895). "Notes on Some Southern Lepidoptera." *The Canadian Entomologist* 27 (9): 242–247.
———. (1895). " On the Larvae of Some Nematoid and Other Saw-flies from the Northern Atlantic States." *Transactions of the American Entomological Society* 22: 301–312.
———. (1896). "The Life-Histories of the New York Slug Caterpillars—III-VI." *Journal of the New York Entomological Society* 4 (4): 167–190.
———. (1896). "A New *Gloveria*." *Journal of the New York Entomological Society* 4 (1): 22–26.
———. (1896). "Note on the Head Setae of Lepidopterous Larvae, with Special Reference to the Appendages of *Perophora melsheimerii*." *Journal of the New York Entomological Society* 4 (2): 92–93.
———. (1896). "Notes on the Phylogeny of Saturnians." *The Canadian Entomologist* 28 (12): 303–305.
———. (1896). "Notes on Sawfly Larvae." *The Canadian Entomologist* 28 (9): 235–239.
———. (1896). "On the Larvae of the Higher Bombyces (Agrotides Grote)." *Proceedings of the Boston Society of Natural History* 27: 127–147.
———. (1896). "On the Probable Origin of the Pericopidae: *Composia fidelissima* H.-S." *Journal of the New York Entomological Society* 4: 68–73.
———. (1896). "Recent Notes on Bacteria." *Transactions of the New York Academy of Sciences* 15: 148–153.
———. (1897). "The Life-Histories of the New York Slug Caterpillars.-X-XI." *Journal of the New York Entomological Society* 5 (2): 57–66.
———. (1897). "On the Larvae of Certain Saw-flies (Tenthredinidae)." *Journal of the New York Entomological Society* 5: 18–30.
———. (1897). "On the White Eucleidae and the Larva of *Calybia slossoniae* (Packard)." *Journal of the New York Entomological Society* 5: 121–126.
———. (1898). "The Life-Histories of the New York Slug Caterpillars. XIII-XIV." *Journal of the New York Entomological Society* 6: 1–9.
———. (1898). "A New *Parasa*, with a Preliminary Table of the Species of the Genus." *Psyche* 8: 273–276.
———. (1899). "The Life-Histories of the New York Slug Caterpillars. (Conclusion)." *Journal of the New York Entomological Society* 7 (12): 234–253.
———. (1899). "Life Histories of North American Geometridae.—I." *Psyche* 8: 310–312.
———. (1899). "Note on the Secondary Abdominal Legs in the Megalopygidae." *Journal of the New York Entomological Society* 7 (2): 69–70.
———. (1900). "Notes on Some North American Species of Tineidae." *The Canadian Entomologist* 32: 305–311.
———. (1900). "Notes on the Winter Lepidoptera of Lake Worth, Florida." *Proceedings of the Entomological Society of Washington* 4: 446–485.
———. (1900). "Papers from the Harriman Alaska Expedition. XII. Entomological Results (6): Lepidoptera." *Proceedings of the Washington Academy of Sciences* 2: 487–501.
———. (1901). "Descriptions of the Larvae of Three Mosquitoes." *Journal of the New York Entomological Society* 9: 177–179.
———. (1901). "Descriptions of Some Pyralid Larvae from Southern Florida." *Journal of the New York Entomological Society* 9: 19–24.

———. (1901). "Further about the Types of *Acronycta*." *The Canadian Entomologist* 33: 191–192.

———. (1901). "Life Histories of Some North American Moths." *Proceedings of the United States National Museum* 23 (1209): 255–284.

———. (1901). "Notes on the Genitalia of *Halisidota harrisii*, Walsh." *The Canadian Entomologist* 33: 30.

———. (1901). "On Certain Identifications in the Genus *Acronycta*." *The Canadian Entomologist* 33 (4): 122.

———. (1902). "Annual Address of the President. The Collection of Lepidoptera in the National Museum." *Proceedings of the Entomological Society of Washington* 5: 61–72.

———. (1902). "A Generic Subdivision of the Genus *Plusia*." *Journal of the New York Entomological Society* 10: 79–82.

———. (1902). "Illustrations of the Larvae of North American Culicidae.—II." *Journal of the New York Entomological Society* 10: 194–201.

———. (1902). "The Life History of a Second Epiplemid (*Callizzia amorata* Pack.)." *Proceedings of the Entomological Society of Washington* 5 (2): 131–133.

———. (1902). "Notes on Mosquitoes in New Hampshire." *Proceedings of the Entomological Society of Washington* 5: 140–148

———. (1902). "Review of the Genus *Ethmia*." *Journal of the New York Entomological Society* 10: 202–208.

———. (1903). "Annual Address of the President: Some Recent Work in North American Lepidoptera." *Proceedings of the Entomological Society of Washington* 5: 167–173.

———. (1903). "Descriptions of the Larvae of Some Moths from Colorado." *Proceedings of the United States National Museum* 25: 369–412.

———. (1903). "List of Lepidoptera taken at Williams, Arizona. By Messrs. Schwarz and Barber. I. Papilionoidea, Sphingoidea, Bombycoidea, Tineioidea (in part)." *Proceedings of the Entomological Society of Washington* 5 (3): 223–232.

———. (1903). "A List of the North American Lepidoptera and Key to the Literature of this Order of Insects." *Bulletin of the United States National Museum* 52: 1–723.

———. (1903). "New North American Lepidoptera with Notes on Larvae." *Proceedings of the Entomological Society of Washington* 5 (4): 290–298.

———. (1903). "Notes on Mosquitoes on Long Island, New York." *Proceedings of the Entomological Society of Washington* 5: 45–53.

———. 1903. *A Preliminary Genealogy of the Dyar Family*. Washington, D.C.: Gibson Bros.

———. (1904). "Additions to the List of North American Lepidoptera, No. 1." *Proceedings of the Entomological Society of Washington* 6: 62–65.

———. (1904). "Brief Notes on Mosquito Larvae." *Journal of the New York Entomological Society* 12: 172–174.

———. (1904). "The Lepidoptera of the Kootenai District of British Columbia." *Proceedings of the United States National Museum* 27 (1376): 779–938.

———. (1904). "Mosquitoes of British Columbia." *Proceedings of the Entomological Society of Washington* 6 (1): 37–41.

———. (1905). "A Descriptive List of a Collection of Early Stages of Japanese Lepidoptera." *Proceedings of the United States National Museum* 28 (1412): 937–956.

———. (1905). "A Few Notes on the Strecker Collection." *Proceedings of the Entomological Society of Washington* 7: 92–94.

———. (1905). "New Facts That Are Not New." *Entomological News* 16: 310.

———. (1905). "On the Classification of the Culicidae." *Proceedings of the Entomological Society of Washington* 7 (4): 188–191.

———. (1905). "A Review of the Hesperiidae of the United States." *Journal of the New York Entomological Society* 8 (3): 111–142.

———. (1905). "Review of 'Synonymic Catalogue of the North American Rhopalocera. Supplement No. 1, by Henry Skinner.'" *Journal of the New York Entomological Society* 13: 217

———. (1906). "Life Histories of North American Geometridae.—LXVII." *Psyche* 13 (5): 117–118.

———. 1906. *The Life-History of a Cochlidian Moth—Adoneta bicaudata Dyar. Biological Studies by the Pupils of William Thompson Sedgwick*. Boston: Printed at the University of Chicago Press.

———. (1906). "A List of American Cochlidian Moths, with Descriptions of New Genera and Species." *Proceedings of the United States National Museum* 29 (1423): 359–396

———. (1906). "The North American Nymphulinae and Scopariinae." *Journal of the New York Entomological Society* 14: 77–107.

———. (1907). "The Life Histories of the New York Slug-Caterpillars.—XIX." *Journal of the New York Entomological Society* 15 (4): 219–225, Plate II.

———. (1907). "Notes on Some Species of Notodontidae in the Collection of the United States National Museum, with Descriptions of New Genera and Species." *Proceedings of the Entomological Society of Washington* 9: 45–69.

———. (1907). "Report on the Mosquitoes of the Coast Region of California, with Descriptions of New Species." *Proceedings of the United States National Museum* 32 (1516): 121–129

———. (1908). "Book Notice: Mosquito Life." *The Canadian Entomologist* 40 (2): 75–76.

———. (1908). "Descriptions of Eleven New North American Pyralidae, with Notes on a Few Others." *Proceedings of the Entomological Society of Washington* 10 (1–2): 112–118.

———. (1909). "The Life History of an Oriental Species of Cochlidiidae Introduced into Massachusetts (*Cnidocampa flavescens* Walk.)." *Proceedings of the Entomological Society of Washington* 11 (4): 162–170, pl. 114.

———. (1909). "A Report on Mosquitoes at Dublin, New Hampshire, Particularly on the Occurrence of *Mansonia perturbans* Walker [Diptera, Culicidae]." *Proceedings of the Entomological Society of Washington* 11: 145–149

———. (1910). "Concerning Dr. John B. Smith at Dublin, New Hampshire." *Entomological News* 21: 17–18.

———. (1910). "Some Moths from Claremont, California, with Notes on Certain Allied Species." *Pomona College Journal of Entomology* 2: 375–378.

———. (1911). "Descriptions of Some New Species and Genera of Lepidoptera from Mexico." *Proceedings of the United States National Museum* 38 (1742): 229–273.

———. (1912). "Descriptions of New Species and Genera of Lepidoptera, Chiefly from Mexico." *Proceedings of the United States National Museum* 42 (1885): 39–106.

———. (1912). "Distribution of Mosquitoes in North America." *Proceedings of the Seventh International Zoological Congress, Boston, 19–24 August, 1907*, pp. 956–957.

———. (1913). "Descriptions of New Species of Saturnian Moths in the Collection of the United States National Museum." *Proceedings of the United States National Museum* 44 (1947): 121–134.

———. (1913). "Notes on Cotton Moths." *Insecutor Inscitiae Menstruus* 1 (1).

———. (1913). "Results of the Yale Peruvian Expedition of 1911. Lepidoptera." *Proceedings of the United States National Museum* 45: 627–649.

———. (1914). "Descriptions of New Species and Genera of Lepidoptera from Mexico." *Proceedings of the United States National Museum* 47: 365–409.

———. (1914). "The Life Histories of the New York Slug-Caterpillars. XX." *Journal of the New York Entomological Society* 22 (3): 223–229.

———. (1914). "Report on the Lepidoptera of the Smithsonian Biological Survey of the Panama Canal Zone." *Proceedings of the United States National Museum* 47: 139–350.

———. (1915). "New American Lepidoptera Chiefly from Mexico." *Insecutor Inscitiae Menstruus* 3 (5-7): 79–85.

———. (1915). "Pyralidae of Bermuda." *Insecutor Inscitiae Menstruus* 3 (5–7): 86–89.

———. (1916). "Mosquitoes at San Diego, California." *Insecutor Inscitiae Menstruus* 6 (4-6): 46–51.

———. (1916). "New *Aedes* from the Mountains of California." *Insecutor Inscitiae Menstruus* 4 (7–9): 80–90.

———. (1917). "The Barnes & McDunnough 'List' (Lepidoptera)." *Insecutor Inscitiae Menstruus* 5: 41–44.

———. (1917). "Descriptions of Some Lepidopterous Larvae from Mexico." *Insecutor Inscitiae Menstruus* 5 (7–9): 128–132.

——— (1917). "The Mosquitoes of the Mountains of California." *Insecutor Inscitiae Menstruus* 5: 11–21.

———. (1918). "Descriptions of New Lepidoptera from Mexico." *Proceedings of the United States National Museum* 54 (2239): 335–372.

———. (1918). "The Male Genitalia of *Aedes* as Indicative of Natural Affinities (Diptera, Culicidae)." *Insecutor Inscitiae Menstruus* 6: 71–86.

———. (1919). "A Revision of the American Sabethini of the *Sabethes* Group by the Male Genitalia (Diptera, Culicidae)." *Insecutor Inscitiae Menstruus* 7 (7–9): 114–142.

———. (1919). "Some Tropical American Phycitinae (Lepidoptera, Pyralidae)." *Insecutor Inscitiae Menstruus* 7 (1–3): 40–63.

———. (1920). "The Mosquitoes of British Columbia and Yukon Territory, Canada (Diptera, Culicidae)." *Insecutor Inscitiae Menstruus* 8 (1–3): 1–27; pl. 21.

———. (1921). "Mosquitoes of Canada." *Transactions of the Royal Canadian Institute* 13: 71–130.

———. (1922). "Mosquitoes of the United States." *Proceedings of the United States National Museum* 62 (2447): 1–119.

———. (1922). "Note on the Male Genitalia of *Culex coronator* and Allied Forms (Diptera, Culicidae)." *Insecutor Inscitiae Menstruus* 10: 18–19.

———. (1923). "The Mosquitoes of the Yellowstone National Park (Diptera, Culicidae)." *Insecutor Inscitiae Menstruus* 11 (1–3): 36–46.

———. (1924). "The Mosquitoes of Colorado." *Insecutor Inscitiae Menstruus* 12: 39–46.

———. (1925). "The Mosquitoes of Panama." *Insecutor Inscitiae Menstruus* 13 (7–9): 101–195.

———. 1926. *Diamonds Going and Coming*. Boston: The Stratford Company.

———. (1926). "Three Psychodids from the Glacier National Park." *Insecutor Inscitiae Menstruus* 14 (7–9): 103–106.

———. (1927). "An Anecdote of the Law-a Story." *Reality* 13 (April): 35–38; (May): 21–24.

———. (1927). "New Species of American Lepidoptera of the Families Limacodidae and Dalceridae." *Journal of the New York Academy of Sciences* 17: 544–550.

———. (1928). "Descriptions of Four South American Moths." *Proceedings of the Entomological Society of Washington* 30 (1): 9–10.

———. (1928). *The Mosquitoes of the Americas*. Carnegie Institution of Washington, publication no. 387. Washington, D.C.: Carnegie Institution of Washington.

———. (1929). "A New Species of Mosquito from Montana, with Annotated List of the Species Known from the State." *Proceedings of the United States National Museum* 75 (2794): 1–8.

———. (1929). "Notes and New Species of American Moths of the Genus *Scoparia* Haworth." *Proceedings of the United States National Museum* 74 (no. 2769, art. 24): 1–9.

———. 1935 [1936–1937]. "Limacodidae." In *The Macrolepidoptera of the World. The American Bombyces and Sphinges*, edited by A. Seitz. Vol. 6.

Dyar, H.G., and C. Heinrich. (1927). "The American Moths of the Genus *Diatraea* and Allies." *Proceedings of the United States National Museum* 71 (2691): 1–48.

Dyar, H.G., and F. Knab. (1904). "Diverse Mosquito Larvae That Produce Similar Adults." *Proceedings of the Entomological Society of Washington* 6: 143–144.

———. (1906). "The Larvae of Culicidae as Independent Organisms." *Journal of the New York Entomological Society* 14 (4): 169–230.

———. (1907). "Book Notice: A Monograph of the Culicidae of the World by F. V. Theobald: London, 1907. Volume IV." *Journal of the New York Entomological Society* 15 (4): 239–248.

———. (1908). "Descriptions of Some New Mosquitoes from Tropical America." *Proceedings of the United States National Museum* 35 (1632): 53–70.

———. (1908). "Notes on Mosquito Work." *The Canadian Entomologist* 40 (9): 309–312.

———. (1910). "Descriptions of Some New Species and a New Genus of American Mosquitoes." *Smithsonian Miscellaneous Collections* 52: 253–266.

———. (1915). "Notes on the Species of *Culex* of the Bahamas." *Insecutor Inscitiae Menstruus*.

Dyar, H.G., and C.S. Ludlow. (1921). "A Note on Two Panama Mosquitoes." *Military Surgeon* 48: 677–680.

———. (1921). "Two New American Mosquitoes." *Insecutor Inscitiae Menstruus* 9: 46–50.

Dyar, H.G., and E.L. Morton. (1895). "The Life-Histories of the New York Slug Caterpillars.-I-II." *Journal of the New York Entomological Society* 3 (4): 145–157; 4(1): 1–9.

———. (1922). "Notes on American Mosquitoes." *Military Surgeon* 50: 61–64.

Dyar, H.G., and R.C. Shannon. (1924). "The American Chaoborinae (Diptera: Culicidae)." *Insecutor Inscitiae Menstruus* 12: 213.

———. (1927). "The North American Two-Winged Flies of the Family Simuliidae." *Proceedings of the United States National Museum* 69 (2636): 1–54.

Engle, R.L. 2003. *In the Light of the Mountain Moon: An Illustrated History of Skyland*. Shenandoah National Park Association.

Epstein, M.E. (1995 [1997]). "Evolution of Locomotion in Slug Caterpillars (Lepidoptera: Zygaenoidea: Limacodid group)." *Journal of Research on the Lepidoptera* 34: 1–13.

———. (1996). "Revision and Phylogeny of the Limacodid Group Families, with Evolutionary Studies on Slug Caterpillars (Lepidoptera: Zygaenoidea)." *Smithsonian Contributions to Zoology* (Number 582): 1–102.

Epstein, M.E., and P.M. Henson. (1992). "Digging for Dyar: The Man Behind the Myth." *American Entomologist* 38: 148–169.

Essig, E.O. 1931. *A. History of Entomology*. New York: The Macmillan Company.

Evenhuis, N.L. (2008). "The 'Mihi Itch'—A Brief History." *Zootaxa* 1890: 59–68.

Felt, E.P. (1917). "Book Review. The Mosquitoes of North and Central America and the West Indies by L. O. Howard, H. G. Dyar and Frederick Knab." *Psyche* 24: 161–163.

Forbes, W.T.M. (1927). "Exit the Tentamen, But...." *Science* 66 (1713): 396–397.

———. (1929). "Obituary. Harrison Gray Dyar." *Entomological News* 40: 165–167.

Gay, F.P. (1939). "A Half Century of Bacteriology at Columbia." *Columbia University Quarterly* (June & Sept.): 125–129; 204–206.

Gossel, P.P. (1988). "The Emergence of American Bacteriology, 1875–1900." Ph.D. dissertation, Johns Hopkins University.

Graf, J.E., and D.W. Graf. (1959). *Leland Ossian Howard 1857–1950: A Biographical Memoir*. Washington D.C.: National Academy of Sciences.

Grote, A.R. (1874). "List of the Noctuidae of North America." *Bulletin of the Buffalo Society of Natural Sciences.*
———. (1881). "New Moths from Arizona, with Remarks on *Catocala* and *Heliothis.*" *Papilio* 1 (10): 153–168.
———. (1888). "The Classification of the Bombycidae (Third Paper)." *The Canadian Entomologist* 20 (10): 181–185.
Grote, A.R., and C.T. Robinson. 1868. *List of the Lepidoptera of North America.* Philadelphia: American Entomological Society.
Gunder, J.D. (1929). "North American Institutions Featuring Lepidoptera." *Entomological News* 40: 244–252; 280–286, pl. 21.
Gurney, A.B. (1976). "A Short History of the Entomological Society of Washington." *Proceedings of the Entomological Society of Washington* 78: 225–239.
Hadley, P. (1927). "Microbic Dissociation: The Instability of Bacterial Species with Special Reference to Active Dissociation and Transmissible Autolysis." *Journal of Infectious Diseases* 40: 1–312.
Harbach, R.E. (2007). "The Culicidae (Diptera): A Review of Taxonomy, Classification and Phylogeny." *Linnaeus Tercentenary: Progress in Invertebrate Taxonomy*, edited by Z.-Q Zhang and W.A. Shear. Special issue. *Zootaxa* 1668: 1–63.
Harrower, C.S. (1875). *In Memoriam: Harrison Gray Dyar.* New York: Jones.
Heinrich, C., and E.A. Chapin. (1942). "William Schaus." *Proceedings of the Entomological Society of Washington* 44 (9): 189–195.
Himmelman, J. (2002). *Discovering Moths: Nighttime Jewels in Your Own Backyard.* Camden, ME: Down East Books.
Holland, W.J. (1927). "Exit Huebner's Tentamen!" *Science* 66 (1696): 4–6.
Hollinger, R. 1996. *Abdu'l-Bahá in America: Agnes Parsons' Diary.* Los Angeles: Kalimat Press.
Howard, L.O. (1901). *Mosquitoes; How They Live; How They Carry Disease; How They Are Classified; How They May Be Destroyed.* New York: McClure, Phillips & Co.
———. (1909). "The Entomological Society of Washington." *Proceedings of the Entomological Society of Washington* 11: 8–18.
———. (1929). "Harrison Gray Dyar." *Science* 69 (8 Feb.): 151.
———. (1933). *Fighting the Insects: The Story of an Entomologist, Telling of the Life and Experiences of the Writer.* New York: The Macmillan Company.
Howard, L.O., and A. Busck. (1936). "In Memorium: Andrew Nelson Caudell." *Proceedings of the Entomological Society of Washington* 38 (3): 34–47.
Howard, L.O., H.G. Dyar, and F. Knab. (1912). *The Mosquitoes of North and Central America and the West Indies.* Vol. 1 (159). Washington, D.C.: Carnegie Institution of Washington.
———. 1915. *The Mosquitoes of North and Central America and the West Indies. Systematic Description.* Vol. 3 (part 1). Washington, D.C.: Carnegie Institution of Washington.
———. 1917. *The Mosquitoes of North and Central America and the West Indies. Systematic Description.* Vol. 4 (part II). Washington, D.C.: Carnegie Institution of Washington.
Hudson, G.H. (1893). "A New Form of *Prionia*, and Notes on *Platypteryx arcuata* and *P. genicula.*" *The Canadian Entomologist* 25: 24.
Hulst, G.D. (1881). "Some Remarks upon the Catocalae, in Reply to Mr. A.R. Grote." *Papilio* 1 (11): 215–218.
Jones, A. (1852). *Historical Sketch of the Electric Telegraph Including Its Rise and Progress in the United States.* New York: George P. Putnam.
Kelly, J. (2002). "A Final Look at D.C.'s Tunnel-Digging Bug Man." *Washington Post*, Nov. 7.
Kelly, N.V. (2009). "Linwood Hill: New Information about an Eccentric Resident." Rhinebeck Historical Society newsletter. *RHS* (winter): 1–3.

———. (2009). *Rhinebeck's Historic Architecture*. Charleston, SC: The History Press.
Kitzmiller, J.B. (1982). *Anopheline Names: Their Derivations and Histories*. The Thomas Say Foundation Monographs. Vol. 8. College Park, MD: Entomological Society of America.
———. (1987). "Biography of Clara Southmayd Ludlow." *Mosquito Systematics* 19 (3): 251–255.
Knight, K.L., and R.B. Pugh. (1974). "A Bibliography of Mosquito Writings of H.G. Dyar and Frederick Knab." *Mosquito Systematics* 6: 11–26.
Leach, W. 2012. *Butterfly People: An American Encounter with the Beauty of the World*. New York: Pantheon Books.
Leonard, J.W., ed. (1914). "Evelyn G. Mitchell." *Woman's Who's Who of America: A Biographical Dictionary of Contemporary Women of the United States and Canada, 1914-1915*. New York: American Commonwealth Co.
Lill, J.T., R.J. Marquis, R.E. Forkner, J. Le Corff, N. Holmberg, and N.A. Barber. (2006). "Leaf Pubescence Affects Distribution and Abundance of Generalist Slug Caterpillars (Lepidoptera: Limacodidae)." *Environmental Entomology* 35 (3): 797–806.
Lillie, F.R. (1988). "The Woods Hole Marine Biological Laboratory." *Biological Bulletin. Marine Biological Laboratory* 174, Supplement: i-vii; 1–191; 193–269; 271–284.
Lindroth, C. (1973). "Systematics Specializes between Fabricius and Darwin: 1800–1859." In *The History of Entomology*, edited by Smith, Mittler and Smith. Annual Review Inc. and The Entomological Society of America.
Ludlow, C.S. (1909). "Mosquito Comment." *The Canadian Entomologist* 41 (1): 21–24.
Mallis, A. 1971. *American Entomologists*. New Brunswick, NJ: Rutgers University Press.
March, H.O., F.H. Chittenden, and H.G. Dyar. (1911). "Papers on Insects Affecting Vegetables: The Hawaiian Beet Webworm (*Hymenia fascialis* Cram.)." *Bulletin of the U. S. Dept. of Agriculture. Bureau of Entomology* (109, pt 1): 15.
Marlatt, C.L. (1896). "Some New Nematids." *The Canadian Entomologist* 28 (10): 251–258.
Maynard, N.C. (1891). *Was Abraham Lincoln a Spiritualist? Or, Curious Revelations from the Life of a Trance Medium. Together with Portraits, Letters, and Poems. Illustrated with Engravings, and Frontispiece of Lincoln, from Carpenter's Portrait from Life. 'After All, It Is the Old Old Story, Truth Is Stranger Than Fiction.'* Philadelphia: Rufus C. Hartranft.
McDunnough, J. (1938). "Check List of the Lepidoptera of Canada and the United States of America. Part 1. Macrolepidoptera." *Memoirs of the Southern California Academy of Sciences* 1: 174.
McPherson, R.S. (2001). *Navajo Land, Navajo Culture: The Utah Experience in the Twentieth Century*. Norman: University of Oklahoma Press.
Mitchell, E.G. (1907). *Mosquito Life: The Habits and Life Cycles of the Known Mosquitoes of the United States; Methods for Their Control; and Keys for Easy Identification of the Species in Their Various Stages*. Putnam.
———. (1907). "Validity of the Culicid Subfamily Deinoceritinae." *Psyche* 14: 11–13.
———. (1908). "A Reply to Dr. Dyar." *The Canadian Entomologist* 40 (3): 93–98.
Morris, J.G. 1860. *Catalogue of the Described Lepidoptera of North America*. Washington, D.C.: Smithsonian Institution.
Morton, E.L. (1892). "Notes from New Windsor. *Isa textula* H.-S." *Entomological News* 3: 1–3.
Munroe, A. (1902). "Concord and the Telegraph." *Concord Antiquarian Society Papers* 4: 1–22.
Murphy, S., J. Lill, and M.E. Epstein. (2011). "Natural History of Limacodidae of the Washington, D.C. Region." *Journal of the Lepidopterists' Society* 65 (3): 137–152.

Neumoegen, B. (1893). "Description of a Peculiar New Liparid Genus from Maine." *The Canadian Entomologist* 25: 213–215.

Neumoegen, B., and H.G. Dyar. (1893). "Descriptions of Certain New Forms of Lepidoptera." *The Canadian Entomologist* 25 (5): 121–126.

———. (1893). "New Species and Varieties of Bombyces." *Journal of the New York Entomological Society* 1 (1): 29–35.

———. (1893–1894). "Preliminary Revision of the Bombyces of America North of Mexico." *Journal of the New York Entomological Society* 1: 97–180; 182; 181–186; 147–173.

———. (1894). "A Preliminary Revision of the Lepidopterous Family Notodontidae." *Transactions of the American Entomological Society* 21: 179–208.

Ogilvie, M.B., and J.D. Harvey, eds. (2000). "Evelyn Groesbeeck Mitchell." In *The Biographical Dictionary of Women in Science: L-Z*.

Oppewall, J. (1979). "From the Gold Rush to the Hollywood Rushes: Some Notes on the History of Butterfly Farming in California." *Terra* 30–35.

Ottolengui, R. (1902). "*Plusia* and Allied Genera with Descriptions of New Species." *Journal of the New York Entomological Society* 10: 57–79, pls. 56–59.

Packard, A.S. (1895). " First Memoir on the Bombycine Moths." *Memoirs of the National Academy of Sciences* 7: 1–83.

———. 1914. "Monograph of the Bombycine Moths of North America, Including Their Transformations and Origin of the Larval Markings and Armature. Part III. Families Ceratocampidae (exclusive of Ceratocampinae), Saturniidae, Hemileucidae, and Brahmaeidae." Edited by T.D.A. Cockerell. *Memoirs of the National Academy of Sciences* 12: 516.

Pape, T., and F.C. Thompson. (2010). "Systema Dipterorum: The Biosystematic Database of World Diptera [Version 1.0]." http://www.diptera.org/. [Accessed 8/9/2011.]

Patterson, G. 2009. *The Mosquito Crusades. A History of the American Anti-Mosquito Movement from the Reed Commission to the First Earth Day*. New Brunswick, NJ: Rutgers University Press.

Pollock, G.F. (1920). *Skyland (the Eaton Ranch of the East) Situated on High Plateau in the Blue Ridge Near Grand Old Stony Man Peak, Overlooking Famous Shenandoah Valley*. Roanoke, VA: Stone Printing and Mfg.

Rathbun, R. (1904). "Annual Report of the Board of Regents of the Smithsonian Institution Showing the Operations, Expeditions, and Condition of the Institution." *Report of the U.S. National Museum, year ending June 30, 1902*.

Ravenel, W.D. (1920). "Report on the Progress and Condition of the United States National Museum for the Year Ending June 30, 1919."

Resh, V.H., and R.T. Cardé. (2009). *Encyclopedia of Insects*. Amsterdam: Academic Press.

Rindge, F.H. (1955). "The Type Material in the J. B. Smith and G. D. Hulst Collections of Lepidoptera in the American Museum of Natural History." *American Museum of Natural History Bulletin* 106 (2): 95–172.

Robinson, L. (1929). "Our Editor." *Reality* 17: 32–33.

Roth, V.L. (1981). "Constancy in the Size Ratios of Sympatric Species." *The American Naturalist* 118 (3): 394–404.

Schwarz, E.A. (1897). "Martin Larsson Linell." *Proceedings of the Entomological Society of Washington* 4: 177–180.

Showalter, W.J. (1927). "Strange Habits of Familiar Moths and Butterflies." *National Geographic Magazine*, July, 76–126.

Simpson, A. "Guide to the Hudson Family Papers 95.3 (revised 2009 by D. Kimok)" Special Collections, Benjamin F. Feinberg Library, Plattsburgh State University College, Plattsburgh, N.Y.

Skinner, H. (1896). "Impressions Received from a study of our North American Rhopalocera." *Journal of the New York Entomological Society* 4: 107–118.

———. (1905). "A Review of a Review." *Entomological News* 16: 316–317.

———. (1906). "On Dr. Dyar's Review of the Hesperidae." *Entomological News* 17 (4): 110–112.

———. (1910). "On an Episode: The Charge." *Entomological News* 21 (2): 85.

Smith, D.R. (1986). "The Sawfly Work of H.G. Dyar (Hymenoptera: Symphyta)." *Transactions of the American Entomological Society* 112: 369–396.

Smith, J.B. (1891). *List of the Lepidoptera of Boreal America*. Philadelphia: American Entomological Society.

———. (1892). "Correspondence. Prof. J.B. Smith's List of Lepidoptera." *The Canadian Entomologist* 24 (4): 103–104.

———. (1901). "*Acronycta* and Types." *The Canadian Entomologist* 33 (8): 232–234.

———. (1901). "Types and Synonymy." *The Canadian Entomologist* 33 (5): 146–148.

———. (1903). "New Noctuids for 1903, No. 4, with Notes on Certain Described Species." *Transactions of the American Entomological Society* 29: 214.

———. 1903. *Checklist of the Lepidoptera of Boreal America*. Philadelphia: American Entomological Society.

———. (1909). "Concerning *Culex perturbans* at Dublin, New Hampshire." *Entomological News* 20: 425–427.

———. (1910). "Letter: Once More *Culex perturbans* in Notes and News." *Entomological News* 21 (2): 84.

———. (1911). "New Species of Noctuidæ for 1911. No. 1." *Journal of the New York Entomological Society* 19 (3): 133–151.

Smith, J.B., and H.G. Dyar. (1898). "Contributions toward a Monograph of the Lepidopterous Family Noctuidae of Boreal North America. A Revision of the Species of *Acronycta* (Ochsenheimer) and of Certain Allied Genera." *Proceedings of the United States National Museum* 21: 1–194.

Smith, P. 1984. "Reality Magazine: Editorship and Ownership of an American Bahá'í Periodical." In *Studies in Babi and Bahá'í History. Vol. 2, From Iran East and West*, edited by J. R. Cole and M. Momen. Los Angeles: Kalimat.

———. 1996. *The Bahá'í Faith: A Short History*. Oxford: Oneworld.

Sorensen, W.C. (1995). *Brethren of the Net: American Entomology, 1840-1880* Tuscaloosa: University of Alabama.

Spilman, T.J. (1984). "Vignettes of 100 years of the Entomological Society of Washington." *Proceedings of the Entomological Society of Washington* 86: 1–10.

Stratton-Porter, G. (1912). *Moths of the Limberlost*. Garden City, NY: Doubleday, Page & Company.

Strecker, H. 1872. *Lepidoptera, Rhopaloceres and Heteroceres, Indigenous and Exotic; with Descriptions and Colored Illustrations*. Reading, PA: Owen's Steam Book and Job Printing Office.

Todd, E.L. (1982). "The Noctuid Type Material of John B. Smith (Lepidoptera)." *U.S. Department of Agriculture Technical Bulletin* (1645).

Wade, J.S. (1936). "The Officers of Our Society for Fifty Years (1884-1934)." *Proceedings of the Entomological Society of Washington* 38: 99–145.

Wagner, D.L., D.F. Schweitzer, J.B. Sullivan, and R.C. Reardon. (2011). *Owlet Caterpillars of Eastern North America*. Princeton, NJ: Princeton University Press.

Wainwright, M. (1997). "Extreme Pleomorphism and the Bacterial Life Cycle: A Forgotten Controversy." *Perspectives in Biology and Medicine*: 40,407–414.

Walton, W.R. (1921). "Entomological Drawings and Draughtsmen: Their Relation to the Development of Economic Entomology in the United States." *Proceedings of the Entomological Society of Washington* 23 (4): 69–99.

Weismann, A. (1882). *Studies in the Theory of Descent*. Vols. 1 & 2. Translated and edited by Raphael Meldola, F.C.S. and Prefatory notice by Charles Darwin. London: Sampson, Low, Marston, Searle, & Rivington.

Yochelson, E.L., and M. Jarrett. 1985. *The National Museum of Natural History: 75 Years in the Natural History Building*. Published on the occasion of the Diamond Jubilee of the Natural History Building (1910–1985) for the National Museum of Natural History, Smithsonian Institution. Washington, D.C.: Smithsonian Institution Press.

# INDEX

Abba Bahá'u'lláh, 117, 121
Abbott, Charles Greeley, 206
Abbott, William L., 66–67
'Abdu'l Bahá,
    Bahá'í, taught as scientific religion
        of universal ideas in the
        West, 196
    bigamy, views on, 199
    birthday, 65th, 266
    stones, monogrammed, given to
        Wellesca, 121–122, Fig. 14.3c
    Persian names given to Wellesca's
        children, 127, 130, 268
    photograph, Fig. 14.1b
    tablets, hand-written correspondence
        in Persian on parchment,
        117
        Wellesca, to, began in 1902, 117
        Wellesca and Dyar, to, in
            response to problems with
            fellow Washington, D.C.
            Bahá'ís, 197
    Washington Bahá'í community in
        1912, visit to, 268
    Wellesca requests prayer from to have
        each child, 118, 129
    Wellesca's hope for him to stay at B
        Street during U.S. visit, 127
Academy of Natural Sciences
    Philadelphia, The, 29, 69
Acca, Palestine. Home of 'Abdu'l Bahá,
    117, 118, 122, 233
Adami, J.G., 44
*Aeneid, The*, by Virgil, 188
Alaska, 31, 174, 179–180, 192, 197, 276
    Cape Fanshaw, 180
    Juneau, 180
    Ketchikan, 180, Fig. 21.3a
    Sitka, 21
    Skagway, 180
Alberta,
    Banff, 90
    Calgary, 92
    Field, 90
    Lake Louise, 92
    Medicine Hat, 92
Aldrich, John Merton, 161, 167, 170,
    172–73, 175, 206
    photograph, Fig. 20.2c
Allen children, 127, 130, 151, 180
    Dyar 1922 adoption of, 193
    names, 130, 268
Allen v. Allen,
    Reno, 147–149, Fig. 18.2
    Wellesca gives wrong marriage date, 148
Allen, Cebas, 151
Allen, Ephraim Williams, 151
Allen, Golshan. *See* Dyar, Harrison Golshan
Allen, Jr., Wilfred. *See* Dyar, Roshan
Allen, Mrs. *See* Wellesca
Allen, Nathaniel T., 151
Allen, Roshan Wilfred. *See* Dyar, Roshan
Allen, Stephen M., 151
Allen, Wallace Parsons. *See* Dyar, Wallace
    Joshan
Allen, Wilfred P., 117–118, 127, 130,
    143–144, 146–155, 192,
    196, 272
    affidavit with signature, 153. *See also*
        Houston, David F.
    census, Fig. 15.1c
    newspaper articles, Fig. 18.2
    marriage license, Fig. 19.1
    letters to George F. Pollock, 'Abdu'l
        Bahá, 148–149
    physical description, 147

( 298 )   Index

American Association for the
    Advancement of Science
    (AAAS), 32, 76, 256
American chestnut, 48
American Literary Bureau, The, 133
American Museum of Natural History, 22,
    35, 71, 95
American Security and Trust co., 122
Anderson, Larz, Mansion, 59
Angelman, J.B., 47
Appleby, Frances (Kaufmann), Pl. 13
Arizona,
    Grand Canyon, 148
    Salt River near Phoenix, 26
Armstrong, Eleanor, 170–172,
    Fig. 20.2d, Pl. 13
Armstrong, Fedora, 272
Army Medical Museum, 176–178,
    Fig. 21.2b
Arts and Industries Building, Fig. 22.2d,
    Map 6. *See* National
    Museum, Old
Aseyeh, Wellesca's name in Baháʼí
    community, 117, 122, 130,
    132, 145, 194- 195, 197–198,
    226, 259, 267–268, 279
    photograph, Fig. 14.1a
Asher family,
    Catherine, George, Louisa, Philip, 242
    Mrs., 8, 16, Pl. 3a
    Norman, 16
Ashmead, William Harris, 33, 49, 56, 67,
    112, 252, 265
    photograph, Fig. 9.1b
Asimov, Isaac, 143
Ayer, Eugene G., 12
Ayer, Frederick, 52
Ayer, Marcellus Seth, 12, Fig. 2.4c

B Street SW, 804
    dinner at, 194
    images, Fig. 15.1b,d, Fig. 22.2,
        Fig. 23.2b
    home of H.G. Dyar, 127, 128, 130, 131,
        132, 137, 177, 218, 267
    life at, by children, 192–194,
    borders, "Baháʼí House" or "Baháʼí
        Hospice," 127, 196. *See also*
        ʻAbduʼl Bahá
    piano, Dyar playing Rachmaninoff's
        Symphonies, 193, 194, 332

purchase and construction, 125, 126
ransacking or raid of house by Zella's
    agents, 139, 142
stamp collection, 193
B Street Tunnel. *See* Tunnel, B Street
bacteria,
    *Bacillus hudsonii*, named for G.H.
        Hudson, 44
    *Bacillus sarracenicolus*, named for
        pitcher plant, 44
    *Bacillus vacuolatus*, named for
        carnivorous bladderwort, 44
    collection at Columbia, living
        species, 43
    critique of Frankland's identification
        tables, 44
    cultures, planting in countryside, 43
    dissertation, "On Certain Bacteria
        from the Air of New York
        City," 43
    fifty new species, named by Dyar, 43
    laboratory in the College of Physicians
        and Surgeons of Columbia
        College, 43
    pleomorphists, 44
    *Serratia plymuthica*, mistakenly
        attributed as Dyar species, 249
    specimens from Kral's laboratory in
        Prague, 43
    variation, reflection of life stages and
        environmental factors, 44
Bahá Abbas, ʻAbduʼl. *See* ʻAbduʼl Bahá
Baháʼí, 117, 118, 126, 148
    B Street, 804, "Baháʼí, House" or
        "Baháʼí Hospice," 196
    community, Washington, DC, 118, 149,
        196, 197
    Feast of Declaration of the Bab, 121,
        Fig. 14.3b
    mixed-race meetings, opposition to, in
        D.C., 149
    movement: free from formal Baháʼí
        rules, 196, 198, 199
    name Aseyeh, for Wellesca, 117
    National Spiritual Assembly, 199
    Twelve Basic Principles of ʻAbduʼl
        Bahá, 196
Bahamas,
    Nassau, Royal Victoria Hotel, 85,
        Fig. 11.2
    New Providence, Island, Bahamas, 132

Baker, Amia Louise Pollock, 142, 226
Baker, Emily. "Mrs. Crawford," wife of J.C. Crawford, 135, 158, 226
Baker, John. Husband of Amia L. Pollock, 226
Ballou, Dr. W.H. Myth that disease carried by mosquito larvae through water supply, 208. *See also* mosquito larvae
Banks, Nathan. Serving lemonade and raison cake at Entomological Society of Washington meeting, 72
Barber, Herbert Spencer, Pl. 13
Barber's tent light trap, 123
Barnes and McDunnough,
    "Contributions to the natural history of Lepidoptera of North America" and Dyar's critique, 113–114
    "Checklist of Lepidoptera of Boreal America," 157
    expedition to south Florida, 114
    unpublished "New Catalogue," 157
Barnes, John "Jack." *See* William Schaus
Barnes, William,
    collection,
        photographs, Fig. 13.3b, Fig. 24.1a
        price, included wing of Macon County Hospital Decatur, IL, 205–206
        purchase, attempts at, by Smithsonian or congressional appropriation, 112–113, 201
        contracts to USNM specialists, 157, 165
        early attempts to remove Dyar from position, 113
        loans, unofficial, of USNM specimens by A. Busck, 165–168, 203
        photographs and portraits, Fig. 20.1c, Fig. 24.1a
        rumors of Dyar's dismissal as Honorary Custodian, 154, 156
Barnes, William C., nephew, letter, 205, Fig. 24.1b
Battle of the Names between Smith and Dyar,
    actual names for each other, 76. *See also* *Euclidia dyari* and *Protorthodes smithii*
    apocryphal names for each other: *corpulentis* or *smithiformis* or *dyaria*, 36
Behr, Hans Hermann. San Francisco lepidopterist, 30
Belgian Congo, mosquitoes from F.W. Edwards, 220
Bellport Long Island, NY, Dyar summer cottage, 33, 54, 67, 82, 84, 85, 88, 259
Benjamin, Foster Hendrickson, 201–204, Fig. 24.2a, 206, 208, 224–225. *See also* Tentamen
Benson, Mary Foley, 171, Fig. 20.2e
bigamy, 143, 199
    legitimize children as reason, 150
    Allen marriage rather than Dyar divorce, 150
black flies (Simuliidae), 180, 181
bladderwort, the greater (*Utricularia vulgaris*). *See* Bacteria (found on), *Bacillus vacuolatus*
Blake, Hannah Elizabeth. Wife of Henry J. Hudson, 11
'Bloodletter, Mrs.' *See* Dyar, Jr., short stories
Blue Book(s),
    account books, used by Eleonora and later Zella for bills, 8, 244
    field notebooks, used by Dyar, for rearing Lepidoptera, mosquitoes and sawflies, 8, 9, 16, 17, 25, 32, 47, 160, 250
    images of, Fig. 3.1, Pl. 2, Pl. 3
Bombycine moths for the National Academy of Science. *See* Packard, Alpheus S.
Bonne-Webster, Mrs. Mosquito worker, 175
Booth, Catherine, 142, 268
Boston Society of Natural History,
    dispute over specimens, 173. *See also* Charles Johnson
    Dyar publication in *Proceedings*, 42
    location, 14, Map 2a
    Strecker book, missing, 19
Böving, Adam Giede, Pl. 13
Bowles, Paul. Author and composer, "Nelson Dyar," 195
Boyd, Mrs., 148

Brakeley, Mr. J. T. Mosquito breeding sites at Dublin, NH, 103
Braun, Carl. Collector of *Dyaria singularis*, 37
Bremen, Germany. Grote's home during correspondence with Dyar, 75
British Columbia (1919), 174
British Columbia, collecting along train route, 180
British Columbia,
  Bennett, 180
  Glacier, 90
  Hazelton, 180
  Kaslo, 90
  Kootenay Lake, 90, Fig. 11.4
  Lake Atlin, 180
  Nanaimo, Vancouver, 31
  Prince George, 180
  Prince Rupert, 180
  Revekstoke, 90
  Shawnigan Lake, 90
  Vancouver Island, 90, 92
  Victoria, 31, 90
British Museum (of Natural History), 69, 110, 211, 220, 232
Bronn, Mary J., 209
Brooklyn Entomological Society, 42
Brooklyn Institute. Hulst pledges specimens to, 80
Brooks, William Keith. Brooks-Dyar Rule, 23, 244
Brown and Belford, Zella Dyar's Reno attorneys, 142
*Brown Rosary, The Lay of the*, 8
Brown, F. Martin, xxii, 173, 275
Brown, George Rothwell, Washington "Post-Scripts," 186
Brown, George, *amici curiae* (friend of the court), 147, 148, 149
Brownell, C.L., 54
Browning, Elizabeth Barrett Browning, 8
Brunswick, The, DC, 63
Bryan, Charles, 186, 277, Pl. 14d
Bryan, William Jennings, 186, 198, 277
    *See also* Scopes Monkey trial
Bureau of Entomology (USDA), 82, 105, 136, 146, 153, 174, 178, 209, 214, 218, 221–222
  hired Dyar, but not until 1913, 112
  taxonomy section, 204
Busck, August, 56

  backup for Dyar's moth identification, 87
  correspondence from Dyar in Reno, nerves improving, in spite of the vexations of laws, 88
  Dyar dismissal, aftermath, 154, 156
  Dyar's "List" contributor, 68–69
  financial loans to, from Dyar, 126, 157
  loans, of specimens, unofficial, to Barnes, 165–168, 203
  mail, holding, and watching desk for Dyar, 131–132, 142
  mentorship from Dyar, 68–69, 157
  mosquito collecting, 103
  named Florida moths reared by Dyar, 85
Panama,
  moths collected, 71
  moth publication, 108
  photographs, Fig. 20.1a, Fig. 26.3, Pl. 9a-b
  Ten Commandments, the, telling Dyar to study, 157
  Tortricidae, group of interest, 168
Butterflies,
  anise swallowtail (*Papilio zelicaon*), 31
  aphrodite fritillary (*Speyeria aphrodite*), Pl. 3a
  Becker's white (*Pontia beckeri*), 28
  black swallowtail (*Papilio polyxenes asterius*), 16, 17
  California hairstreak (*Satyrium californica*), 30
  *Callidryas*, 99. *See also* cloudless sulphur
  cloudless sulphur (*Phoebus sennae*), 26
  eastern tiger swallowtail (*Papilio glaucus*), 17
  giant swallowtail (*Papilio cresphontes*), 17, 23, 242, Pl. 3c
  golden hairstreak (*Habrodais grunus*), 28
  great southern white (*Ascia monuste*), 18
  Hawaiian blue (*Udara blackburni*), 30
  madrone butterfly caterpillar, Mexican (*Eucheira socialis*), 34
  map butterfly (*Cyrestis aza*), 67
  melissa arctic (*Oeneis melissa semidea*), 33

monarch (*Danaus plexippus*), 28
mourning cloak (*Nymphalis antiopa*), 16
nokomis fritillary (*Speyeria* [*Argynnis*] *nokomis*), 213, 282
ox-eyed woodnymph (*Cercyonis pegala*), 242
*Parnassius*, 31, 89
pea blue (*Lampides boeticus*), 30
pearl crescent (*Phyciodes tharos*), 17
red admiral (*Vanessa* [*Cynthia*] *atalanta*), 16, 17, Fig. 3.1
regal fritillaries (*Speyeria idalia*), 17
sierra sulphur butterfly (*Colias behrii*), 29
thoas swallowtail (*Papilio thoas*). See giant swallowtail
western white (*Pontia sisymbrii*), 28
*Butterfly Farmer*. See Ximena McGlashan
Button, Charles, 4

California,
   Arroyo Seco, 92
   Plumas County, 276
   Berkeley,
      related to Dyar divorce, 135, 137, 138, 139, 145
      Zella and children's home, 191
   Carpinteria, 92
   Fallen Leaf Lake and Lodge, near Lake Tahoe, 145, 179
   Gardenia, 92
   Gold Lake (Sierra County), 276
   Laguna, 92
   Lake Tahoe, 180, 181, 276
   Long Beach, alternative medicine for Dyar's goiter, 181
   site of Zella's fatal accident, 225
   Los Angeles. Mosquito collecting, 92
   Mt. Shasta, 92
   National City, 92
   Orr's Hot Springs, Mendocino County, 276
   Pasadena,
      Eagle Rock Valley, 15, 26
      Home of Zella Peabody, 15
      Ostrich Farm, 92
   San Diego, 92, 145, 153
   San Francisco, Golden Gate Park, 26
   San Juan Capistrano, 180
   San Onofre, 92
   Sisson (present day Shasta City), 92
   Summit (Sierra County), 276
   Sweetwater Junction, 92
   Tahoe City, 180
   Thrall, 92
   Truckee, 141, 276
   Watsonville, 31
Callender, Major, mosquito control, 178
Calvinism, 3
Cambridge Entomological Club, 22, 34
Camp Bird Gold Mine, Ouray, Colorado, 59
*Canadian Entomologist, The* (*Can. Ent.*), 20, 29, 33, 37, 62, 70, 76–78, 98, 102, 177, Fig. 5.2b
Canadian Rockies, 92
Carlin, NV, 150, 180
Carnegie Institution of Washington, 83, 99, 111
Carnegie Monograph, 83. See also "The Mosquitoes of North and Central America and the West Indies"
Carpenter, Mathilde "Tilly" M., 171–172, Fig. 20.2d, Pl. 13
caterpillars, "cats" to young Dyar siblings, 8, 16
Caudell, Andrew Nelson,
   Dyar assistant, USNM and in field, 56, 62, 67, 85–87, 88, 89–90, 92, 99, 103, 179
   financial loans from Dyar, 126, 165. See also Dyar, Jr. Loans to colleagues and friends
   Penelope, "Poodle", wife of, Fig. 11.3a
   photographs, Fig. 9.1b, Fig. 11.3a, Pl. 9e-f, Pl. 13
   unpublished catalog of food plants, 165, 259
Ceratopogonidae, "no see um" named for Dyar by Coquillett (*Tanypus dyari*), 88
Channing Club. See First Unitarian Church, Berkeley, CA
Chapman, Thomas Algernon, 40, 41
Chase, Isabelle B., 63
Chastleton, The. See Sixteenth Street Mansions
Chautauqua, 61, 110. See also Stony Man Camp
Chesapeake Bay crabs, 59–60

Children's Mission, Tremont St., Boston, Map 2a, 63. *See also* Harriet M. Peabody
Chironomidae, 160, 209
Chittenden, Frank Hurlbut, 254
Christian, Walter, marriage license clerk, 154, Fig. 19.1. *See* Wellesca marriage in Richmond, VA
cicada, periodical
   Marlatt, system for determining broods, 33
   Brood X, 88, Pl. 9b
Civil Service Commission, 135
Civil War, United States, *xxiii*, 6, 185
Clark, B. Preston, 156, 169
Clarke, John F. Gates, of Bellingham, WA, 214
Clough, Dr. Jacob N.M., 12, 13, 50
Clough, ER Dyar. *See* Eleonora Rosella Hannum Dyar
Cockerell, Theodore Dru Alyson,
   complaints about mishandling specimens at USNM, 159
   "Mosquitoes of Colorado, The," 161, 273
   Packard's monograph, co-edited with Dyar, 110–111
   portrait, Fig. 19.2b
   reviewer of,
     "List," Dyar's, 68, 70
     "Mosquitoes of North and Central America," 160–161
Cockle, J. William, collector and proprietor of Kaslo Hotel, 89, 90
Coffin, T.H., found mosquito larvae in Bahamas, 132
Colburn, Aunt Parthenia "Parnie," 9, 18, 50, 212, 238, Pl. 4c-d
Colburn, Nettie,
   Lincoln Séances, 5–6, Fig. 1.2b
   Séance, 1891. Attended by Eleonora Dyar's sisters, 13
Colburn, Parthenia "Parnie" Hannum. *See* Colburn, Aunt Parthenia "Parnie"
Coleman, Judge. Nevada State Supreme Court, 149
Colloday, Mr. Zella's divorce attorney in Berkeley, 146

Colorado, Denver,
   Albany Hotel, 26
   Union Station, 26, 28, Fig. 4.2a
Colorado, trip by Dyar and Zella, 28
Colorado trip, 1901, 85–87
   Grand Junction, Caudell visit to Lookout Mountain. *See* Golden Summit
Colorado, trip with Wellesca, Western slope around Grand Lake, 181
Columbia University,
   alumni card on file, 150, 224, Epi. Fig. 1b,d
   Alumni Federation, questionnaire on file, 224, Epi. Fig. 1a,c
   doctoral student, 43
   faculty reappointment, 54
   Faculty of Pure Science. Program where Dyar got graduate degrees, 43
   master's student, 35
Comstock, John Henry, 36, 38–41, 77
Conger, Rev., First Universalist Church, 15
convergent or parallel evolution of adult characters in Limacodidae, 48
Cook and Peary, 104, 105. *See* also Henry Skinner
Cooper, George S. architect, 119
Cope, Edward Drinker, 74
Coquillett, Daniel William,
   Honorary Custodian of Diptera, 88, 90, 99, 100, 101, 102, Fig. 9.1b
Cornell University, 38, 77, 160, 201, 210, 211
*corpulentis* or *smithiformis*, apocryphal names for species named for Smith by Dyar, 36. *See also* "Battle of the Names"
Crawford, Jr., James Chamberlain,
   burial plot, in, with Dyar and Pollock family, 226
   donation of Dyar's collection through, 154
   editor, *Proceedings of the Entomological Society of Washington*, 106
   husband of Wellesca's niece Emily Baker, 135, 158, 226
   photograph, Fig. 9.1b

specimens, mishandled, USNM's Division of Insects, 158
USNM Associate Curator, 112, 136, 154
Cresphontes Hollow, NY, young Dyar, Jr.'s name for collecting site near Rhinebeck, 17
Crosby, William Otis. M.I.T. geology professor, 12
Curd, Eleonor and Lewis, 228, 252
Currie, Rolla P.,
    Kaslo, B.C. trip, on, 89
    financial loans to by Dyar, 126
    hired as USNM Aid, 56, 252
Curtis, Lemuel, clockmaker, 3
Cushman, Robert Asa, Pl. 13
cyanide jars/bottles. *See* insect collecting supplies

Dak. Allen children's name for Dyar, 188
Darlington, Joseph J., D.C. attorney, 120
Darwinian principles, 31, 36, 38, 96, 98
    struggle for existence, 98
Davis, Joseph of Hotel Arlington, NY, 142
Dean, Bashford, Columbia University, 242
Decatur, IL. William Barnes Collection, original location, 112
DeGarmo, Dr. James M., 11, 240
"Diary of Lepidoptera"(June 17-October 6) 1882, 16, Fig. 3.1. *See also* Blue Book(s)
Dimmock, Anna Katherina, 22
Draudt, Max, 211
Drown, Thomas M., 12, 43
Dupont Circle,
    collecting moths, 88
    residences. *See* Dyar, Jr., District of Columbia residences
Dupree, Dr. James Dr. Louisiana's Surgeon General and Evelyn Mitchell's mentor, 101
Dyar Clough, Mrs. ER. *See* Dyar, Eleonora Rosella
Dyar v. Dyar,
    divorce in Alameda County, CA, 146
    negotiating with Mrs. Peabody, 137, 139, 140
    separation of Dyar and Zella, negotiations and settlement, 133, 135, Fig. 16.1
    testimony,
        Panama Canal trip, 138
        Hotel Arlington, clerk Joseph Davis a witness, 142
Dyar, Charles Warren, 50, 250
Dyar, Daniel Everett, 50, 250
Dyar, Dorothy,
    Columbia University, 1922–1923, religious studies student, 191
    Halloween party in tunnel, 186
    illness, prone to,
        mastoiditis and operation, 91–92
        measles, chicken pox, 123, 125
    mosquito collecting
        adults collected by dropping net over her in B.C., 90
        larvae in pools, finding, at Tuppers Lake, 91
    nickname Bertha, 212
    Peabody, Harriet M., her influence on, according to Dyar, Jr., 128
    photographs, including portraits, Fig. 23.1a, Pl. 10c-d, Pl. 11a
    Unitarian minister, Seattle, Dean of the Tuckerman School, Boston, 191
Dyar, Eleonora Rosella. *See also* Eleonora Rosella Hannum,
    Boston,
        death and burial, 13
        move to, 11
    marriage to Dr. Clough, 13
        medium, trance, 12–13
    images, Fig. 2.4a,b
    marriage to Dyar, Sr., 6
    materialized spirit, as a, 13, 240
    Rhinebeck, Map 1
        robbery, diamonds and pearls, 239
        visiting sisters, 9
        wealthy widow, 8
Dyar, ER. *See* Eleonora Rosella Dyar
Dyar, Gertrude, 212
Dyar, Harrison Golshan, 125, 127, 130, 193, Fig. 23.2a
Dyar, Jeremiah and Susanna Wild. Dyar, Sr.'s parents, 3
Dyar, John Wild. Last survivor of Dyar, Sr.'s siblings, 12, 50, 151
Dyar, Joseph. Dyar, Sr.'s brother, 3
Dyar, Joseph. First "AR" Dyar (b 1719), 229

Dyar, Jr., Harrison G.,
  autobiographic sketch, "The Old New England Stock," 199
  birth on Valentine's Day 1866, 6
  Boston, MA,
    collecting spots,
      Arnold Arboretum, 12
      Columbus Avenue, 12
      Dartmouth Street, 12
      Forest Hills and Franklin Park, 32
    homes,
      Roxbury, 123 Mt. Pleasant Avenue, Dyar and Mother, 11; Forty-nine Winthrop, Dyar and Zella, 34
      West Chester Park or 170 West Chester Park Road, 12–14, 18, Fig. 2.3b-c, Map 2,
  calligraphy, distinct style, 197, 211, Fig. 25.2c
  caterpillar livestock,
    home, in tumblers and backyard flower pots, 48, 250
    travels, while on, 30, 246
  chemistry, bachelor's degree, 12. See also Massachusetts Institute of Technology
  children's names, derived from Hannum relatives, 130
  collecting trips, personal, at his own expense, 67
  collection, personal, 1917 donation to USNM, 154. See also "Lepidoptera, Catalogue of"
  death, 223
    burial, Glenwood Cemetery, Washington D.C.
    death as a celebration, from spiritualist background, 13, 161. See also Dyar, Jr., colleagues' obituaries
  dismissal as government employee, 154
  dissertation. See Dyar, Jr., bacteria
  District of Columbia, residences
    B Street, 804, SW, 126, 132, 142, 187, 15.1b, 15.1d, 22.2a, 22.2c, Map 6, Pl. 15a
    C Street, 110, SE, flat, 252
    Twenty-first Street, 1512 and 1510, Dupont Circle, NW, 59, 63, 131, 134, 186, 189, 252
      backyard gardens, 186, 253, 277. See also Tunnel, Dupont Circle
      images, Fig. 8.1a, Fig. 14.3a, Fig. 22.1a, Fig. 22.1c-d, Map 6
      sale to Harriet Peabody and then back to Zella to regain control of assets, 121
  divorce. See Dyar v. Dyar
  editor, as, or on editorial board,
    Insecutor Inscitiae Menstruus, 107
    Journal of the New York Entomological Society, 95
    Proceedings of the Entomological Society of Washington, 106
    Reality Magazine
  entomological societies, as president
    Entomological Society of Washington, 80
    Cambridge Entomological Society, 34
  eugenic-related beliefs. See Eugenics
  financial loans, to,
    colleagues at USNM, 126, 165
    Wellesca's brother. See George F. Pollock
  financial losses, 1920s, 215
  genealogy
    Dyar family tree, Fig. 6.1b
    number of Dyars of each sex tabulated between 1740 and 1900, 50
    "Preliminary," published, 151
  health issues, 131
  honeymoon, Zella, 26–27
  "Lepidoptera, A list of the North American and key to the literature of this order of insects" of 1903, 41, 67–70, 158, Fig. 9.2
  "Lepidoptera, Catalogue of," 11, 14, 17, 18, 242–243, 267
    number of specimens (1897), 251
    images of, Pl. 4
    title and first entries, Pl. 4a-b
  Linwood Hill, NY,
    collecting, 11, 16, 18
    home, 8, 10
    images, Fig. 2.1, Fig. 2.2b

location, Map 1b
"Woodville Inn," Fig. 2.2a
Ludlow, Clara S., suggestion by Dyar
of becoming coworker under
many preconditions, 178
marriage to Zella Peabody, 15,
Fig. 2.5
sexual relations, ended in the early
1900s, 151
ultimatum to Zella that she be
friends with Wellesca, 127
"Moths of Limberlost," critique, 106
music,
notation, musical, "Sleep Baby
Sleep," 10
piano, 11, 59, 193, 194
nervous breakdown, 158
New York City, rented flats, Map 3b
400 West Fifty-seventh Street,
Windermere, in The, 27
243 West Ninety-ninth Street, 32
76 West Sixty-ninth Street, 35
*New York Entomological Society, Journal of the*, dominant contributor over first seven years, 49
nicknames,
Dak, by Allen children, 188, 193, 279
Dickie, by Wellesca, 193, 223
Harry, among Hannum relatives,
Zella, Perle, and early friends,
8, 50, 61
personality, philosophy,
cantankerous, 65
caustic or critical, 40–41, 95, 97, 101,
158, 190
compulsiveness, Dyar, in work and
pastimes, 190
"materialist, a thoroughgoing,"
195, 239
photographs and portraits, Fig. 5.1, Fig.
5.3a, Fig. 7.2a, Fig. 9.1b, Fig.
10.1a, Fig. 11.1, Fig. 12.1c,
Fig. 13.3a, Fig. 17.1a, Fig.
20.2a, Fig. 21.1a, Fig. 21.4a,
Fig. 25.2a, Fig. 26.2a, Fig.
26.3, Frontispiece, Pl. 10a, Pl.
11a, Pl. 13, Pl. 14b,f,h
productivity, unmatched in one area,
then shifted into another, 190
properties, District of Columbia,

Ashburn Apartments, 119, 120
exchanges, transfers, 216, 221, 266,
Fig. 14.2
Fourteenth Street Apartments, 120
images of, Fig. 14.2
Kalorama Heights, D.C., 119
Winchester, 120, 266
properties, owned with Perle N. Knopf,
New York City, Map 3b
Fifth Avenue, 331, first owned by
Dyar, Sr., 5
Davenport, Albert H., tenant, 51
599 Broadway. Probable location of
Dyar's collection 1890s, 30
Shakers, United Society, New-
Lebanon, sixty-three year
lease, 51
Ayer, Frederick, sold to, 52
West Ninety-fifth Street, 16, Knopf
residence, 51, Fig. 7.1
Rhinebeck, NY
collecting spots, including
Montgomerys and Pells Dock,
243. *See also under* New York
home. *See* Linwood Hill, NY
short stories,
early, from spiritualist
upbringing, 10
male protagonist seeks a marriage
license, 272
Shakespeare, quoting, 157, 194
spiritualist. *See* spiritualist
upbringing
typewriter, portable, 123, 193
Dyar, Lowell, oldest grandson of
Wellesca
collected streetcar license plates, 227.
*See also* Roberta Hill
Wellesca's grandson, 227, 236, 283
Dyar, Nora. Nickname for Eleonora, 212
Dyar, Otis Peabody, 61, 63, 85
Berkeley, graduate in engineering, 192
Colombia trip, possible, with Shannon
and father, 192
disposition "spoiled by too much
hectoring" by grandmother,
mother and sister, 129
Friend's School Intermediate Class,
graduate, 129
measles, 125

Dyar, Otis Peabody (*Cont.*)
  marriage to Catherine Ross and known as "Uncle Oat," 192
  photographs, Fig. 11.1, Fig. 23.1b-c. Pl. 10a, c-e, Pl. 11a
  seeking father at B Street home, when not at museum, 127
Dyar, Perle Nora, 6, 8, 10, 11, 14, 18. *See also* Perle Nora Knopf
Dyar, Roshan, 118, 121, 122, 130, 137, 138, 148, 149, 193, 195, 218, 227, 270
  name derived from Eleonora Rosella Hannum Dyar, 130
  photographs, Fig. 14.3b, Fig. 15.1e, Fig. 23.2a
Dyar, Sr. Harrison G.,
  Ben Franklin's kite experiment, performed, 3
  inventions,
    carbonate of soda production with J. Hemming, 4
    steam carriage, 5
    telegraph, 3, 4
    time division telegraph, 5
    underground dirt removal, invention, 4
    white lead, manufacture, U.S. patent, 4
  mastery, when achieved, passed on to another problem or topic, 190
  photograph, Fig. 1.1
  President of New York and New Haven Railroad, 4, 151
  purchases building at 331 Fifth Avenue, NY, NY, 5
  wealth accrued from monetary reward for dye patents from one of the Royal Societies of France, 4
Dyar, Susanna Wild, mother of Dyar, Sr., 3
Dyar, Wallace Joshan, 130, Fig. 23.2a,c
Dyar, Wallace Joshan. Accounts of his growing up in the Dyar household, including tunnels, 130, 188, 189, 192–195, 216, 222, 224, 228

Dyar, Zella,
  angle sprain at Sierra Madre, 1906, with husband, 84, 92
  birth and ancestry, 14, 63
  choice to remain married for financial stability, rather than stigma of divorce, 133
  description,
    character, "of refinement and high birth," 127
    physical, 14
  Everett School, Boston, MA, 14, Map 2b
  hearing loss/deafness,
    around time of Otis's birth, 128, 253
    ear trumpet, 63
  interred at Forest Lawn Cemetery in Los Angeles, 226
  motherhood, choosing to have two children, as disappointment to Dyar, 128
  paddle caterpillar, description written by Zella, 246
  photograph with husband, after divorce, on Knopf's front steps, 213, Fig. 25.2a
  photographs, Fig. 15.1a, Pl. 10b-d
  pianist, 15, 118
  mosquito collecting, 131
  moths named, genus and several species by husband, 70, 141, 212, 213
  refusal to give up Dupont Circle home/property in property settlement, 133
  struck by a trailer, 226
  typhoid fever, 125
  Wellesca,
    jealousy, 123
    refusal to be friends with, 125, 127
Dyar's Catalogue, numbered records of Lepidoptera specimens in Dyar's collection, obtained through rearing, collecting, exchanges, or purchases, 251, Pl. 4
Dyar's Law (or Rule) of Geometric Growth, 22–25, Fig. 3.2
*dyaria*, apocryphal name for species named for Dyar by Smith, 36. *See also* "Battle of the Names" and Moths, *Dyaria*, the genus

Dyarian (in Dyar's unique style),
    absolution, 213
    riddle, names of Allen children, 130
    critique, 106

Early, Charles. First D.C. realtor of H.G. Dyar, Jr., 120, 266
earthquake, Nevada, 1915, while Dyar was in Reno, 138, 160
East Grand Forks, ND, 181
Edwards, Frederick Wallace, 161, 211, 220
Edwards, Henry, 21, 22, 53, 276
Edwards, William Henry. Number of molts of promethea moth, 22
electric lights and globes, for collecting moths, 11, 12, 14, 34, 37, 89, 240
embryology course, 35, 250, Fig. 5.1. *See also* Woods Hole
embryonic development, viewed from outside egg, 49, 250
Empire State Building, future location across from Dyar, Sr. residence, 5
*Entomologica Americana*. Journal of the Brooklyn Entomological Society, 20
*Entomological News*, 30, 47, 79, 90, 96–100, 103
Entomological Society of Washington,
    account of early meetings, 72
    meetings, 60, 80, 82, 105, 106
    Saengerbrund Hall, 106
eugenics,
    including themes in Dyar's short stories, 35, 143, 151
    woven into Dyar's genealogy, in *Reality Magazine*. "The Old New England Stock," 199
Eugenics, International Congress of, 35
Eureka Secret Canyon Mine. An investment scam, 216
Ewing, Henry Ellsworth, Pl. 13

*Facilis decensus averni*, 188
Fernald, Charles H.,
    Crambidae, encouragement for Dyar to work on, 71
    Dyar's "List," participation in, 68–69
    early correspondence with Dyar, Jr., 19
    moth family names, unstable, 255
    portrait, Pl. 8e
    snafu over *Butalis basilaris* sent unspread for identification, 40, 41
Fernald, Henry T., 93
First Spiritual Temple, Boston, MA, 12, 13, Map 2a
First Unitarian Church, Berkeley, CA. Dorothy Dyar secretary on board of directors and Otis on the same on the youth members' Channing Club, 191
Fisher, W. Samuel, Pl. 13
Florida,
    Altamont Spring, 18, 243
    Jacksonville, 27
    Jupiter, 27
    Key West. The cemetery, 85
    Lake Worth, 27, 244
    Miami River, 85
    Palm Beach, 27, 61, 85, 244. *See also* Lake Worth
'Flossy (ie).' *See* Dyar, Jr., short stories
food plants, caterpillars, 19, 21, 29, 30, 32, 33, 47, 86, 165
Forbes, William Trowbridge Merrifield, "Pink Whiskers," 160, 204, 211, 273
    photograph, Fig. 24.2c
Ford, Mary Hanford. Short-term Editor of *Reality Magazine* under Dyar, Jr., 198
Ford, Richard A. Dyar, Jr.'s attorney, 138, 139, 146
Fox Sisters, in Hydesville, NY, 240
Franclemont, John, 273
Franklin Park, Washington, DC, 120
*Frass: An Occasional Journal of Para-lepidopterology*, xxiii
Freeborn, S.B., as possible successor to Dyar on mosquitoes at USNM, 220
Frenatae or Generalized Frenatae, 38, 40–41
'French, Paul,' *See* Dyar, Jr., short stories, "A Fool and His Friend." *See also* Isaac Asimov
frenulum, 38
Friends School, 129, 227

Gahan, Arthur Burton, Pl. 13
Gangs of New York. *See* The New York Draft Riots of 1863
genealogy,
  *See* Limacodidae caterpillars, phylogeny
  *See* Dyar, Jr. "Preliminary Genealogy"
George Washington University Hospital, place of treatment for F. Knab, 161
Getsinger, Lua Moore, 117, 118
"Ghost Makers, The," 184
ghost stories or attending séances as a youth, 10, 184, 189–190
giant skipper caterpillars (*Megathymus*), 52
Gilded Age children, xxiv
Gillett, Charlotte, wife of William Barnes, and sisters Amaryllis T. and Jessie D., 113, 265
Glenwood Cemetery, Washington, D.C., 223, 226
Glover, Townend and his specimens, 72, 213
Goiter, Dyar medical issue, 181
Golshan. *See* Harrison Golshan Dyar
Gough, Eugene. D.C. realtor and friend of Pollock family, 120
government printing office. Delays in publishing taxonomic works, 108
grasshopper, "the green fool" (*Acrolophitus hirtipes*), 87
Greene, Charles Tull, 175, Pl. 13
Grochek, Nettie, Pl. 13
Grossbeck, John A., 83, 103, 114
Grote, Augustus Radcliffe, 37, 41–2, 69–70, 74–77, 81, 161, 202, Pl. 8f
  Dyar's dedication of *Insecutor* to, 75, 108, Fig. 13.2b
  Issues with Smith and others over *Tentamen*, 75, 76
ground beetles (Carabidae), 20, 244
Guenée, type specimens, 77–78
Gunder, Jeane Daniel, 109

Hadley, Philip, 44
Haimbach, Frank, 159, 222

Hannum siblings. Surviving past 1900, 251
  Barden, Angeline (Angie) Nancy
  Bushnell, Mary Robinson,
  Eastabrook, Helen Sophia
Hannum, Bessie. *See* Elizabeth Hannum
Hannum, Eleonora Rosella, before marriage to H.G. Dyar, Sr., 5, 6
Hannum, Elizabeth (Aunt Bessie), wife of Josiah C., 30, 61
Hannum, Josiah Cushman (Uncle Joe), 6, 17, 30, 50, 61, 130, 229, 242, 251
  Battery Light Artillery of the New York Volunteer Army, 28th, 6, 239
  Civil War artillery sabre, 50
Hannum, Parthenia "Parnie" Robinson, 5–6, Fig. 1.2. *See also* Parthenia Hannum Colburn and Lincoln, séances
*Harpalus pensylvanicus*, 20, 244. *See also* ground beetles
Harriman Alaska Expedition of 1899, 79, Fig. 10.2b
Harrower, Rev. Charles S., 7
Hatch Act of 1887, The, xxiv
Hawaii, Pacific Kingdom of, 30
Heidemann, Otto, 72, Fig. 9.1b
Heinrich, Carl, 160, 165, 167, 168, 201–205
  dissections of *Acharia* and other moths in collaboration with Dyar, Jr., 211, 281
  photograph, Fig. 26.3
Henson, Pamela M., donation of Dyar's collection to keep his job, 274
  matching initials of Wellesca P. and Wilfred P. Allen, 152
Hering, E. Martin, 211
Hesperiidae, 96, 98, 201. *See also* skippers
Higbie, Alfred, 120–121, 133, 134, 216, 217, 266, 282
Hill, Dorothy Dyar. *See* Dorothy Dyar
Hill, Roberta. Zella's granddaughter recounts family lore, 60, 227, 236
Hoff, Louisa (Lou), 18, 48, 243

Holland W.J. Director of the Carnegie Museum, Pittsburgh, and author of "The Butterfly Book" and "The Moth Book", 213
Holland, Caleb, father of Harriet M. Peabody, 63
Holland, Harriet M., 14. *See also* Harriet M. Peabody
Holland, John M., brother of Harriet M. Peabody, 191, Pl. 10e
Holland, William Jacob. President of Cambridge Entomological Society, 34
Holley, Horace. On editorial board of *Reality Magazine*, 198
Hommon, Harry B. In charge of mosquito control at Yosemite, 208
Honorary Custodian of Diptera, USNM, 90, 160, 161. *See also* Coquillett, Knab, and Aldrich
Honorary Custodian of Lepidoptera, USNM, 112, 114, 135, 213
  appointment, 65
  keeping position after scandal, 272
Hopp, Walter, 211
Horton, J.H. Collector owlet moth specimen Dyar named, *Hortonius euenemus*, 208
Hotel Mahackemo, Norwalk, CT. Place of Dyar's recovery from nervous breakdown, 158, Fig. 19.2a
"houses of transformations," caterpillar rearing chambers of Dyar's youth, xxiv
Houston, David F., 153. *See also* USDA, Secretary of Agriculture
Howard, Buell V., 226
Howard, Leland Ossian,
  compensation, early efforts to obtain for Dyar, 65, 67, 112
  diplomacy, attempts at, between Dyar and others, 69, 113–114, 136, 156–157, 166–170, 176
  Dyar's arrival at USNM, prior to, 21, 40, 52–56
  Entomological Society of Washington, description of meetings, 72
  portraits and other photographs, Fig. 7.2b-c, Fig. 26.3, Pl. 13
Howard, Mrs. Leland O., 63

Hübner names. *See Tentamen*
Hudson, Arthur, 14
Hudson, Charles, 240
Hudson, George Henry, 18, 36, 44, 76, 94, Pl. 8b
Hudson, Lucy Ann, 8, 11, 17, 240
Hudson, Lucy, grandniece of Lucy A. Hudson, 240
Hudson, Rev. Henry James, 11, 240
Hulst, Rev. George D., 42, 68–69, 71, 74, 76, 79–81, 83
  About Dyar being stood up for Brooklyn Entomological Society meeting, 42
  Dyar's harsh postmortem critiques, 79–80
  photograph, Fig. 10.2a
Humboldt, NV, 26
Hustings Court, Richmond, Virginia, 153

Idaho, Sandpoint, 179
Independence Avenue, formerly B Street, 228, Fig. 22.2d
insect collecting supplies,
  pins, card points, 160
  cyanide jars/bottles, 89, 132, 160
  Schmidt boxes, wood, to ship specimens, 160
  *See also* moth collecting
*Insect Life*, 21. *See also* C.V. Riley
*Insecutor Inscitiae Menstruus*,
  Dyar's entomology journal, 75, 108, 132, 157, Fig. 13.2a-b
  Knab handling operations, 132
  meaning of name, 107
  "secular city metros" to Dyar's second family, 193
  valedictory of, 225
intelligent design, Dyar, Sr., 7
International Congress of Entomology, 2$^{nd}$, 1912, 129
invasive species. Dyar's attempt at introducing a limacodid to D.C., 93
Island House. *See* Wachapreague, VA

Johnson, Charles, 173. *See also* USNM, policy on keeping specimens from identifications

Jones, Frank Morton. Dyar's willingness to collaborate, 204, 205, Fig. 24.2b. *See also* Psychidae
*Journal of the New York Entomological Society*,
  cover, Fig. 12.1a
  *See* Dyar, Jr. editor and dominant contributor
*Journal of the Washington Academy of Sciences*, 211
Joutel, Louis H., 111, 265, Fig. 10.3b
Jugatae, 38–40
jugum, 38, 40

katydid, 'ventriloquist.' *Cyphoderris piperi* (now *C. monstrosa piperi*), 92–93
katydid, brown (*Atlanticus davisii*), 124
Kearfott, W.D., 114
Khayyam, Omar. "Rubaiyat, The", 6, 105
Kheiralla, Dr. Ibrahim G., 64
Kidder, Louise and Zinie. Educators and Zella's California cousins, 137, 270, 281
Kinzel, F., 248
Knab, Frederick, 78, 82, 91, 99–102, 108, 111, 119, 132, 159–161, 173, 176–177, 210–211
  collecting mosquitoes with Busck, 159
  Leishmaniasis, visceral form, 161
  letter, image, Fig. 19.3b
  managing *Insecutor*, 159
  photograph, Fig. 19.3a
  *See* mosquito larvae as separate organisms
Knopf, Dr. Siegmund Adolphus and Perle Nora,
  advice on venue for divorce, 138–140
  residence, 16 W 95th Street, New York, 51, Fig. 7.1, Map 3
Knopf, Dr. Siegmund (later "Sigard") Adolphus, 14, 15, 43, 51–52, 61, 131, 191, 203
Knopf, Perle Nora, 15, 40, 48. *See also* Dyar, Perle Nora
Koebele, Albert, 209
Kyser, Janice, Pl. 13

labial palpi, 48. *See also* Lintner, Dyar's apprenticeship

Lampton, James J., 217, 281
Langley, Samuel Pierpont, 65
Leech, Coray. Husband of Maud Leech, 137, 142
Leech, Laurel. Young son of Maud Leech, who travelled to Reno with his mother, 138
Leech, Maud,
  affidavit, convinced Zella of her husband's guilt of adultery, 146
  Allen children
    baby sat, while Wellesca went to Bahamas with Dyar, 132, 149
    thought she was their mother, 193
  blackmail towards Dyar and Wellesca, claimed, 138, 193, 220, 270
  false claim that Dyar fabricated Persian names from 'Abdu'l Bahá, 268
  perjury, accused of, by Dyar, 158
  testimony, 142
  *See* Dyar, Jr., short story, "A Fool and His Friend"
Leishmaniasis, visceral form, 161
Lembert, John B., 29, Fig. 4.2b
Leng, Charles W., 95
Lepidoptera classification,
  larval setiferous tubercles (setae), based on, a bridge between ancestral and derived Lepidoptera, 39, 41, Fig. 5.3b-c
  talks on, by Dyar, 41–42
Lepidoptera head setae, homology, 42
Lesca. *See also* Wellesca,
life histories,
  Limacodidae, 45–47, 49, 123
  Lepidoptera, other, sawflies, mosquitoes, 30, 41, 42, 52, 88, 89, 108, 109, 190, 209
Lincoln, President Abraham and Mary Todd,
  séances, 5, 6, Fig. 1.2b. *See also* Nettie Colburn and Parthenia Hannum Colburn
Lindsey, Arthur Ward. Curator for William Barnes, 201
Linell, Martin. Aid at USNM before Dyar's tenure, 36, 54, 55
Lintner, Joseph Albert, New York State Entomologist,

Dyar's apprenticeship and correspondence, including request for critique of mouthpart morphology, 18, 19, 20, 21, 22, 26, 30, 34, 35, 36, 55, 65, 161, 190
   photograph, Pl. 8c
Locke, Catherine Virginia "Jennie" Mrs., 142, 203, Fig. 9.1b, Pl. 13
Long Island racetrack, NY, 4. *See also* Dyar, Sr., telegraph experiment
*Lucky Starr*, by Isaac Asimov, 143
Ludlow, Clara Southmayd,
   accusation by Dyar and Knab of mixing mosquito specimen from Philippines with one from Pennsylvania, 177
   rebuttal, stating responsibility of coauthored piece "is jointly shared," 177
   photograph, Fig. 21.2a
Lugger, Otto, 72

"macros" (Macrolepidoptera), 41–42, 165
madrone, 166
Mail, G. Allen, assisted Dyar in 1926 trip to Montana, 281
Maine,
   Bangor, 37. *See also* Carl Braun
   Canton, birthplace of Zella Peabody Dyar and parents, 14, 63
   Dixfield nr. Bangor, 32
   Lincolnville, 84
   Mexico, Oxford County, Maine, 191. *See also* Harriet M. Peabody
   Webb Pond, 32
   Weld, 123
mangroves, leaves, 49, 72, 85
Manitoba,
   Winnipeg Beach, 179
   Winnipeg, 181
Mann Act, 139
Mann, William E., 217
Mann, William M., Pl. 13
Marble, Ruth Banister Hannum, (Aunt Ruth), 9, 13, 50, 212
Marlatt, Charles Lester, 33, 218, 220, 221
marriage, between Dyar and Wellesca. *See* Reno, NV

marriage license. *See* Wilfred P. Allen
Marsh, Orthniel Charles, 74. *See also* Edward Drinker Cope
Martini, E. German mosquito worker, 175–176
Maryland,
   Baltimore, 90, 159
   Chesapeake Beach, 84, 90
   Herzog's Island, Fig. 10.1a
   Hyattsville, water lilies, 93
   Plummers Island, 88, 259, Fig. 10.1b
Massachusetts,
   Boston, Map 2. *See also* Dyar, Jr. collecting spots, homes
   Concord, 3, 4
   West Springfield, Mount Holyoke, Longmeadow, 91
Massachusetts Institute of Technology, M.I.T., 14, 19, 34, 43, Map 2a
Maynard, Nettie Colburn. *See* Nettie Colburn
Maynard, William P. Nettie Colburn's husband, 50
McAdam, Judge David. Mock trial judge Lake Placid, 32
McDunnough, James Halliday,
   Barnes and McDunnough "Checklist," 114, 154
   conflict with Dyar, 280
McFarland, Mrs. William, 63
McGlashan, Ximena. *The Butterfly Farmer* in Truckee, CA, 141
McGrath, Mr., fictitious employer of Wilfred Allen as a typist, 125, 148
McKinley, Rep. William Brown. Barnes lobbied for government purchase, 113
McLean, Edward B. Publisher of *Washington Post*, owner of Hope Diamond, 184
McLean, Wallace. Dyar and Wellesca's confidant and business associate, 130, 138, 139, 146, 151, 192, 207
Meade, Mary Robbins. Author, theosophist, and caregiver to Allen/Dyar children, 195, 239, 279

Megathymidae, 52. *See also* giant skippers
Merrick, Frank, collector from New Brighton, PA, 113
Metropolitan Hotel, Washington, DC. Dyar's official residence after separation, 135
Meyrick, Edward, 109
"micros" (Microlepidoptera) 19, 40–42, 157, 165
Middlebury, VT, 3
Middlesex, England, 4
midge (*Chaoborus trivittatus*), 183
*mihi itch*, 204
Miller, A. Grant. Wellesca's Reno attorney, 146, 149
Minnesota,
 Mayo Clinic, Rochester. Visit by Dyar for goiter, 181
 Moorhead, 181
 Thief River Falls, 181
 Warroad, 181
M.I.T., *See* Massachusetts Institute of Technology
Mitchell, Evelyn Groesbeeck,
 author, mosquito worker, and illustrator, 91, 101
 images, newspaper, Fig. 12.2a–b
 libel lawsuit against Dyar for his attack on *Mosquito Life*, 102, Fig. 12.2
 Dyar critique of *Mosquito Life*, 101–102, Fig. 12.2c
 rebuttal, flattered by Dyar's comparison of her with predatory mosquito larva, 102, Fig. 12.2d
Monever, 'Abdu'l Bahá's daughter, 118, 267
*Monopoly* game, 119. *See also* Dyar, Jr., properties, District of Columbia
Montana,
 Bitter Root Mountains, 213
 forest fires, 207
 Glacier National Park, 129, 181, 197, 207
 Grinnell Glacier, 207, Fig. 25.1
 Swiftcurrent Pass, 207
Montana (1917), 179, 276

Moore, Allen Francis. Congressman in William Barnes's district, 205, 206
Morrill Land Grant Act of 1862, xxiv
Morrison, Harold, 173
Morse, Samuel F. B., 146
Morton, Emily L.,
 co-author at onset of New York Slug Caterpillars series, 46–48, Fig. 6.2a
 incorrect published identification of *Tortricidia pallida* as *Isa textula*, 47
Mosquitoes
 conservation concerns about, in Yosemite, 209
 control of,
  lowering water level "to destroy breeding conditions," 103
  medical concern, of no, "mosquito-killing conspirators," 209–210
  oilers, as exterminators, 207
 Culicidae the family name, 99, 176
 egg rafts (boats), 103, Pl. 12c
 larvae (wrigglers), 82, 90, 208, Fig. 10.3c, Pl. 12a–b
 "independent organisms," treating as, by Dyar and Knab, 82, 99, 100
 lectures at Howard University to medical students, 222
 male genitalia, study, along with larvae by Dyar and Knab changed details of classification, 160
 monographs or major works,
  "Mosquitoes of the Americas, The," 210
  "Mosquitoes of Canada, The," 176
  "Mosquitoes of North and Central America and the West Indies, The," 91, 100, 101, 108, 159, Pl. 12
  "Mosquitoes of Panama," 183
  "Mosquitoes of the United States, The," 176
 National Parks, in, 209. *See also* Yosemite and Yellowstone
 pupae, Pl. 12c

species, common names,
   common house. *See Culex pipiens*
   eastern salt marsh mosquito. *See Aedes sollicitans*
   rock pool mosquito. *See Aedes atropalpus*
   western tree hole. *See Aedes varipalpus*
species, Latin names
   *Aedes atropalpus*, 124
   *Aedes calapus*, 132, 179
   *Aedes cantator Coquillett*, 93
   *Aedes increpitus*. Isolated population in Yosemite exterminated by oiling, 183, 209
   *Aedes nearcticus*, 207
   *Aedes palustris, var. pricei*, 179
   *Aedes sollicitans Walker*, 82, 124, 181, 261, 267
   *Aedes varipalpus*, 90, 276
   *Aedes ventrovittis*, 183, 209
   *Anopheles crucians*, 267
   *Anopheles* larvae, 82
   *Anopheles maculipennis* Meigen, collected by Zella Dyar, 269
   *Anopheles perplexens*, 177
   *Anopheles*, as malaria carrier, 90, 103, 161
   *Culex*, 82, 88, 134
   *Culex aseyehae*, 132, 269
   *Culex confinis*, 82
   *Culex curriei*, collected by Zella Dyar, 90
   *Culex decorator*, 100
   *Culex derivator*, 100
   *Culex extricator*, 100
   *Culex habilitator*, 100
   *Culex lactator*, 100
   *Culex lamentator*, 100
   *Culex mitchellae*, 91, Fig. 11.5
   *Culex mortificator*, 100
   *Culex mutator*, 100
   *Culex pipiens*, 90, 159, Fig. 21.1b
   *Culex quinquefasciatus*, 132
   *Culex regulator*, 100
   *Culex signifer*, 83
   *Culex similis*, 269
   *Culex sphinx*, 269
   *Culex sylvestris*, 82
   *Culex territans*, 82
   *Culex*, "ator" endings named by Dyar and Knab, 99
   *Culiseta alaskaensis*, Fig. 21.3b
   *Culiseta incidens*, 90
   *Melanoconion humilis*, 100
   *Mochlostyrax caudelli*, species named from larva, 99
   *Psorophora*, predatory mosquito larvae, 102. *See also* Evelyn G. Mitchell
   *Uranotaenia coquilletti*. Named by Dyar and Knab for Coquillett, who misidentified, 100
   swarms of, 180, 207
   workers on Carnegie monograph as "scientific Aedids of Washington," 102
Moths,
   Acronicta, 72, 77–79, 98
      *Acronicta funeralis* (paddle), 32, 42, 246, Pl. 5a-b
      *Acronicta impleta*, 78
      *Acronicta luteicoma*, 78
      *Acronicta morula*, 246
      *Acronicta oblinata* (smeared dagger), Pl. 5c
      Smith and Dyar's monograph on Acronicta, 72, 77
   *Acronycta*. *See Acronicta*
   *Agriopodes fallax* (green marvel), Pl. 5d
   *Alabama argillacea* (Hübner), cotton worm, 108
   *Apatela* (= *Acronicta*), 98
   Arctiinae (Tiger Moths),
      *Arctia docta*. *See Notarctia proxima*
      *Euchaetes zella*, named for Zella Dyar, 70, Fig. 9.2b
      *Haploa confusa*, 73
      *Haploa triangularis*, 73
      *Haploa*, variation in wing pattern, 167
      *Hyphantria cunea* (fall webworm), 23, 244, Fig. 3.2
      *Notarctia proxima*, 27, 70
      virgin tiger moth (*Grammia virgo*), 18–19, 238, Plate 4c-d
      woolly bear (*Pyrrharctia isabella*), 23
   *Zellatilla*, 213
   aspen leaf miner, common. *See Phyllocnystis populiella*

Moths (*Cont.*)
  *Blastobasis guilandinae*. Larvae found on nicker bean, *Guilandina bonducella* (now *Caesalpina bonduc*) in Florida, 258
  Bombyces, 35–37, 41, 71
  Bombycidae, 22, 36, 41. *See also* Bombyces
  burnet. *See* Zygaenidae
  *Butalis basilaris*. Larva important in Dyar's classification of Lepidoptera, 248
  *Callizzia amorata*. *See* Epiplemidae
  *Callosamia promethea* (promethea), 22, Pl. 7b
  case bearers (Psychidae), 204
  *Catocala ultronia*, 18
  ceanothus silk moth. *See Hyalophora euryalis*
  *Charandra derideris* (the laugher), Pl. 5f
  *Chrysodeixis*, 30
  *Citheronia regalis* (the regal moth), 17, 106
  *Clisiocampa*. *See* tent caterpillar, western (*Malacosoma californica*),
  clothes moths (Tineidae), 204
  *Coenodomus hockingii* Walsingham. Synonym of *Dyaria singularis*. Obscuring name but not the myth, 247
  *Composia fidelissima* (faithful beauty), 42, 248, Pl. 7d
  cotton worm. *See Alabama argillacea* (Hübner)
  Crambidae, many species named by Dyar between 1904 and 1919, 71
  *Cucullia laetifica* (camphorweed paint), Pl. 5h
  Dalceridae, 211
  definite tussock moth. *See Orgyia definita*
  *Dendrolimus*, 53
  *Dendrolimus howardi*. *See Gloveria*
  *Dryopteryx*. *See Oreta*
  *Dyaria singularis* Neumoegen, 36, 37, 247, Fig. 5.2b. *See also Coenodomus hockingii* Walsingham

eight-spotted forester (*Alypia octomaculata*), Pl. 5e
Epiplemidae, gray scoop wing or crenulate moth (*Callizzia amorata*), 61, 87–88
Erebidae, 248
*Euclidia dyari*, 76
*Euplexia lucipara*, 20
*Euscirrhopterus gloveri*, 208
*Eutrapela clemataria* (curved toothed geometer), Pl. 7i-j
*Euxoa*, 201
fall webworm. *See* Arctiinae, *Hyphantria cunea*
flannel moth. *See* Megalopygidae, *Gelechia ocellella*, 87
Gelechiidae, 68
Geometridae (inchworm moths), Pl. 7g-j
ghost moths. *See* Hepialidae
*Gloveria arizonensis*, 53
*Gloveria howardi*, 53
gypsy moth (*Lymantria dispar*), 37
*Gyros muirii*, (Muir's crambid moth), 180–181. *See also* Crater Lake
*Hadena devastatrix*, 18
*Harrisimemna trisignata* (Harris' three spot), 36, 97
Heliozelidae (fairy moths), 12
*Hemeroplanis cumulalis*, 104–105
Hepialidae, 40
*Hepialus lembertii*, 29
herald, the. *See Scoliopteryx libatrix*
*Hortonius euenemus*. *See Euscirrhopterus gloveri*
*Hyalophora euryalis* (ceanothus silk moth), 28
*Hylesia*. Photographed Latin American species for Packard monograph, 111
io moth (*Automeris io*), 16, 17, Pl. 3c
lappet moth (*Phyllodesma americana*), 17, Pl. 3b, Pl. 7a
Lasiocampidae. *See* tent caterpillars
leaf miner on sour gum (*Nyssa sylvatica*), 36
leaf rollers (Tortricidae), 68
*Leuculodes dianaria*. Short published description, 109

Limacodidae (Slug Moth),
   caterpillars:
      early encounters, 16, Pl. 3d
      first-instar, 49
      mangrove leaves, on, 72
      nettles, gelatines, or hairys, 45
      "Slug Caterpillars,
         The life-histories of the
         New York," 46-49, Fig. 6.1a,
         Fig. 6.2b-c
      biogeography, land bridges, 49
      Phylogeny/genealogy, divides by
         caterpillar types: Tropic hairy,
         Tropic spiny, Tropic smooth,
         and Palearctic smooth, 45,
         Fig. 6.1a
   cocoon, oval, 46
   family name, nomenclatural
      instability, 249
   parasitoid, (Crypturus dyari), 49
   pet group, 'Codes, Dyar's "most
      cherished treasures,"
      45, 242
   species
      *Acharia gertrudens*, named for
         cousin, 212
      *Acharia norans*, named for
         mother, 212
      *Acharia stimulea* (saddleback), 248,
         250, Pl. 1c
      *Acharia ximenans*, 141, 212. See
         also Ximena MacGlashan
      *Acharia zellans*, named for Zella,
         141, 212, Fig. 25.2b
      *Alarodia (Calybia) slossoniae*, 49,
         250. See also embryonic
         development
      *Apoda y-inversa*, Fig. 6.2c
      *Euclea delphinii* (spiny oak),
         84, Pl. 1c
      *Euclea zygia* and *Metraga costilinea*,
         synonyms, 110
      *Euphobetron*. See *Phobetron dyari*
      *Isa textula* (crown slug), 20, 267,
         Fig. 14.4, Pl. 1b
      *Kronea minuta*, 65, 254
      *Lithacodes fiskeana*, 91, 98
      *Monema flavescens* (oriental
         limacodid), 93. See also
         invasive species
      *Monoleuca semifascia* (pin-striped
         slug), 123, 267, Pl. 1d
      *Parasa indetermina* (rose slug), 84,
         250, Pl. 1f
      *Parasa chloris* (smaller Parasa), 17,
         Pl. 1e, Pl. 2d-e
      *Parasa prasina*, 62, 251
      *Parasa wellesca*, 62, 251. See also
         *Parasa prasina*
      *Phobetron dyari*, 203
      *Phobetron pithecium* (hag moth),
         13, 49, 250, Pl. 1g
      *Prolimacodes badia* (skiff moth
         caterpillar), 250, Pl. 1a, Pl. 2c
      *Sibine*. See *Acharia*
      *Semyra zinie*, named for Zella's
         cousin Zinie Kidder, 281
      *Tortricidia flexuosa* (red-cross button slug),
         48, Pl. 2f
      *Tortricidia pallida* (red-cross slug),
         47, Pl. 1h
      *Venadicodia ruthaea*, named for
         aunt, 212
luna moth (*Actias luna*), 16
Lymantriinae (formerly Liparidae),
   37, 42, 109, 248, Pl. 7e
*Malacosoma*, 53. See also moths, tent
   caterpillar
Megalopygidae (flannel moths),
   associated with 'Bombyces,' 71
   caterpillars have extra
      prolegs, 49
   *Megalopyge crispata*. In
      transformation series of
      larval tubercles, 248
   *Megalopyge opercularis*. Mistaken
      for limacodid by Caudell, 88
   *Megalopyge partheniata*. Named for
      Aunt Parnie, 212
*Mesoleuca simulata* Hübner var. otisi
   (now *Thera otisi*), 90
Mexican tiger moth. See Arctiinae,
   *Notarctia proxima*
Micropterygidae, *Micropteryx*, 39
*Noctua phyrophiloides* var. *peabodyae*. See
   *Pronoctua peabodyae*
Noctuidae (owlet moths),
   genealogical tree, 42
   many species described by Dyar
      between 1904 and 1919, 71

Tortricidia flexuosa (Cont.)
  Nola clethrae. Named by Dyar for sweet pepper bush (*Clethra*), the larval food plant, 85
  Nola sexmaculata, 21
Nolidae, Dyar described over 60 species, 20
Notodontidae (Prominent Moth),
  Cerura. See also *Furcula*, 21, 76
  Cerura scitiscripta, black-etched prominent, Pl. 6g-h
  Datana californica, first species named by H.G. Dyar, 20. See also Riley, C.V.
  Datana contracta, 19
  Datana drexelii (Drexel's datana), Pl. 6e-f
  Datana floridana, 20
  Datana integerrima, 19
  Datana major, 19, 20
  Datana ministra, 19
  Datana palmii, 20
  Datana perspicua, 21
  Datana robusta, 19
  Ellida gelida [now *caniplaga*], 22
  Furcula cinerea and *F. cineriodes* caterpillars, 21
  Heterocampa guttivitta (saddled prominent), Pl. 6c-d
  Heterocampa unicolor, 22
  Macrurocampa dorothea, named for daughter Dorothy, Dyar, 51
  Nadata. Rare example of Dyar's Law used in own work after original study, 25
  Oligocentra lignicolor (lace-capped caterpillar), 20, Pl. 6a-b
  Symmerista albifrons (white-headed prominent), 28
Omiodes blackburni, on coconut palm, 30
Oreta rosea. Life history described in Dyar's first scientific paper, 20
Orgyia,
  cana (western tussock), 30
  definita (definite tussock), 23, 35, Pl. 7e
  leucostigma (white-marked tussock), 17, Pl. 3c
  vetusta (western tussock), 26
paddle caterpillar. See *Acronicta funeralis*
Pericopinae, 42, 248. See also *Composia fidelissima*
Phryganidia californica (California oak moth), 15
Phyllocnystis populiella (common aspen leaf miner), Pl. 7f
pink-striped oakworm (*Anisota virginiensis*), 18, 247
Pleonectyptera. See *Hemeroplanis cumulalis*
plume moth. See Pterophoridae
Polloccia alticolalis, 263. See also Wellesca, species named for her
polyphemus (*Antheraea polyphemus*), stemmed or unstemmed cocoons, 97
promethea (*Callosamia promethea*), 22, Pl. 7b
prominents. See Notodonidae
Pronoctua peabodyae. Named by Dyar for Harriet Peabody, 64
Psychidae (case bearers), 204–205. See also Frank Morton Jones
Pterophoridae (plume moths), 40, 248
Pyralidae (Aquatic Snout Moths),
  Elophila gyralis, 260
  Elophila obliteralis, 243
  Paraponynx badiusalis, 243
Pyralidae (Snout Moths)
  Epipaschiinae, 37. See also *Dyaria singularis*
  Philodema (Sarata) rhoiella, 87
  Sarata (Megasis) caudellella, 87
Pyraloidea (Snout Moths), 37, 68
  many species named by Dyar between 1904 and 1919, 71
rose hooktip moth. See *Oreta rosea*
Saturniidae (giant silk moths), 16, 22–23, 42, 71, 111, Pl. 7b
  larval versus adult phylogeny by Grote, 42
Scoliopteryx libatrix (the herald), including cultured bacteria on, 44, Pl. 5g

Sphingidae (sphinx moths or hornworms),
    *Agrius convolvuli*, 30
    *Pachysphinx imperator*, western poplar sphinx caterpillar. Drowned in irrigation ditches, 27
    *Sphinx kalmiae* (laurel sphinx), Pl. 7c
*Spodoptera*, 30
*Synchlora aerata* (wavy-lined emerald), Pl. 7g-h
*Syntomeida epilais* (spotted oleander caterpillar), 21
tent caterpillars, 34, 52
    phylogeny, 72
    western (*Malacosoma californica*), 53
*Thyatira pudens*, 20
Tortricidae, 68, 168, 250. *See also* Fernald, C.H.
velvet bean caterpillar (*Anticarsia gemmatalis*), 132
woolly bear. *See* Arctiinae (tiger moths)
Zygaenidae, 41, 248
    grape leaf skeletonizer (*Harrisina americana*), 50
    net-winged beetle mimicry, 72
    Procridinae (once Pyromorphidae), 50
moth collecting. *See also* electric lights and globes
    paper envelops and rubber stamps, 29–30
    pocketed canvas apron, 89
    sugaring (sugar baits), 17, 26, 28, 89
Müller, Roberto, 109
Munsey Co., Frank A. Publishing house Dyar attempted to publish his pulp fiction at, 133
Myers, Eunice, Pl. 13

National Geographic Society, 171, 275
National Mall, 56, 88, 178, 192, 194
    photographs and figures, Fig. 13.1, Fig. 15.1d, Fig. 23.2b, Map 6, Pl. 9b, Pl. 13
National Museum of Natural History, 214, Fig. 13.1
National Museum building, Old. 125, 192, Fig. 9.1a, Fig. 22.2c. *See also* Arts and Industries Building

Neumoegen, Berthold, 35–38, 43, 49, 51, Fig. 5.2a
Nevada residency, purchasing a home in Reno, 146
Nevada Supreme Court, Allen v Allen,
    appeal, 149, Fig. 18.2c
    final judgment, 149
New England Peabody Home for Crippled Children, Newtown, MA. *See* Harriet M. Peabody
New Hampshire,
    Adirondacks, 30, 61
    Adirondack House, The, 32
    Center Harbor, 88
    Dublin, 84, 93, 103
        Chemical and Pathological Laboratory, 102–103
    Durham, 88
    Fabyan, 32
    Moultonborough, 88
    Mt. Washington, 32, 88
    Profile House Hotel, 88
    Red Hill, 88
    White Mountains,
        Bethlehem, 12
        Campton Village, 12
        Jefferson (Highlands), 12
        North Woodstock, 12
New Jersey,
    Fort Lee, 33
    Summit, 84
New York Draft Riots of 1863, The, 6. *See also* Josiah Cushman Hannum
New York Entomological Society
    auctions, 51, 242, 251
    editor of journal, Dyar's selection, 95
    lecture at meeting, 49
    meeting, 68
New York State Collection of Insects, Albany, NY, 19, 21
New York,
    Bronx Park, 48
    Central Park, 35
    Dutchess County, 12, 22
    Elizabethtown, 18, 91
    Esopus Creek, 243
    Hussey Mt. Spring, 18
    Kaaterskill Falls, 243
    Keene Valley, 32, Map 5
    Lake Champlain, 90

New York (*Cont.*)
  Linwood Cove, 18
  New Windsor, home of Emily
      L. Morton, 46
  Onteora Club in the Catskills, 54, 251
  Pelham Manor, 33
  Plattsburgh, 11, 14, 32, 43–44, 84, 91, 94
    collections from globe fixtures by
      G.H. Hudson, 18
  Poughkeepsie, moths at electric lights,
      18–21, 27, 34
  Raven Pass, 18
  Red Hook (Fraleigh's Hook), 12
  Rhinebeck, 12
  Rhinecliff, 6
  Rouse's Point, 32
  Saratoga Springs, 18
  Shaupeneek Ridge, 243
  Staten Island. Home of Josiah
      Cushman Hannum, 17, 30
  Syracuse, 16, 242
  Tupper[s] Lake and Moody, 90, 91
  Ulster County, 12
  Valcor Island, 94
  Van Cortlandt Park, 32
  Watkins Glen, 180, 195
*North American Entomologist* (Grote's
    journal), 75, 108
North Carolina, Tryon, 84, 91, 93, 123,
    267
Norwalk, CT. Hotel Mahackemo, 158,
    Fig. 19.2a
Noyes, Francis, Fig. 26.3

O'Conner, Jeremiah, 206
O'Leary, Eugene B., 142
oak, emory (*Quercus emoryi*), 53. *See also*
    *Gloveria howardi*
obituary, no record of Dyar writing one
    for colleague, 161
oiling ponds, 183, 209. *See* also mosquito
    control
Ontario,
  Dryden, 179
  Gravenhurst, 90
  Kenora, 179
  Nipigan, 179
  White River, 179
Oregon,
  Albina, 31

  Crater Lake, 180–181
  Medford, 181
  Mt. Tabor, 31
  Portland, 31
Osborn, Henry Fairfield, 35, 49, 247
Oslar, Ernest, 87
otosclerosis. *See* Zella Dyar, hearing loss/
    deafness
Ottolengui, Rodrigues, 32, 204

P Street, Dyar's rear gardens, 59, 60,
    186
Packard, Alpheus Spring,
    classification,
      belief that Limacodidae were related
        to Saturniidae, 41, 111
      use of pupae, 41
    last monograph completed by Cockerell
      and Dyar, 111
    Neolamarckian School, 38
    photograph, Pl. 8d
Packard, Elizabeth Walcott, 111
Panama and the Canal,
    Biological Survey of Canal Zone, 108
    collecting with Shannon, 182
    construction by United States and
      concern about mosquito-
      borne disease, 99
    Dyar's trip through canal, New York to
      San Francisco, 137, Fig. 17.1b.
      *See also* ships, SS Finland
*Papilio* (H. Edwards's journal). Devoted to
    Lepidoptera, 21, 26
Paris, France. *See* Dyar, Sr.,
Parsons, Agnes and Arthur, 268
Parsons, Dr. Mary, 130, 131, 132
Peabody, Harriet M.
    Bahá'í religion, 117, 191, 196
    death, 191
    Dyar's mother-in-law, 59, 91, 92, 126,
      129, 133–134, 137–141, 143,
      150, 191, 196
    Ethical Club of the 20th Century in Los
      Angeles, California, 63
    Knopf, Dr. S. Adolphus,
      correspondence, 191
    "Little Louise Mine" investment,
      Leadville, CO, 63
    Navajo school in Aneth, Utah,
      organized Navajo fair, 63, 252

New England Peabody Home for
Crippled Children in
Newtown, MA, 63, 253–254
photographs, Fig. 17.1c, Pl. 11b
Poland, AZ, starts reading
room in, 64
See Moths, *Pronoctua peabodyae*, named
for her by Dyar
Peabody, Philo, 14
Peabody, Zella M.,
See also Zella Dyar
Wedding in Los Angeles, Fig. 2.5
collecting and rearing caterpillars for
future husband, 14
Pelham Courts Apartments in Dupont
Circle, 184, Fig. 22.1c
pendulum. See Dyar, Sr., telegraph
Pergande, Theordor, 53, 72
Persian names from 'Abdu'l Bahá for
Wellesca and her children,
117, 118, 127, 130
Roushan and Joushan became
Roshan and Joshan, perhaps
connection to Hannums, 130
Pierce Arrow automobile, owned by H.G.
Dyar, Jr., 194
Pierce, Ulysses Grant Baker, Rev., 283
Pilgrim's March. Played by Zella Dyar on
piano, 118
pitcher plant (*Sarracenia purpurea*). See
Bacteria (found on), *Bacillus
sarracenicolus*
Poe, Edgar Allen, 217. See Dyar's short
story "Annabel Lee"
political cartoon, "Fighting Bob" La
Follette's presidential
campaign, 186, Pl. 14d
Pollock, Addie, 150
Pollock, George Freeman, 61, 148, 195,
226, 259, 284
ashes scattered at Skyland, 226
celebrating news of Wellesca's
'marriage' to Dyar, 150, 271
early photograph, Fig. 8.2a
financial loans from Dyar, 119,
121, 221
Wellesca's children staying with at
Skyland, 195
Pollock, George H., 284, Fig. 8.2a
remains at Glenwood Cemetery, 226

Pollock, Loue, 61
Pollock, Louise Plessner, 61, 151, 283
Pollock, Uila, 61, 283
Pollock, Wellesca. See Wellesca
Pomona College, Claremont, California, 104
Pope, C. H. (Munsey Trust Co.). Creditor
seeking payment from Dyar,
Jr., 217
Porter, Gene Stratton. Author, 106. See
also "Moths of Limberlost"
Pratt, Miss. Dyar's requests her as
specimen preparator, 157
Price, William W. Proprietor of Fallen
Leaf Lodge, 179
*Proceedings of the United States National
Museum*,
Outlet for large Dyar works during
publication of *Insecutor*,
108–109, 264, Fig. 13.2c
See Smith and Dyar Monograph on
*Acronicta*
prohibition, L.O. Howard's view against
and W.J. Bryan for, 72, 198
prolegs, pseudo legs on caterpillar
abdomen, 33, 45, 49
extra in flannel moth caterpillars. See
Moths, Megalopygidae
Prudden, Theophil Mitchell, 43–44
*Psyche* (the journal). Dyar's inchworm
series, 70
*Psyche* (the journal). Publisher of Dyar's
early works, 20, 34
pulp detective fiction, 133

Quaintance, Altus Lacy, 207
quarantine laws, U.S. not enacted in early
twentieth century, 93
quarreling (quarrels),
Grote, A.R., in poem, 74
lack of, at Entomological Society of
Washington meetings, 72
Smith, J.B., "more interested
in mosquitoes than
quarreling," 83
quick divorce, failures of
California, Dyar and Zella's first plan to
avoid bad publicity, 137

Rabbání, Shoghí Effendí, 200, 342
Rathbun, Richard, 112

Ravenel, William de Chastignier, 136, 176
Reading, PA. Home of H. Strecker, 19, 27
*Reality Magazine*, 191, 198–199, 216–217, Fig. 23.3b
   Deuth, Eugene and Wandeyne, original publishers, 198
   Mr. Bryan and the Monkeys, 198
Rehoboth Beach, DE, 181
reincarnation, 143, 239
reinstatement, Dyar's pursuit of. *See* Dyar, Jr.
Reno, NV,
   divorce, Dyar attempts, 138–142, 144–146, Fig. 18.1
   Dyar's move to, via Panama Canal, 137, Fig. 17.1b
   'marriage' on April 26, 1921 between Dyar and Wellesca, 150
   *See also* Allen v. Allen
Riley, Charles Valentine,
   *Acronicta* notes received from his widow, 77
   correspondence with Dyar
      *Datana californica* and to publish in *Insect Life*, 20, 21
      beetles at electric light globes, 20
   grave, 226, Epi. Fig. 2b
   photograph, Pl. 8a
   resignation as Division Chief, Bicycle accident and death, 40
   Riley insect collection oldest at USNM in late 1920s, 213
   snafu over Dyar's visit. 36, *See also* Linell
Roberts, C.H., President of N.Y. Entomological Society, 95
Robinson, Herold Sweetser. Publisher of *Reality Magazine*, 198, 216–217
Rock Creek Park, Washington, DC, 88
Rockefeller, Jr., John D. Smithsonian appeal to purchase Barnes Collection, 206
Rohwer, Sievert A., 203, 204, Pl. 13
Roosevelt, President Franklin D., 121
Roosevelt, President Theodore. Wellesca compares Dyar with the president's belief in the "refined, higher class," 148

Root, F.M., mosquito worker, 210
Roxbury Latin School,
   Images, Map 2a, Fig. 2.3a
   Latin and Greek courses, 11, 12
      Prescribed Mathematics, Prescribed Physics, and French: received college credit, 12
*Rubayyat*, 105. *See also* Omar Khayyam

Saint Mark's Lutheran Church, 187, Fig. 22.2a,c, Fig. 23.2b
salt marsh mosquito, eastern. *See Aedes sollicitans*
Sanitary Corps, 178–179
Saskatchewan, 92
Sawfly,
   *Ameteastegia pallipes*, Fig. 4.3a
   Caterpillars also make plant galls and leaf mines, 31
   *Craesus latitarsus*, Fig. 4.3b
   Dyar names sawfly for F.C. Pratt (*Lophyrus pratti*), 72
   Dyar's significant contribution to life histories stands alone, 33
   *Nematus hudsonii magnus*, 44
Schaus, William, 154, 156, 159, 165–167, 181, 201–202, 205, 211, 213–214, 251
   Barnes, John "Jack," 71, 170, Pl. 9d
   Barnes, William, and confusion about "Schaus and John Barnes" specimens in National Museum, 274
   biographical sketch, 71
   donation of collection to USNM, early interest, 55
   honorary titles, problem, 168–171, 203
   mosquitoes, Dyar netting over Zella, 90
   portraits and photographs of, Fig. 20.1b, Fig. 20.2b, Fig. 26.3, Pl. 9c-d, Pl. 13
   species, of Lepidoptera named, 109
   tenure at USNM, including Dyar's support, 169, 255, 274
   tiffs, with Dyar and staff, 170–171
   types, labeling practices, 110
   visit by Dyar and Zella, England, 156
Schmidt boxes to ship insect specimens, 160

Schoenberger, Paul ranger of the National Park Service, 207
Schwarz, Eugene Amadeus, 66, 70, 72, 88, 210, Fig. 9.1b, Pl. 13
Scopes Monkey Trial, 198
Scott, Austin. President of Rutgers University, 81
Scudder, Samuel H., 34, 69
séances. *See* Spiritualist and Lincoln
Searle Mr. [the printer], 132, 159,
Secretary of Agriculture. *See* USDA
Secretary of the Smithsonian. *See* Smithsonian Institution
Sedgwick, William Thompson, 34, 43
Seitz, Adalbert, "The Macrolepidoptera of the World," 211
Sellins, Miss, Pl. 13
sex pheromones, 35
Shannon. Raymond Corbett,
  Colombia, possible trip with Otis and Dyar, 192
  independence from Dyar on mosquito faunas of Chile and Argentina, 210
  meal at Dyar home, 194
  Panama, 181–182, 210. *See also* Panama and the Canal
  photographs, Fig. 9.1b, Fig. 21.4b, Pl. 13
  possible successor to Dyar on mosquitoes at USNM, 220
Shenandoah National Park. *See* Skyland and Stony Man Mountain
Sherman, Jr., Franklin, 97
steam ships, SS
  *Carrillo*, 182
  *Finland*, 137, 142, Fig. 17.1b
  *Toloa*, 182
Sixteenth Street Mansions, 215, 216, Fig. 26.1
Skinner, Henry,
  battle between Philadelphia (Skinner) and New York (Dyar) entomology editors, 95–98
  butterfly catalogue, 69
  disagreement with Skinner over the value of caterpillars in classification and with typographical errors prior to being editor of *JNYES*, 42, 98
  Dyar's wager with him over typographical errors, 99
  editorial battles with Dyar and Knab over naming mosquito species from larval stage, 100
  identification of reared Yosemite hairstreak (*S. californicum*), 30
  photograph, Fig. 12.1b
  poking fun at Smith and Dyar's battle over who discovered mosquito breeding grounds in Dublin, NH, writes poem about them as Cook and Peary, 103–105
skippers,
  southern cloudy wing (*Thorybes bathyllus*), 242
  *See* Hesperiidae
Smith, John Bernhardt,
  *Acronicta*, conflicts with Dyar over names, 77–79
  *Acronicta*, monograph with Dyar on, 77
  Assistant to C.V. Riley, 20
  battle of the names. *See* death from Bright's disease and Dyar's "The Man Who Did Return" (main character John Smith), 105
  *Haploa triangularis* an actual species, or according to Dyar a southern form of *confusa*, 73
  Hulst, problems with Dyar's of, 79
  "Lepidoptera of Boreal [North] America, List of," 20, 76
  "Lepidoptera Checklist" updated in 1903, soon after Dyar's "List," 70
  mosquito control movement, leader, in New Jersey, 82
  mosquitoes, report on New Jersey, Fig. 10.3a,b
  New Jersey State Entomologist, 42, 77, 79
  photograph, Fig. 10.1b
  Schmidt, Johann, original name of, 72
  unauthorized additions to his publications by Dyar, including *Proceedings of the U.S.N.M.*, 79

Smithsonian Institution,
  Castle, Fire of 1865, destroyed all the insect specimens, 213
  Endowment Fund. Possible source of funds for Barnes Collection, 206
  Secretary of the Smithsonian, 65, 112, 206
  Undersecretary, assigned to National Museum, 159
Soule, Caroline Gray, 96, 97
Southern Pacific Railway. Collecting mosquitoes along, 92
spiritualist,
  community, Boston, 12, 14, 241. *See* also Working Union of Progressive Spiritualists (WUPS)
  medium, trance and physical, 13, 61, 240
  upbringing, Dyar, Jr., 9, 10, 139, 161
spruce, injurious Lepidoptera on, 174
Staudinger's European catalogue, 70
steam carriage, inventors along with Dyar, Sr., 5
Stejneger, Leonard, 167–169, 172
Stockberger, Warner W. USDA Director of Personnel, 215, 218, 220–221
Stone, Alan, 224
Stoneman House. *See* Yosemite
Stony Man, VA,
  Camp, 61, 123, 263, 267
    Cake Walk and Carnival, 61
  location in Blue Ridge Mountains near Luray, VA, 61
  Mountain, 61, 84, 87
  *See also* Virginia, Skyland
Stover, M. Alma, nurse for birth of Wallace Dyar, 130, 142
Stowell, E.C., Dublin Chemical and Pathological Laboratory, 102, 103
Strecker, Ferdinand Heinrich Herman, 19–20, 27, 29, 36–37, 61, 65, 67, 74–75, 88, 161, 254, Pl. 8g–h
sugaring. *See* Moths, collecting
Susie, collector in Los Angeles, 14, Pl. 4d. *See also* Arthur Hudson

Swasey, Mrs. John P., Wife of representative from Maine, 63

Tabanidae (horse flies). Requested by Knab for Dyar to send, 160
*Technical World Magazine.* Dyar called "A New Type of Patriot," 112
telegraph. *See* Dyar Sr.
*Tentamen,* of Hübner, 75, 76, 256
Teuton War Spies, 184
Texas, Brownsville, 197
  Kerrville, 180
  San Benito, 180
Theobald, Frederick V., 101, 161, 211
Theosophy, 195
Thompson, Miss, pianist/violist, Chevy Chase, MD, 63
tiger beetle. *Cincindela cupracens*. Typo in *JNYES* missed by Dyar, 98
Tijuana, Mexico. Dyar's only known visit to Mexico, 92
Toumey, James W., 53
Tovar, M. Nunez, 210
Townsend, C.H. Tyler, 132, 159
trilobite, nickname for Roshan, according to evidence and testimony in Allen v. Allen, 149
Truckee, California, 212
Tunnel, B Street, 187–188, 277–278
  air-raid shelter proposed, 228
  darkness inside compared with Luray Caverns, VA, 189
  detailed illustration in Modern Mechanix in 1932, 17
  gathering bricks for archways and ceiling with children from the brickyard, 188
  images, Fig. 22.2b, Map 6, Pl. 15b
  inspected by District for widening Independence Avenue, 228
  newspaper articles, Pl. 14e, g
  proposed experiments growing mushrooms, 228
tunnel, Dupont Circle,
  collapses (1917, 1924, 1958), 184–187, 189, 227,
  images and photographs, Fig. 22.1b, Pl. 14c, Pl. 16

newspaper articles, Pl. 14a-d
  shaft, ladder of pipes set in
    concrete, 186
  dirt from tunnel, dumped on vacant lot
    behind home, 186
  'Dr. Otto von Golph,' German
    chemist, purported to be
    digger, 185
  German language newspapers, inside,
    186
  hollyhocks, digging bed of, onset of
    tunnel, 185
tunnels,
  after Dyar's death, 227
  "Great oaks from little acorns
    grow," 190
  Twenty-second Street 1950
    collapse, 227
  Why Dyar dug them, 189
Turkish Empire, 117
type specimens,
  destroying, Dyar's jest to Smith
    77–79
  loaning, 44, 53, 69, 105, 110–114, 156,
    167, 178, 202, 204, 211, 257
  photographs, Fig. 11.5, Fig. 25.2b

U.S. Army,
  mosquito identifications, 176
  Sanitary Officers' Reserve Corps, 215
U.S. Congress, 93, 113, 120, 205, 206
Undersecretary, assigned to National
    Museum. *See* Smithsonian
    Institution
University of Colorado, Boulder, CO, 158
University of Massachusetts, Amherst,
    19, 40
USDA, United States Department of
    Agriculture, 20, 31, 49, 52
  administrative leave, Dyar's, 145
  Civil Service, 135
  Division of Entomology (becomes
    Bureau of Entomology),
    33, 40, 56
  entomology examination, 112
  insect collections in Agriculture
    Building, 36
  compared Dyar's handwriting to
    Wilfred Allen's, 153–154

Horrigan, Mr., investigator from the
    Solicitor's Office, 153
Plant Quarantine Division, 204
reinstatement, attempts, 158, 174,
    215, 218, 220, 221, 223,
    Fig. 26.2b-c
Secretary of Agriculture,
  detectives, 175
  L.O. Howard's refusal to approach for
    funding Dyar trip, 174–175
  rebuffed Dyar's attempts for
    rehire, 176
USNM and USDA collections,
  in separate building during Dyar's first
    visit, 36
  need to consolidate them, 55, 56
  Pergande instructed to document tent
    caterpillars in each, 53
USNM (United States National Museum),
    Smithsonian Institution, 65
  Division of Insects, 56, 112, 158, 168
  Lepidoptera collection, the national,
    67, 71, 78, 81, 86, 109, 111,
    113, 114
    arrangement during Dyar's tenure
      as Honorary Custodian,
      "duplicates" separate from
      main collection, 156,
      213, 274
  purchase of Dognin collection, 169
  lepidopterists, 165–166, 173,
    201–206, 214, 222, 225
  hierarchy, 202
  division of duties, 165, 168
  policy on keeping specimens in
    exchange for identification,
    173, 214
U.S. National Museum Building, Fig. 9.1a
Utah,
  Garfield, 180
  Salt Lake City and the Great Salt Lake,
    26, 28, 180

Van Ingen, Gilbert, 49
Vanderburgh, Federal, 230, 239, Fig. 2.1
Vaupel, Karl, 142. *See also* Steamship
    *Finland*
Vermont, Middlebury, 3. *See also*
    Middlebury College

Virginia,
  Alexandria, 6, 188
  Chain Bridge, 124
  Grassymead near Mount Vernon, 90
  Paramore's Island, 124
  Richmond. Marriage between Wellesca and Wilfred P. Allen, 117
  Skyland, 111, 127, 148, 195, 226, Fig. 14.4
  Virginia Beach, 127, 267
  Wachapreague, 131, 267

Walcott, Charles D. Smithsonian Secretary, 112, 169, 206
Walker, Charles. Dyar, Sr.'s legal counsel and Morse's brother-in-law, 4
Walker, Thomas, type specimens, 77–78
Walsh, Benjamin. Description of *Aplodes mimosaria*, 71
Walsh, Thomas. Dupont Circle neighbor, 59, Fig. 8.1b
Walsingham, Lord, 41, 211, 222
Wardman, Harry. D.C. architect and real estate developer, 189, 215
*Washington Post*, 59, 138, 144, 184
Washington State,
  Ashford, 179
  Bellingham, 179
  Bremerton, 183
  Glacier, 179
  Hoodsport, 183, 276
  Hoquiam, 179
  Lake Cushman, 276
  Longmire's Springs, base of Mt. Rainier, 92, 179
  Oroville, 276
  Seattle, 179, 191, 197, 282
  Spokane, 197
  Sumas, 276
Weekapaug, RI, 84, 90
Weeks, Archibald C., 42
Weismann, August, "Studies in the Theory of Descent," 36, 38
Wellesca,
  admission to fictitious marriage, 224, Epi. Fig. 1
  Aseyeh Allen (Dyar), as, "The Teaching of Abdul Baha. Short Talks on the Practical Application of the Bahai Revelation [sic]," 145, 197, Fig. 23.3a
  death and final resting place, 226, 283, Epi. Fig. 2a. *See also* Glenwood Cemetery
  divorce in Maryland, 150
  driving and accident, 194
  fictional marriage in Reno and fictional divorce in Maryland, 150
  flat with Mr. Allen at 1228 B Street, SW, 125, Fig. 15.1c
  food shopping for family at Central Market, 194
  gossip about getting rich at Dyar's expense through real estate deals, 118, 125
  heart disease, 218
  marriage to Wilfred P. Allen in Richmond, VA, 117, 148, 153, 154, 224, Fig. 19.1
  miscarriage, 10, 125, 144
  nicknames, 259
  nurse, Dyar's, 132, 135, 137
  photographs of, Fig. 8.2b, Fig. 11.3b, Fig. 14.1a, Fig. 21.4a
  shunned by Washington, D.C. Bahá'ís, 196
  species or genera named for, 62, 132, 263
  testimony that she had operation to bear children, 151
  travel with Dyar (as nurse) kept secret to avoid gossip, 133
  real estate purchases, proxy, 121, 134
  visit to 'Abdu'l Bahá in Acca, Palestine, 117–118
  Zella, letter to,
    about retaliatory return of the stones from 'Abdu'l Bahá, 122
    pressuring to stop the divorce proceedings in D.C., 141–142
    response to Zella about her anxiety over Dyar's generous gifts to the Allen family, 127
Westchester Park Rd, 170, Boston, MA, Map 2. *See also* Dyar, H.G., Jr. home and West Chester Park
Wetmore, Alexander, 172, 206
White Sulphur Springs, WV. Knab assists with mosquito control, 160
White, Robinson. Zella Dyar's D.C. attorney, 138, 139, 142
Whitman, Charles Otis, 35. *See also* Wood's Hole

Wilfred, Sir. Character in Dyar short story "The Three Green Sisters" and perhaps in choice of 'Wilfred Allen' as alter ego, 10, 152
Wilson, President Woodrow, 153
Wolfe, Frank C. Marlborough Hotel Co, 216. *See also* Sixteenth Street Mansions
women's suffrage, 102, 191, 196
   Dyar's "Taming of a Suffragette," 102
Woods, Grace, 16
Woods Hole Biological Station, 32, 34–35, 231, 250, Fig. 5.1
Working Union of Progressive Spiritualists (WUPS), 12. *See also* spiritualist
World's Fair, San Francisco 1915, 138
World War I, 156, 160, 184, 186, 189
Worthington, A.S., 102

Yates, Mrs., Pl. 13
Yellowstone National Park, 31, 129, 181
   bear dump, evening at, 181
   black fly (*Simulium*) bites, 209
   Mammoth Hot Springs. Springs not a factor into mosquito breeding conditions, 31, 181
   Old Faithful, 181
   superintendent, gave Dyars housing in exchange for mosquito control advice, 181
   tourist complaints of "mosquitoes," 209
   Yellowstone Canyon, 181
Yosemite National Park,
   Anderson's Trail, 28
   Big Trees Mariposa Grove, CA, 26
   butterfly survey, 1891, most intensive to date, 28, 29
   Cloud's Rest Peak, 28
   final trip, 1924. 183. *See also Aedes increpitus*
   Inspiration Point, 28
   "Loco," person said to have destroyed Dyar's caterpillar livestock, 29
   Miwok names for places in, 30
   Mt. Lyell, 29
   Stoneman House, 26, 29, Fig. 4.1
   Tenieya Canyon, 28
   Yosemite Valley, 28
Yukon, 174
   Carcross, 180
   Dawson, 180
   Tahkini ("Tahkeena") River, 180
   White Horse, 180

Zetek, James, 182, 276–277
Zoological Congress, Seventh International in Boston, 112